Camera-Aided
Robot Calibration

Hanqi Zhuang
Zvi S. Roth

Department of Electrical Engineering
Florida Atlantic University
Boca Raton, Florida

CRC Press

Taylor & Francis Group
Boca Raton London New York

CRC Press is an imprint of the
Taylor & Francis Group, an **informa** business

CRC Press
Taylor & Francis Group
6000 Broken Sound Parkway NW, Suite 300
Boca Raton, FL 33487-2742

© 1996 by Taylor & Francis Group, LLC
CRC Press is an imprint of Taylor & Francis Group, an Informa business

First issued in paperback 2019

No claim to original U.S. Government works

ISBN-13: 978-0-367-44853-0 (pbk)
ISBN-13: 978-0-8493-9407-2 (hbk)

Visit the Taylor & Francis Web site at
http://www.taylorandfrancis.com

and the CRC Press Web site at
http://www.crcpress.com

Library of Congress Card Number 96-19093

Library of Congress Cataloging-in-Publication Data

Zhuang, Hanqi.
　　Camera-aided robot calibration / by Hanqi Zhuang and Zvi S. Roth.
　　　　p.　cm.
　　Includes bibliographical references and index.
　　ISBN 0-8493-9407-4 (alk. paper)
　　1. Robots--Calibration.　　I. Roth, Zvi S.　　II. Title.
　　TJ211.419.Z48　1996
　　670.42'72--dc20　　　　　　　　　　　　　　　　　　　　　96-19093
　　　　　　　　　　　　　　　　　　　　　　　　　　　　　　　　CIP

PREFACE

Many of today's industrial robots are still programmed by a teach pendant: The robot is guided by a human operator to the desired application locations. These motions are recorded and are later edited, within the robotic language residing in the robot controller, and played back, for the robot to be able to repetitively perform its task. Examples of typical robotic applications, for which such a strategy is more than sufficient, include pick-and-place of objects and automobile painting. For a successful run of such applications it is required that the robot be repeatable and that its work environment be unchanged, i.e., all parts and tools must be in well-defined fixed locations. Slight inaccuracies, while being tolerable in applications such as spray-gun painting, are not allowed for precision assembly, such as placement of surface mount electronic components on a circuit board.

Modern automation trends, on the other hand, have placed an increasing emphasis on sensor-guided robots and off-line programming. In the first, sensors such as vision are often employed to detect differences between actual and desired part locations. These offsets are then communicated to the robot controllers, so that the robot can correct its preprogrammed path. The robot motion commands are generated off-line from a CAD system. The programming task, in such a case, can be greatly simplified with the aid of interactive computer graphics, to simulate the effects of planned motions, without actually running the robot. Such software programs utilize data on the work piece and the robot that already exist in CAD databases. For a successful accomplishment of these more advanced tasks the robots need to be not only repeatable but also accurate. One of the leading sources of robot imperfect accuracy are differences between the geometric model of the robot as exists in its "blueprint drawings" and its actual geometry, as results from its construction tolerances. Robot calibration is a process by which the accuracy of a robot manipulator is enhanced, sometimes by orders of magnitude, through modification of the robot control software. To be able to better tune the robot geometric model through calibration, a sufficient amount of precision measurement data must be collected. Such data include the measured internal robot joint positions, and the coordinates of one or more points on the robot with respect to a designated reference frame.

These and many more fundamental ideas on robot accuracy and calibration are nicely laid out in several earlier references, most noteworthy is the book *Fundamentals of Manipulator Calibration*, by Mooring, Roth, and Driels (1991). Other books include Stone (1987) and Bernhardt and Albright (1993). Robot calibration, being somewhat a "tough sell" to industry users and robotics researchers of the 1980's, has evolved in the 1990's into an active mainstream robotics research area, as evidenced by numerous publications on the subject and many significant recent contributions.

Our book may, on one hand, be viewed as a compilation of our own research results, developed mostly in the early 1990's. In that sense, the book complements earlier books that focused on developments during the 1980's. This book is not intended, on the other hand, as a comprehensive review of all new results in the area of robot calibration. It is aimed at robotics researchers and practitioners as a handy and self-contained reference in the, narrower yet critically important, area of "robot calibration using computer vision." The key issue is no longer - "should we calibrate?", but rather "How to do it fast and cheap?". Shifting the burden of calibration from the robot manufacturer to the robot user raises issues of measurement rate, total calibration time, automated operation, user-friendliness, non-invasiveness and total cost.

Robot calibration consists of four steps: selection of a suitable robot kinematic model, measurement of robot end-effector's pose (i.e. position and orientation) in the world coordinate system, estimation of the robot model parameters, and compensation of robot pose errors. The measurement phase is unquestionably the most critical step towards a successful robot calibration.

Many robot pose measurement techniques discussed in the Robot Calibration literature are still far from being attractive tools for robot users who need to calibrate their robots on the manufacturing floor. Calibration measurement instruments such as theodolites, laser tracking systems, and coordinate measuring machines are either too slow or overly expensive or both.

Cameras and vision systems have become standard automation components. A typical robotic cell may feature an integrated multiple-camera system for product inspection, part presentation, and real-time monitoring of assembly accuracy. Some of these cameras may be fixed in the robot cell area, whereas others may be permanently attached to the moving robot arm to assist in on-line component alignment. Implementation of a robot calibration system using cameras may require little additional hardware, and only a modest amount of additional software.

Calibration by a camera system is potentially fast, automated, non-invasive and user-friendly. Cameras can also provide full pose measuring capability. There are two typical setups for vision-based robot pose measurement. The first is to fix the cameras in the robot environment so that while the robot changes its configuration the cameras can "view" a calibration fixture mounted on the robot end-effector. The second setup is to mount a camera or a pair of cameras on the end-effector (hand) of the manipulator.

If the cameras in the system are calibrated in advance, the locations of the calibration fixture in world coordinates for various robot measurement configurations can be computed by the vision system. The stationary-camera setup is non-invasive, as the cameras are often placed outside the robot workspace, and need not be removed after robot calibration. The major problem existing in all stationary camera setups is that in order to have a large field-of-view for the cameras, one has to sacrifice measurement accuracy. By using higher resolution cameras, the cost of the system and in particular its image processing part may increase dramatically.

The moving camera approach can resolve the conflict between high accuracy and large field-of-view of the cameras. The cameras need only perform local measurements, whereas the global information on the robot end-effector pose is provided by a stationary calibration fixture.

Methods for robot calibration using hand-mounted cameras can be further classified into "two-stage" and "single-stage" methods. In a two-stage approach, the cameras are calibrated in advance. The calibrated cameras are then used to perform robot pose measurements. In a single-stage approach, the parameters of the manipulator and those of the camera are jointly and simultaneously estimated. Depending upon the number of cameras mounted on the robot hand, these methods can be further divided into stereo-camera and monocular-camera methods. In the stereo-camera case, two cameras that have the same nominal optical characteristics are mounted on the robot hand. In the monocular case, only one camera is used.

This book, being the first on the topic of robot calibration using computer vision technology, covers the entire process of vision-based robot calibration, including kinematic modeling, pose measurement, error parameter identification, and compensation. It also addresses the issue of hand/eye calibration.

It is assumed that the reader is familiar with the basic theory of and practical approach to cameras, lenses, and image processing algorithms such as image preprocessing and segmentation. Moreover, even though most basic definitions are provided, it is assumed that this book is not the reader's first exposure to robotics. Sufficient familiarity with robots at the practical level (i.e. programming) and an introductory course on Robotics would be very helpful.

The book starts with an overview that emphasizes the authors' personal perspective on the history of robot calibration and of available techniques with focus on vision-based methods. It addresses some standing issues related to kinematic modeling, pose measuring, kinematic identification, camera calibration and autonomous calibration.

Chapter 2 covers the review of camera calibration techniques that are relevant to the robot calibration problem. It starts with the distortion-free pin-hole camera model to introduce the concept of camera calibration. By using a lens distortion model, a number of camera calibration techniques which are suitable for camera-aided robot calibration are presented. The chapter also addresses relevant issues such as the estimation of the image center and the compensation for perspective projection distortion. Finally, camera calibration simulation and experimental results are given that demonstrate the effectiveness of the camera calibration techniques outlined in this chapter.

Chapter 3 studies the properties of kinematic modeling techniques that are suitable for robot calibration. It summarizes the well-known Denavit-Hartenberg (D-H) modeling convention and points out the well-known drawbacks of the D-H model for robot calibration. After the presentation of a modified D-H model, the chapter then develops the Complete and Parametrically Continuous (CPC) model and the modified CPC model, both designed to overcome the D-H model singularities.

Pose measurement is a key element in a successful robot calibration task. If cameras are mounted on the robot hand, poses of the robot end-effector can be measured by a single camera or by a pair of stereo cameras. On the other hand, if stationary cameras are used, at least two cameras must be used. Chapter 4 discusses in great detail various vision-based pose measurement techniques. Methods for the identification of the relationship between the robot tool coordinate frame and the camera coordinate frame are also discussed.

Kinematic identification is a central element in a robot calibration task. Chapters 5 to 10 address this issue from different perspectives. Chapter 5 concentrates on error-model-based kinematic identification, while Chapter 6 presents linear solution approaches under the assumption that the robot measurement configurations follow a certain pattern.

Autonomous calibration of robot and camera systems is important in certain applications. In Chapter 7, a procedure for simultaneous calibration of a robot-camera system is developed.

Although hand/eye calibration is a highly practical problem, this issue is not addressed in existing books. Chapter 8 is devoted to this particular area. The chapter starts with a review of quaternion algebra, a key mathematical tool. Linear solution approaches for estimating the unknown rotation matrix are then covered. These methods are fast, but less accurate, comparing nonlinear approaches, which are also discussed in this chapter. This chapter demonstrates the pros and cons of the various approaches.

The geometric relationships in a robot system that most frequently need to be calibrated are the base and tool transformations. Chapters 9 and 10 deal with this aspect of calibration. Whenever the entire pose of the robot can be measured, the calibration of the base transformation becomes very simple. However when only point measurements are available, the task is more complex. Chapter 9 presents a linear approach that solves for the base transformation using point measurements only.

The final stage of a robot calibration task is accuracy compensation, using the identified kinematic error parameters. Chapter 11 presents a number of accuracy compensation algorithms, including the intuitive task-point redefinition algorithm and the linear quadratic regulator algorithm. The first is accurate and fast, provided that the robot is not in its singularity points. On the other hand, the latter is more robust. A simple bilinear interpolation method suitable for 2D compensation is also given in this chapter.

Off-line optimal selection of measurement configurations can significantly improve the accuracy of kinematic identification. In Chapter 12, a number of procedures that are designed for robot measurement configuration selection are outlined.

In Chapter 13, we present experimental results which were obtained by calibrating two industrial robots. Practical considerations important for conducting robot calibration experiments are also given in this chapter.

A brief appendix to the book provides readers additional mathematical background.

We acknowledge the support provided by the National Science Foundation, Motorola, National Radio Astronomy Observatory, and the Florida Atlantic University (FAU) Robotics Center for our robotics research. We would like to thank our graduate students whose thesis results are spread all over the book - Dr. Shoupu Chen, Dr. Kuanchih (Luke) Wang, Dr. Xangdong Xie, Ms. Xuan (Sharon) Xu, and Mr. Wen Chang Wu; and to other faculty, staff and students of the FAU Robotics Center for useful suggestions and discussions, in particular to Dr. Oren Masory, Dr. Ming Huang, Dr. Daniel Raviv, Dr. Jian Wang, Mr. Baiyuan Li, Mr. Yan Jiahua, and Mr. Zeer Gershgoren. Our special thanks are given to Mr. Roy Smollett for his wonderful Engineering support. Last but not least to our families for their patience, understanding and moral support.

Hanqi Zhuang
Zvi S. Roth

Boca Raton, Florida
January 1996

TABLE OF CONTENTS

1. OVERVIEW OF ROBOT CALIBRATION   1

  I.   The Motivation   1
  II.   Historical Perspective   3

2. CAMERA CALIBRATION TECHNIQUES   11

  I.   Introduction   11
  II.   Camera Models   11
     A.   A distortion-free camera model   11
     B.   Camera calibration: Basic concepts   14
     C.   Lens distortion model   16
  III.   Tsai's RAC-Based Camera Calibration Algorithm   20
     A.   Stage 1: computation of the rotation matrix R and the translation parameters t_x and t_y 21
     B.   Stage 2: computation of t_z, k, f_x and f_y 24
  IV.   A Fast RAC-Based Algorithm   24
     A.   Stage 1: computation of the rotation matrix R and the translation parameters t_x and t_y 25
     B.   Stage 2: computation of t_z, k, f_x and f_y 26
  V.   Optical Axis Perpendicular to the Calibration Board   27
     A.   Modification of the camera model   28
     B.   A calibration algorithm   29
  VI.   Nonlinear Least-Squares Approach   30
     A.   A linear estimation procedure   31
     B.   A nonlinear estimation procedure   34
  VII.   Estimation of the Ratio of Scale Factors   35
     A.   Single camera method   35
     B.   Stereo cameras method   38
  VIII.   Estimation of the Image Center   39
  IX.   Perspective Projection Distortion of Circular Calibration Points   41
     A.   A special case - one dimensional distortion   41
     B.   The general case   46
  X.   Simulation and Experimental Results   50
     A.   Simulation study of image center estimation   50

B. Simulation study for the estimation of the ratio of scale factors ... 52
C. Experimental study of the RAC-based camera calibration ... 53
D. Experimental results from a near singular camera configuration ... 57
XI. Summary and References ... 58

3. KINEMATIC MODELING FOR ROBOT CALIBRATION
63

I. Introduction ... 63
II. Basic Concepts in Kinematics ... 65
III. The Denavit-Hartenberg Model and Its Modification ... 68
A. The Denavit-Hartenberg modeling convention ... 68
B. Completeness, proportionality and shortcomings of the D-H model for robot calibration ... 75
C. Modification to the D-H model ... 78
IV. The CPC and MCPC Models ... 80
A. A singularity-free line representation ... 80
B. The CPC model ... 82
C. The MCPC model ... 87
D. Examples ... 89
V. Relationship between the CPC Model and Other Kinematic Models ... 92
A. Extraction of CPC link parameters from link homogeneous transformations ... 92
B. Mapping from the D-H model to the CPC model ... 93
C. Mapping from the CPC model to the D-H model ... 94
VI. Parametric Continuity: General Treatment ... 97
VII. Singularities of the MCPC model ... 99
VIII. Discussion and References ... 103

4. POSE MEASUREMENT WITH CAMERAS
107

I. Introduction ... 107
II. System Configurations ... 110
A. Stationary camera configurations ... 110
B. Hand-mounted camera configurations ... 113
III. Pose Measurement with Moving Cameras ... 114
A. Coordinate system assignment ... 115
B. The stereo-camera case ... 116
C. The monocular-camera case ... 118
IV. Identification of the Relationship between Robot End-Effector and Camera ... 119
A. Methods for stereo cameras ... 120

| | B. | A method for monocular camera with two views | 128 |
| V. | | Summary and References | 130 |

5 ERROR-MODEL-BASED KINEMATIC IDENTIFICATION   **133**

I.	Introduction	133
II.	Differential Transformations	134
III.	Finite Difference Approximation to Kinematic Error Models	139
IV.	Generic Linearized Kinematic Error Models	141
	A. Linear mappings relating end-effector Cartesian errors to Cartesian errors of individual links	142
	B. Linear mapping relating Cartesian errors to link parameter errors	144
	C. Elimination of redundant parameters	145
	D. Observability of kinematic parameters	147
V.	The D-H Error Model	153
	A. Linear mapping relating Cartesian errors to D-H parameter errors of individual links	153
	B. The linearized D-H error model	154
VI.	The CPC Error Model	156
	A. Linear mapping relating Cartesian errors to independent CPC parameter errors of individual links	156
	B. The linearized CPC error model	161
	C. Linearized *BASE* and *TOOL* error models	163
VII.	The MCPC Error Model	166
	A. Linear mapping relating Cartesian errors to MCPC parameter errors of individual links	166
	B. The linearized MCPC Error Model	166
VIII.	Summary and References	168

6. KINEMATIC IDENTIFICATION: LINEAR SOLUTION APPROACHES   **171**

I.	Introduction	171
II.	Problem Formulation and a Solution Strategy	172
III.	A Hybrid Linear Solution Method for All-Revolute Manipulators	174
	A. Solution for the orientation parameters	176
	B. Solution for the translation parameters	178
IV.	An All- Recursive Linear Solution Approach for General Serial Manipulators	181
	A. Problem reformulation	181
	B. Calibration of a prismatic joint	183

C. Calibration of a revolute joint 185

D. Determination of the World-to-Base (*BASE*)
transformation matrix 187

V. Extension of the Hybrid Linear Solution to General Serial
Manipulators 188

A. Solution for orientation parameters 188

B. Solution for translation parameters 188

VI. Numerical Studies 189

A. The effect of static errors in the measuring system
on pose measurement errors 191

B. The effect of random errors in the measuring system
on pose measurement errors 193

C. The effect of propagation errors in the orientation
parameter estimation due to the recursive nature of the
algorithm 194

D. Comparison between an error model based identification
technique and the linear solution method 195

VII. Summary and References 199

7. SIMULTANEOUS CALIBRATION OF A ROBOT AND
A HAND-MOUNTED CAMERA 201

I. Introduction 201

II. Kinematic Model, Cost Function and Solution Strategy 202

III. The Identification Jacobian 206

IV. Implementation Issues 210

A. Camera parameters 210

B. Robot parameters 211

C. Change of reference coordinate system 211

D. Observability of the unknown parameters 212

E. Verification of the calibration results 213

V. Extension to Stereo-Camera Case 214

VI. Discussion and References 215

8. ROBOTIC HAND/EYE CALIBRATION 217

I. Introduction 217

II. Review of Quaternion Algebra 220

A. Quaternions 220

B. Quaternion algebra 221

III. A Linear Solution 222

A. Solution for the rotation matrix 223

B. Solution for the translation vector 227

IV. A Nonlinear Iterative Solution 229

	A.	An alternative mathematical formulation of the hand/eye calibration problem	229
	B.	The cost function	231
	C.	The identification Jacobian	232
	D.	Observability issues	233
V.		Simulation Results	236
VI.		Discussion and References	241

9. ROBOTIC *BASE* CALIBRATION — 245

I.		Introduction	245
II.		Problem Statement	245
III.		Estimation of the Base Orientation	247
	A.	Quaternion-based algorithms	247
	B.	SVD-based method	251
IV.		Estimation of the Base Position	251
V.		Experimental Results	252
VI.		Summary and References	254

10. SIMULTANEOUS CALIBRATION OF ROBOTIC *BASE* AND *TOOL* — 255

I.		Introduction	255
II.		Problem Statement	256
III.		A Linear Solution	257
	A.	Solution for the rotation matrices	258
	B.	Solution for the position vectors	263
IV.		Simulation Studies	266
	A.	Number of pose measurements required for calibration	267
	B.	Calibration effectiveness under different measurement noise levels	267
	C.	Calibration effectiveness when the nominal robot geometry deviates from its actual one and joint readings are not perfect	269
V.		Summary and References	270

11. ROBOT ACCURACY COMPENSATION — 273

I.		Introduction	273
II.		Workspace-Mapping Method	273
	A.	System setups	274
	B.	Accuracy compensation sub-tasks	274
	C.	Bilinear interpolation	276
III.		Model-Based Pose-Redefinition Algorithm	277

IV. Gradient-Based Algorithms   278
    A. Solution strategy   278
    B. Derivation of the manipulator Jacobian   279
    C. A Newton-Raphson compensation algorithm   282
    D. DLS and LQR algorithms   282
    E. Simulation results   285
VI. Summary and References   289

12. SELECTION OF ROBOT MEASUREMENT CONFIGURATIONS   291

I.   Introduction   291
II. Problem Statement   291
    A. Performance measures   291
    B. A general problem statement   292
    C. A more restricted problem statement   292
III. Two Simple Search Algorithms   293
    A. Uniform random search   293
    B. Pairwise exchange   293
IV. Configuration Selection Using Simulated Annealing   294
    A. The SA algorithm   294
    B. Selection of robot measurement configurations   295
    C. Design of a practical cooling schedule   297
    D. Simulation studies   299
    E. Experimental results   300
V.   Summary and References   305

13. PRACTICAL CONSIDERATIONS AND CASE STUDIES   307

I.   Introduction   307
II. Practical Considerations   307
III. Calibration of a PUMA Arm   312
    A. The system setup   312
    B. PUMA calibration using a hand-mounted stereo cameras 313
    C. PUMA calibration using a hand-mounted monocular
       camera   317
IV. Calibration of a SCARA Arm   323
    A. The system setup   323
    B. Calibration of an Intelledex robot   324
V.   Summary and References   328

REFERENCES   331

APPENDICES   341
   I.   Summary of Basic Concepts in Matrix Theory   341
       A.   Eigenvalues and eigenvectors   341
       B.   Vector and matrix norms   341
       C.   Singular value decomposition   342
   II.   Least Squares Techniques   343
       A.   Linear least squares   343
       B.   Nonlinear least squares   343
  III.   Sensitivity Analysis   344

INDEX   347

Chapter 1

OVERVIEW OF ROBOT CALIBRATION

I. THE MOTIVATION

Robot Calibration is the process of enhancing the accuracy of a robot manipulator through modification of the robot control software. Humble as the name "calibration" may sound, it encompasses four distinct actions, none of which is trivial:

Step 1: Determination of a mathematical model that represents the robot geometry and its motion (*Kinematic modeling*).

Step 2: Measurement of the position and orientation of the robot end-effector in world coordinates (*Pose Measurement*).

Step 3: Identification of the relationship between joint angles and end-point positions (*Kinematic Identification*).

Step 4: Modification of control commands to allow a successful completion of a programmed task (*Kinematic Compensation*).

The need for robot calibration arises in many applications that necessitate off-line programming and situations that require multiple robots to share the same application software. Examples of the first are assembly operations, in which costly hard-automation (such as the use of accurate *x-y* positioners for the assembly part) to compensate for robot inaccuracies may be avoided through the use of calibration, as shown in Figure 1.1.1. An example of the latter is robot replacement, where calibration is an alternative to robot reprogramming, as shown in Figure 1.1.2.

Without calibration, robots which share application programs may experience significant accuracy degradation. The need to reprogram the machine upon replacement (or upon other maintenance actions that may cause permanent changes in the machine geometry) may result in a significant process down-time. Robots should be calibrated in a time period which is a fraction of the reprogramming time for calibration to be economically justifiable.

The growing importance of robot calibration as a research area has been evidenced by a large number of publications in recent years, including books and survey papers. Readers interested in surveys of robot calibration and detailed reference lists are referred to the book by Mooring, Roth, and Driels (1991) and a survey paper by Hollerbach (1988).

2   Overview of robot calibration

This book is not intended to be one more comprehensive survey of calibration. It focuses on camera-based techniques for robot calibration utilizing a unified modeling formalism developed and refined by us over recent years. We naturally chose to put main emphasis on our own research results; but, of course, these results were not developed in "empty space", as will be shown in the next section.

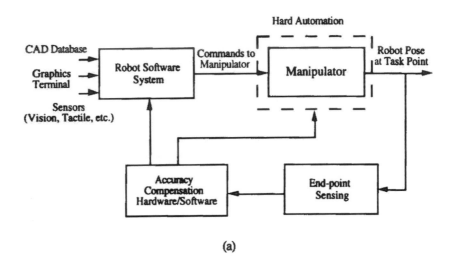

(a)

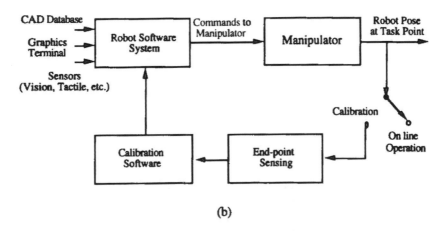

(b)

Figure 1.1.1. Assembly Operations: (a) Without robot calibration
(b) With robot calibration

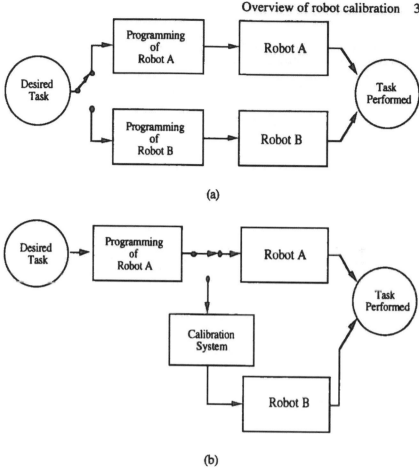

(a)

(b)

Figure 1.1.2. Robot replacement: (a) Individual programming of each robot
(b) With calibration

II.   HISTORICAL PERSPECTIVE

The following brief historical review of Robot Calibration research and practice portrays our own subjective view of key references and achievements that were most related to our own work and have had the biggest influence on us.

The booming growth of Robotics research in the late 1970's and early 1980's was a direct result of successful application of robot manipulators to automated manufacturing, particularly in the automotive industry and parallel to that the rapid growth in the computer industry. The predominant method of robot programming, suitable for the applications at that time, was "Teaching by Doing"; that is, physically moving the manipulator to each task point, recording and later replaying the joint-space description at these joints. Manipulators were designed to be highly repeatable and most applications involved a relatively low number of task points with minimal interaction between the robot and external sensors.

Richard Paul's book (1981) has been a major influence on all robotics researchers of the 1980's. His systematic use of homogeneous transformations and the Denavit-Hartenberg (D-H) mechanism kinematic modeling formulation for robot path planning and control quickly became universal standards. Many manipulator controllers were subsequently built featuring D-H forward and inverse models. The theoretical tractability of Task-Space robot path planning and the advent of sophisticated sensors such as solid state cameras, force sensors, and various types of proximity sensors and the personal computer revolution all fueled the high expectations that robot manipulators would be soon used to implement fully automated "factories of the future", and be key elements in many sophisticated multi-step applications involving task-space description and on-line interaction with large magnitudes of sensory data.

Such applications require, in principle, repeated use of the robot inverse kinematic model. In addition there is a need to program the robots off-line to move to task points never visited before by the robot. Such off-line programmed robots must be designed not only to be repeatable but, more importantly, to be accurate. The accuracy of a manipulator depends strongly on the accuracy of the robot geometric model implanted within its controller software. Robotics researchers and practitioners from academia and industry began to study the effects of joint offsets, joint axis misalignment and other accuracy error sources on the manipulator end-effector position and orientation errors.

The first major discovery, found independently by Mooring (1983) and Hayati (1983), which in retrospect established Robot Calibration as a new research area, was that the D-H model is singular for robots that possess parallel joint axes. More specifically, the common normal and offset distance parameters may undergo large changes when consecutive joint axes change from parallel to almost-parallel. Both researchers offered alternative robot kinematic models: Hayati introduced a modification to the D-H model which gained popularity and was subsequently adopted by many other researchers. Mooring advocated a model introduced earlier in the mechanism kinematics literature (for instance, refer to Suh and Radcliffe (1978)) based on the classical Rodrigues equation. Since most industrial manipulators are designed to be "simple", that is, to have parallel or perpendicular consecutive joint axes, this singularity problem is a major issue from a practical view point.

During the 1980's many robot calibration researchers came with their own model versions. In fact, the number of models almost equaled the number of researchers. An excellent survey of this flood of models, categorized into 4-, 5- and 6-link paıameters models is Hollerbach's paper (1988). The survey concluded with the following comment:

One issue that should be settled in the future is the choice of coordinate system representation. One strong alternative seems to be the Hayati modification of the Denavit-Hartenberg representation. It is not clear at this point what advantage the six-parameter representations would have for modeling lower-order kinematic pairs, while they have the disadvantage of redundancy.

To be able to improve the accuracy of the robot one needs to be able to measure the world coordinates of the robot end-effector at different robot joint-space configurations and record the joint positions at such configurations. If end-effector position and orientation, as predicted by the robot nominal forward kinematic model by plugging in the joint readings at the selected measurement configurations, differ from the actual end-effector pose measurement, the robot model needs to be suitably adjusted. That is what Robot Calibration is all about.

From a data collection point of view Robot Calibration is not different from Robot Performance Evaluation (in particular for assessing repeatability and accuracy). Much work on evaluation of machine tools and robot manipulators was performed during the 1980's at the National Bureau of Standards and one excellent review compiling many such testing techniques is the book chapter by Lau, Dagalakis and Myers (1988). Another beautiful survey of robot end-joint sensing techniques is the paper by Jiang, Black, and Duraisamy (1988). One of the first major reports of actual robot calibration experiments was the paper by Whitney, Lozinsky and Rourke (1986). Data collection was performed by Whitney and his co-investigators using theodolites.

Many Calibration or Robot Testing studies during the 1980's were done using a variety of measurement techniques ranging from expensive Coordinate Measuring Machines (CMM) and Tracking Laser Interferometer Systems to ones that employed inexpensive customized fixtures. The "heart of the matter" is the measurement in "world coordinates" of one point on the robot end-effector. World coordinates are often defined by the calibration measurement equipment itself. The measured point represents the end-effector position. The measuring of the coordinates of three or more non-colinear end-effector points provides the full pose (position and orientation) of the end-effector. Some measurement devices are capable of measuring the full 6-dimensional pose, some can measure only the 3D position and others, such as single theodolite, measure even less than that.

A major contribution to the Kinematic Identification phase of Robot Calibration was the paper by Wu (1984) in which the Identification Jacobian, a matrix relating end-effector pose errors to robot kinematic parameters errors, is systematically derived. This mathematical tool is very useful for both machine accuracy analysis and machine calibration. Another contribution by Wu and his co-authors was the paper (Veitschegger and Wu (1988)) that introduced two techniques for Accuracy Compensation.

Casting the full robot calibration problem as a four-step problem – modeling, measurement, identification and compensation, was featured in the survey paper by Roth, Mooring and Ravani (1987) and expanded into a full scope book (Mooring, Roth and Driels (1991)). That book was indeed a comprehensive survey of all phases of manipulator calibration as evidenced from research done mostly during the 1980's. The book is a simply written tutorial to many of the fundamental concepts and methods and is highly recommended for first-time robot calibration practitioners.

In addition to Hollerbach's "standing question" regarding calibration models, many other open research issues have lingered, such as:

1. What is the relative importance of robot geometric errors compared to non-geometric errors?
2. How is the calibration quality related to the resolution and accuracy of the calibration instrumentation and the method of calibration?
3. How should robot measurement configurations be optimally chosen?
4. How is observability of robot kinematic error parameters related to the selection of calibration configurations and method?

Some of these problems, even today, are not yet fully answered. Practical implementation questions were even more acute:

1. How can robot calibration be done "fast" and "cheap"?
2. Should calibration be done primarily by the robot manufacturer or can the calibration load be shifted to the robot user?
3. Should robots be designed differently from a hardware and software point of view to accommodate on-line calibration capability?
4. What current technology will make robot calibration, performed "on the manufacturing floor", economically feasible?

Starting with some of the research issues, we believe that our Complete and Parametrically Continuous (CPC) type models, as introduced in the thesis by Zhuang (1989), the paper by Zhuang, Roth and Hamano (1992) and explained in detail in this book, are a step forward toward answering Hollerbach's question. The CPC model was inspired by a paper by Roberts (1988) in the Computer Vision literature, which discussed a very useful line representation with respect to a local coordinate frame using the directional cosines of the line. In the case of robot modeling a joint axis directional vector is represented in terms of a coordinate frame located on the previous joint axis. The CPC model is a natural evolution of Hayati's model (1983), Mooring and Tang's model (1984) and Sheth and Vicker's model (1972). Hayati's model utilizes a plane perpendicular to one of the joint axes. Mooring's model also represented joint axes, however with respect to the world frame. Sheth and Vicker introduced the concepts of Motion and Shape

Matrices. Readers are also referred to Broderick and Cipra (1988).

Important observability issues of kinematic error parameters can be addressed through a generic link-by-link error model, originally introduced by Everett and Suryohadiprojo (1988). These are explained in detail in this book.

The CPC models and error-models apply uniformly to manipulator internal links as well as the *BASE* and *TOOL* transformation. This makes the model highly convenient to robot partial calibration. Ideas of progressive calibration – starting from *BASE* only through *BASE* and joint offsets only, to full scale calibration, were first pursued by Mooring and Padavala (1989).

It is important to fully recognize that the kinematic identification and accuracy compensation processes are merely least squares fittings of a suitable number of design parameters to improve on the overall accuracy. Calibration can be done with any number of available design parameters depending on the actual physical set up. Some robot software systems do not allow the user access to all the coefficients of the robot kinematic model. For instance, in some commercial SCARA arms a user is allowed to modify only the joint variable offsets and the link length parameters. In this light the question about relative importance of non-geometric errors may not be fully meaningful as the least squares fitting is done based on noisy data affected by both geometric and nongeometric sources.

Breaking robot calibration into different levels and focusing on specific partial calibration problems such as Hand-Eye Coordination and Robot Localization is one of the central themes of this book.

With regard to the implementation issues, the "fast and cheap" guideline automatically rules out expensive instrumentation such as Laser Tracking Systems or methods that are highly invasive such as placing contract calibration fixtures within cluttered and application dependent robot work environments. From a calibration cost viewpoint the use of cameras and vision systems is extremely beneficial as these already exist as integral components of most industrial robotic cells. For instance, electronic assembly operations often require the use of a multiple camera setup, one that is attached to the robot end-effector which monitors "fiducial" points on the circuit boards and transmits data which are used to finely adjust the end-effector location and another that may be located within the conveyor system monitoring from below the relative alignment between the robot and the assembly area. Hence, calibration implementation may involve, at most, additional camera calibration boards and specialized calibration software at a cost which is a tiny fraction of the total robotic cell cost. Data acquisition systems using cameras are non-invasive, very fast (potentially) and, in principle there is no increase in the level of difficulty in monitoring more than one point on the robot. In other words, full pose measuring ability is easily feasible.

The major stumbling block that prevented widespread use of cameras for machine tool and robot metrology has been camera resolution, most importantly image sampling due to nonzero pixel size. It can be shown, for instance, that with typical off-the-shelf stereo cameras arranged in stationary locations, the coordinates of a target point located on the moving robot and viewed from a few feet away may have a diameter of uncertainty of nearly 1 mm. Taking directly end-effector position measurements using common stereo cameras is of course totally unacceptable for most robotic applications that typically may require repeatability in the order of magnitude of 0.001-0.1 mm. Sklar (1988) experimented with robot calibration using 1024x1024 pixel camera and customized vision software. With today's technology such a solution increases dramatically the overall cost of the calibration system.

The question "can low resolution sensors be used to provide (indirectly) accurate end-effort position readings ?" received an interesting answer in three different independent studies done in the 1980's. Stone (1987) used triangulation of very noisy acoustic sensors tracing a spark-generating target. These noisy points were obtained through moving each robot joint one at a time. For a revolute joint the collection of measured points is used to fit circles in 3D space. These circles define accurately the plane of rotation and center of rotation for the respective joint. For a prismatic joint the measured points are used to fit 3D lines which describe the respective joint axis direction. Similar studies using stereo cameras were performed by Barker (1983) and Sklar (1988). Another good reference is a thesis by Chen (1987). The reader should keep in mind that any robot kinematic modeling starts by specifying the manipulator's joint axes, in an arbitrary configuration, followed by set-up of link coordinate frames and construction of the link transformations using any convention for selection of link parameters. Joint axis identification provides a solution to robot calibration that is radically different from the method of identification of error parameters using linearized accuracy error models. This solution, which is potentially "cheap", is most certainly not "fast". Accurate joint axis identification requires the measurement of a large number of target points along each joint travel. When using stationary stereo cameras to track a light source mounted on the robot, the issue of target visibility arises, complicating further the calibration process.

Resolving the difficult tradeoff between camera position measurement resolution and the camera field of view, necessitates that the camera(s) always remain close to the moving robot target. In other words, it is necessary that the cameras, used to calibrate the robot, move together with the robot. This idea may appear somewhat counterintuitive to metrology practitioners. After all, position measurement using stationary stereo cameras first requires careful calibration of the cameras. These calibrated cameras subsequently define the robot "world coordinate frame". Presumably any intentional or accidental displacement of these cameras may take the entire measurement system out of calibration.

Reported experimental results by Puskorius and Feldkamp (1987) and by Zhuang, Roth and K. Wang (1994) showed that robot calibration using moving cameras is feasible. The point is that cameras attached to the robot hand continue to remain calibrated with respect to a fictitious camera calibration fixture that moves together with the moving robot.

The field of Camera Calibration has been a rich area of research. Readers are referred to Tsai (1987) and Weng, Cohen and Herniou (1992) and others for references. Camera models could range from the simplest distortion-free "pinhole" model that leads to the well-known perspective transformation to more involved models that take into account lens distortion effects. Of particular interest to robot calibration are camera models that contain an explicit description of the camera pose (position and orientation) with respect to the viewed object.

Key references are Tsai's paper (1987), proposing a method for calibration of the camera pose and relevant camera internal parameters using a single flat calibration board, and the paper by Lenz and Tsai (1989) describing the following interesting idea for robot calibration: Since the camera model includes the pose of the camera and since the camera is rigidly attached to the robot arm, the pose of the camera represents the pose of the manipulator. In other words, the robot calibration measurement phase is done by recalibrating the camera at each robot joint space measurement configuration.

The above references opened up the entire research area of Camera-Aided Robot Calibration in which methods based on moving cameras play a central role. Scanning this issue is what this book is all about.

In recent years Robot Calibration became a very popular research area as evidenced by many recent publications and special sessions in Robotics Conferences. Camera-Aided Robot Calibration is only a minor representative of the current research directions in this area.

The inherent challenge and the critical importance of robot pose measuring in the world coordinate frame prompted several researchers to strongly consider the idea of "Autonomous Calibration". In autonomous calibration, kinematic identification is to be performed based only on data obtained internally within the robot. Most prominent of this new wave is Hollerbach and his co-researchers (Bennett and Hollerbach (1989, 1991)). The idea in its basic form was to create a closed kinematic chain by addition of several links that connect the robot end-effector to the ground. This type of kinematic redundancy created more kinematic unknowns offset by a larger number of equations, which all-in-all enabled the identification of the robot kinematics as well as the additional parameters of the calibration system.

A natural development of this idea was that rather than closing the loop mechanically, one can close the loop optically. Bennett, Geiger and Hollerbach (1991) reported autonomous calibration using servo actuated stereo cameras attached to the robot hand. The camera readings were conveniently considered "internal sensory data".

Recently, Zhuang, Wang and Roth (1995) explored simultaneous calibration of a robot and a single camera attached rigidly to the robot hand. This idea is described in detail later in this book.

Much attention has been given recently to the calibration of parallel manipulators and other multiple degrees of freedom systems such as machine tools and laser tracking systems (Zhuang, Li, Roth, and Xie (1992)). In all the concept of autonomous calibration is central. The level of accuracy required in machine tools and laser tracking systems prohibits the use of off-the-shelf cameras. Cameras can play an important role in the calibration of large parallel robots such as Stewart platforms, but this has not yet been fully explored experimentally and we chose not to include this topic in the book. Most of the techniques that are presented in this book have been verified experimentally.

It is important to stress that although much work has been done in recent years, Camera Aided Robot Calibration as a research area is far from being closed. One of the book's purposes is to encourage robotics researchers to find solutions to many of the numerous open problems listed throughout the book. The goal of "cheap and fast" calibration, to allow off-line accurate, yet user-friendly, robot programming right on the manufacturing floor remains one of the key goals in manufacturing and automation. Successful attainment of it carries a potential enormous economic pay-back. We hope that this book contributes in documenting the current knowledge.

Chapter 2

CAMERA CALIBRATION TECHNIQUES

I. INTRODUCTION

It is appropriate to discuss first camera models and camera calibration. The main intent is not to provide an exhaustive and comprehensive review of this rather rich area, but to focus on those techniques that have practical relevance to the subsequent problem of robot calibration.

This chapter is organized as follows. Relevant camera models as well as the so-called perspective transformation matrix (PTM) method are overviewed in Section II. Tsai's radial alignment constraint (RAC) method for camera calibration is described in Section III. A simplified RAC-based algorithm is given in Section IV. A procedure that handles a near singular case is presented in Section V. Weng's two-phase nonlinear optimization approach is outlined in Section VI.   Methods for determining the ratio of scale factors and estimating the image center are derived in Sections VII and VIII. An analysis of distortion of the centroid of a circular point due to perspective projection is brought in Section IX. Calibration simulation and experimental results are presented in Section X. The chapter concludes with references and discussion.

II. CAMERA MODELS

A. A DISTORTION-FREE CAMERA MODEL

The purpose of the model is to relate the image coordinates of an object point visible by the camera, to the coordinates of this point in a reference coordinate system. Let $\{x_w, y_w, z_w\}$ denote the world coordinate system; $\{x, y, z\}$ denote the camera coordinate system, whose origin is at the optical center point O, and whose z axis coincides with the optical axis; and $\{X, Y\}$ denote the image coordinate system centered at O_I (the intersection of the optical axis z and the image plane) (refer to Figure 2.2.1). $\{X, Y\}$ lies on a plane parallel to the x and y axes.

The transformation from the world coordinates $\{x_w, y_w, z_w\}$ to the camera coordinates (x, y, z) is

$$\begin{bmatrix} x \\ y \\ z \end{bmatrix} = R \begin{bmatrix} x_w \\ y_w \\ z_w \end{bmatrix} + t \tag{2.2.1}$$

where the rotation matrix R and translation vector t are written as

$$R = \begin{bmatrix} r_1 & r_2 & r_3 \\ r_4 & r_5 & r_6 \\ r_7 & r_8 & r_9 \end{bmatrix}$$

and

$$t = [t_x \; t_y \; t_z]^T$$

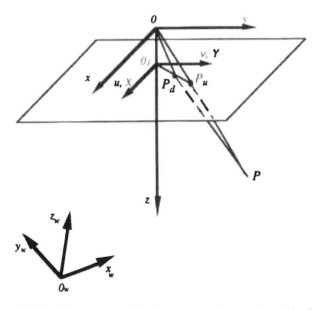

Figure 2.2.1   Camera coordinate system assignment and the Radial
Alignment Constraint

The standard distortion-free "pin-hole" model assumption is that every real object point is connected to its corresponding image point through a straight line that passes through the focal point of the camera lens.   Thus, the following perspective equations result:

$$u = f\frac{X}{z} \qquad (2.2.2)$$

$$v = f\frac{y}{z} \qquad (2.2.3)$$

where f is the (effective) focal length of the camera and (u, v) are the analog coordinates of the object point in the image plane.   The image coordinates (X, Y) are related to (u, v) by the following equations,

$$X = s_u u \qquad (2.2.4)$$

$$Y = s_v v \qquad (2.2.5)$$

The scale factors, s_u and s_v, not only account for TV scanning and timing effects, but also perform units conversion from the camera coordinates (u, v), the units of which are meters or inches, to image coordinates (X, Y) measured in pixels.

A distinction is made between *extrinsic* parameters (the elements of R, t) which convey the information about the camera position and orientation with respect to the world coordinate system, and the *intrinsic* parameters (such as s_u, s_v, f and distortion coefficients that will be discussed later) that convey the internal information about the camera components and about the interface of the camera to the vision system.

Since there are only two independent parameters in the set of intrinsic parameters s_u, s_v and f, it is convenient to define

$$f_x \equiv f s_u \qquad (2.2.6)$$

$$f_y \equiv f s_v \qquad (2.2.7)$$

Combining the above equations with (2.2.1) yields the undistorted camera model that relates the world coordinate system $\{x_w, y_w, z_w\}$ to the image coordinate system $\{X, Y\}$

$$X = f_x \frac{r_1 x_w + r_2 y_w + r_3 z_w + t_x}{r_7 x_w + r_8 y_w + r_9 z_w + t_z} \qquad (2.2.8)$$

$$Y = f_y \frac{r_4 y_w + r_5 y_w + r_6 z_w + t_y}{r_7 x_w + r_8 y_w + r_9 z_w + t_z} \qquad (2.2.9)$$

Note that the computed image coordinates stored in the computer memory of the vision system are generally not equal to the image coordinates (X, Y). Let (X_f, Y_f) be the computed image coordinates for an arbitrary point, and let

(C_x, C_y) be the computed image coordinates for the center O_I in the image plane. (C_x, C_y) is referred to as the *image center*. (X, Y) is then related to (X_f, Y_f) by the following equation,

$$X = X_f - C_x$$
$$Y = Y_f - C_y$$

The ideal values of C_x and C_y can be obtained from the image size of a given vision system. For instance, when the image size is 512 by 512, then C_x and C_y are both equal to 256. However, in common off-the-shelf image processing systems, the amount of fixed uncertainties in the image center coordinates may reach 10-20 pixels.

B.   CAMERA CALIBRATION: BASIC CONCEPTS

In the basic camera model defined by (2.2.8)-(2.2.9), the camera parameters to be calibrated are the three independent extrinsic parameters of R, the three extrinsic parameters of t, as well as the intrinsic parameters f_x, f_y, C_x and C_y. The calibration of the camera is done by taking a set of m object points which: (a) have world coordinates $\{x_{w,i}, y_{w,i}, z_{w,i}\}$, $i = 1, ..., m$ that are known with sufficient precision, (b) are within the field of view of the camera. These points, referred to as "calibration points" are detected each on the camera image at the respective image coordinates, $\{X_i, Y_i\}$.

The camera calibration problem is to identify the unknown camera model coefficients given the above known/measured data. Identification of the model (2.2.8)-(2.2.9) provides explicitly the pose of the camera in world coordinates. This is an important by-product of the camera calibration process which will be shown to have much relevance to the problem of camera-aided robot calibration.

The most basic camera calibration strategy involves linear least squares identification of the *Perspective Transformation Matrix*.

The model (2.2.8)-(2.2.9) can be rewritten as

$$X = \frac{a_{11}x_w + a_{12}y_w + a_{13}z_w + a_{14}}{a_{31}x_w + a_{32}y_w + a_{33}z_w + a_{34}} \qquad (2.2.10)$$

$$Y = \frac{a_{21}x_w + a_{22}y_w + a_{23}z_w + a_{24}}{a_{31}x_w + a_{32}y_w + a_{33}z_w + a_{34}} \qquad (2.2.11)$$

where we set $a_{34} = 1$ since the scaling of the coefficients $a_{11}, ..., a_{34}$ does not change the values of X and Y. The coefficients $a_{11}, ..., a_{34}$ correspond to what is called the "Perspective Transformation Matrix". Equations (2.2.10) and (2.2.11) can now be combined into the following identification model:

$$
\begin{bmatrix}
x_w & y_w & z_w & 1 & 0 & 0 & 0 & 0 & -Xx_w & -Xy_w & -Xz_w \\
0 & 0 & 0 & 0 & x_w & y_w & z_w & 1 & -Yx_w & -Yy_w & -Yz_w
\end{bmatrix}
\begin{bmatrix} a_{11} \\ \vdots \\ a_{33} \end{bmatrix}
=
\begin{bmatrix} X \\ Y \end{bmatrix}
$$

$$(2.2.12)$$

The eleven unknown coefficients can be obtained using linear least squares. The minimum number of calibration points is six. Each data point pair $\{(x_{w,i}, y_{w,i}, z_{w,i}), (X_i, Y_i)\}$ contributes two algebraic equations for the unknown coefficient vector $a = [a_{11}, ..., a_{33}]^T$ of the type shown in (2.2.12). One can further show that the calibration points must not be all coplanar. To illustrate this point, consider, for instance, the case of a planar calibration board characterized by $z_{w,i} = C = $ constant, for any i. In this case the coefficient matrix obtained on the left hand side of the "stacked" equations (2.2.12) is singular since columns 3 and 4 as well as columns 7 and 8 are linearly dependent.

Another point, that needs to be fully understood, is that the linear least squares solution of the equation (2.2.12) is, in general, not equal to the global optimal solution of the camera calibration problem, due to the fact that the model does not take into consideration any lens distortions.

Yet another point is that we are unable to uniquely recover the rotation matrix and the translation vectors from the coefficients $a_{11}, ..., a_{34}$. As will be seen in later chapters, this inhibits the use of the perspective transformation matrix method to calibrate a robot with a hand-mounted monocular camera.

In summary, for these three reasons: (a) camera pose is not directly available, (b) a 3D camera calibration fixture as accurate as a 2D calibration board, is significantly more expensive, and (c) the model obtained is only suboptimal, this method is unattractive for Robot Calibration.

The method does however provide quick, albeit somewhat inaccurate, means to conduct stereo measurements of target points: Given two cameras A and B, which have been calibrated to obtain two sets of coefficient vectors a^A and a^B, one can compute the unknown coordinates $\{x_w, y_w, z_w\}$ of any object point, visible by both cameras, using the point's image coordinates. For the unknown object point, the corresponding image points are $\{X^A, Y^A\}$ and $\{X^B, Y^B\}$, from which one is able to construct two equations of the type

$$\begin{bmatrix} a_{11} - a_{31}X & a_{12} - a_{32}X & a_{13} - a_{33}X \\ a_{21} - a_{31}Y & a_{22} - a_{32}Y & a_{23} - a_{33}Y \end{bmatrix} \begin{bmatrix} x_w \\ y_w \\ z_w \end{bmatrix} = \begin{bmatrix} X - a_{14} \\ Y - a_{24} \end{bmatrix}$$

Two pairs of equations, one for camera A and the other for camera B, provide in the least squares sense the *3D stereo measurement* of a point from its measured image planar coordinates.

C.   LENS DISTORTION MODEL

It is well-known that actual cameras and lenses sustain a variety of aberrations and thus do not obey the above perfect model. The main error sources are: a) Image spatial resolution defined by spatial digitization is relatively low (e.g., a typical CCD sensing array has about 512X480 pixels). b) Video camera lenses are nonmetric. These low-cost "off-the-shelf" lenses sustain a substantial amount of distortion. c) Camera assembly involves a considerable amount of internal misalignment (e.g., the center of the CCD sensing array may not be coincident with the optical principal point, that is, the intersection of the optical axis with the image plane; and the CCD array may not be parallel to the lens). d) Hardware timing mismatches between the image acquisition hardware and the camera scanning hardware, or the imprecision of the TV scanning timing itself.

Geometric distortion has to do with the position of image points on the image plane. As a result of several types of imperfections in the design and assembly of lenses residing within the camera optical system, the distortion-free pin-hole model does not hold and has to be replaced by models that take into account positional errors due to distortion,

$$u' = u + D_u(u, v) \tag{2.2.13}$$
$$v' = v + Dv(u, v) \tag{2.2.14}$$

where u and v are the unobservable distortion-free image coordinates, and u' and v' are the corresponding coordinates taking distortion into account.

Lens distortion effects can be classified into *radial* and *tangential distortions*, as shown in Figures 2.2.2 and 2.2.3. Radial distortion causes an inward or outward displacement of a given image point from its ideal location. This type of distortion is mainly caused by flawed radial curvature of the lens elements. A negative radial displacement of image points is referred to as *barrel distortion*.   It causes outer points to be increasingly crowded together and the scale to decrease. A positive radial displacement is referred to as a *pincushion distortion*.   It causes outer points to spread and the scale to increase. This type of distortion is strictly symmetric about the optical axis.

Actual optical systems are also subject to various degrees of decentering; that is, the optical centers of lens elements may not be strictly collinear. This defect introduces what is called *decentering distortion*. This type of lens distortion has both radial and tangential components.

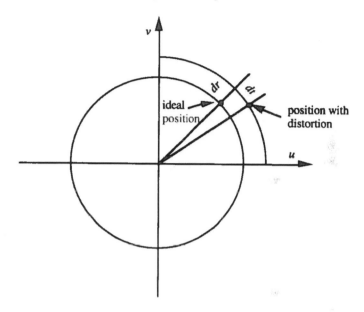

Figure 2.2.2 Radial and tangential distortion effects
on the location of a single image point

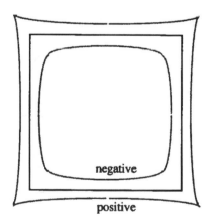

Figure 2.2.3 Effect of radial distortion. Solid lines: no distortion;
other lines: with radial distortion

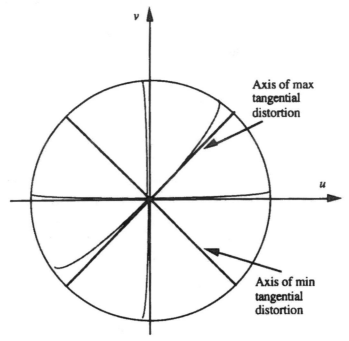

Figure 2.2.4 Effect of tangential distortion. Solid lines: no distortion;
other lines: with tangential distortion

Camera calibration researchers argued and experimentally verified that radial distortion is the dominant distortion effect. If as a good approximation, only radial distortion is to be considered, then

$$D_u(u, v) = ku(u^2 + v^2) + O[(u, v)^5] \qquad (2.2.15)$$
$$D_v(u, v) = kv(u^2 + v^2) + O[(u, v)^5]. \qquad (2.2.16)$$

The higher order terms can for all practical purposes be dropped. Substituting the above into (2.2.13)-(2.2.14) yields

$$u' = u(1 + k'r^2)$$
$$v' = v(1 + k'r^2)$$

where

$$r^2 = u^2 + v^2$$

Because the undistorted image coordinates u and v are unknown, it is desirable to replace these by measurable image coordinates of X and Y. Thus,

$$r^2 = (X/s_u)^2 + (Y/s_v)^2$$

Define $k \equiv k's_v{}^2$, the *radial distortion coefficient*, then

$$\mu \equiv \frac{f_y}{f_x} = \frac{s_v}{s_u} \qquad (2.2.17)$$

and

$$r^2 \equiv \mu^2 X^2 + Y^2 \qquad (2.2.18)$$

where μ is termed *the ratio of scale factors*. With the above modification, one obtains the following camera model that takes into account small radial distortion effects:

$$X(1 + kr^2) = f_x \frac{r_1 x_w + r_2 y_w + r_3 z_w + t_x}{r_7 x_w + r_8 y_w + r_9 z_w + t_z} \qquad (2.2.19)$$

$$Y(1 + kr^2) = f_y \frac{r_4 y_w + r_5 y_w + r_6 z_w + t_y}{r_7 x_w + r_8 y_w + r_9 z_w + t_z} \qquad (2.2.20)$$

A mathematical "trick", that will prove to be useful in finding a linear least squares solution to the camera calibration problem when radial distortion is considered, is to use a variation of the above model,

$$\frac{X}{1 - kr^2} \cong f_x \frac{r_1 x_w + r_2 y_w + r_3 z_w + t_x}{r_7 x_w + r_8 y_w + r_9 z_w + t_z} \qquad (2.2.21)$$

$$\frac{Y}{1 - kr^2} \cong f_y \frac{r_4 y_w + r_5 y_w + r_6 z_w + t_y}{r_7 x_w + r_8 y_w + r_9 z_w + t_z} \qquad (2.2.22)$$

The approximation is reasonable under the assumption that $kr^2 \ll 1$.

Note that among the parameters f_x, f_y and μ, only two are independent.

The parameters to be calibrated in this case are the *extrinsic* parameters of R and t, and the *intrinsic* parameters f_x, f_y (or μ), and k. Whenever all distortion effects other than radial lens distortion are zero, a *radial alignment constraint* (RAC) equation is maintained:

$$\frac{X}{Y} = \mu^{-1} \frac{r_1 x_w + r_2 y_w + r_3 z_w + t_x}{r_4 y_w + r_5 y_w + r_6 z_w + t_y} \qquad (2.2.23)$$

This equation can also be written as

$$X_d : Y_d = x : y \qquad (2.2.24)$$

where $X_d = f_x X$ and $Y_d = f_y Y$.

It is clear that the RAC is independent of the lens distortion coefficient k and the focal distance f. This can also be observed from Figure 2.2.1, in which P_u and P_d are respectively undistorted and distorted image points. The direction of the vector $O_I P_d$ is independent of the effective focal length f. As only radial lens distortion is assumed to exist, the direction of the vector $O_I P_d$ remains unchanged, and only the magnitude of the vector changes if the effective focal length changes. Furthermore, translation in the z direction does not alter the direction of $O_I P_d$.

III. TSAI'S RAC-BASED CAMERA CALIBRATION ALGORITHM

Recall that the camera calibration problem is to identify the set of extrinsic parameters (camera location and orientation in world coordinates) and intrinsic parameters (such as focal length, scale factors, distortion coefficients, etc.) of the camera using a set of points known both in world coordinates and image coordinates.

Let us initially assume that the image center coordinates (C_x, C_y) and the ratio of scale factors μ are known. Estimation methods for μ, C_x and C_y will be discussed in detail in later sections. The problem, addressed in this section, is the computing of the camera extrinsic parameters, i.e. the elements of R and t, and the modified intrinsic parameters f_x, f_y and k, based on a set of *coplanar* calibration points, the object coordinates of which in the world coordinate system are known, and the computer image coordinates of which are measured.

The camera calibration algorithm consists of two stages. In the first stage, the rotation matrix R, and the translational parameters t_x and t_y are computed. In the second stage, the remaining camera parameters are estimated using the results of the first stage.

The rotation matrix has only three independent parameters. If one attempts to represent the rotation matrix in terms of any three independent parameters, such as three Euler angles, the elements of the rotation matrix then become nonlinear functions of these parameters. Consequently, the RAC equation becomes a nonlinear relationship of these parameters. Nevertheless, an interesting discovery made by Tsai is that one can still define sets of *independent* intermediate parameters, compounded in terms of the physical camera parameters, for which the RAC equation can be manipulated into a set of linear equations in terms of these intermediate parameters. Because these intermediate parameters are independent, a unique solution under certain conditions, and the rotation and translation parameters can subsequently be recovered from these intermediate parameters, without the need to invoke measurement data again.

A.   STAGE 1: COMPUTATION OF THE ROTATION MATRIX R, AND THE TRANSLATION PARAMETERS t_x AND t_y

1.   Computation of   the image coordinates $(X_i,\ Y_i)$

Let N be the number of calibration points. Then for i = 1, 2, ..., N

$$X_i = X_{f,i} - C_x$$
$$Y_i = Y_{f,i} - C_y$$

where $X_{f,i}$ and $Y_{f,i}$ are the computer representation of the image coordinates for the ith calibration point.

2.   Computation of the intermediate parameters $\{v_1, v_2, v_3, v_4, v_5\}$

Since RAC is independent of the lens distortion coefficient k and the focal distance f, the rotation matrix R and the two translation components t_x and t_y may be solved from the RAC equation.
Define

$$\{v_1, v_2, v_3, v_4, v_5\} \equiv \{r_1 t_y^{-1},\ r_2 t_y^{-1},\ t_x t_y^{-1},\ r_4 t_y^{-1},\ r_5 t_y^{-1}\}$$

For the ith calibration point, dividing both sides of (2.2.23) (the RAC equation) by t_y and rearranging the resulting expression, yields

$$\begin{bmatrix} x_{w,i}Y_i & y_{w,i}Y_i & Y_i & -x_{w,i}\mu X_i & -y_{w,i}\mu X_i \end{bmatrix} \begin{bmatrix} v_1 \\ v_2 \\ v_3 \\ v_4 \\ v_5 \end{bmatrix} = \mu X_i \qquad (2.3.1)$$

where $x_{w,i}$ and $y_{w,i}$ are the x and y world coordinates of the ith calibration point. The minimum number of non-colinear calibration points needed to solve this equation is $N = 5$. Practically, N should be larger to create an over-determined system of linear equations for the unknowns $\{v_1, v_2, v_3, v_4, v_5\}$ that can then be solved by a linear least-squares algorithm.

Remark: If t_y is zero and t_x is nonzero, the equation can be rewritten in terms of t_x, rather than t_y. To avoid a situation in which both t_y and t_x are zero, the projection of the origin of the world coordinate system onto the image plane should be arranged to lie off the origin of the image coordinate system. In a fully automated camera calibration process, if the identification process happens to result in $t_x = t_y = 0$, the calibration procedure must be repeated after adjustment of the camera position to avoid such a singularity.

3. Computation of R, t_x and t_y

It is now possible to express the unknown parameter t_y in terms of the intermediate parameters v_1 to v_5 obtained in the previous step. After t_y is determined, t_x and some of the elements of the rotation matrix R can be directly computed from t_y as well as the intermediate parameters. The remaining elements of R can then be uniquely determined using the orthonormal condition of R. All these are done as follows:

Step 1. Define C to be a 2x2 matrix as follows

$$C \equiv \begin{bmatrix} v_1 & v_2 \\ v_4 & v_5 \end{bmatrix}$$

If no row or column of C identically vanishes, it can be shown that

$$t_y^2 = \frac{S_r - \sqrt{S_r^2 - 4(v_1 v_5 - v_4 v_2)^2}}{2(v_1 v_5 - v_4 v_2)^2}$$

where $S_r \equiv v_1^2 + v_2^2 + v_4^2 + v_5^2$. Otherwise, the solution is simply

$$t_y^2 = (v_i^2 + v_j^2)^{-1}$$

where v_i and v_j are the nonzero elements in the appropriate row or column of C.

Step 2. Determination of the sign of t_y.

Physically, the signs of x and X as well as y and Y should be consistent, which can be seen from the simple geometry conveyed by Figure 2.2.1. This condition can be utilized to determine the sign of t_y, denoted in short by sign(t_y). Starting with an assumption that $t_y > 0$, we first calculate the parameters

$$r_1 = v_1 t_y$$
$$r_2 = v_2 t_y$$
$$r_4 = v_4 t_y$$
$$r_5 = v_5 t_y$$
$$t_x = v_3 t_y$$

Picking up an arbitrary calibration point, we then compute the following coordinates,

$$x = r_1 x_w + r_2 y_w + t_x$$
$$y = r_4 x_w + r_5 y_w + t_y$$

If both sign(x) = sign(X) and sign(y) = sign(Y), then sign(t_y) = 1 is consistent, and r_1, r_2, r_4, r_5 and t_x are retained. Otherwise we let sign(t_y) = -1 and reverse the sign of r_1, r_2, r_4, r_5 and t_x accordingly.

Step 3. Computation of R

There are exactly two possible solutions for a rotation matrix given any of its 2x2 submatrices. These two sets of solutions produce two different values of f_x, one that has a negative sign and the other having a positive sign. Because the sign of f_x must be positive, one can use this condition to eliminate one of the solutions of R. The first attempt could be

$$r_3 = (1 - r_1^2 - r_2^2)^{1/2}$$
$$r_6 = - \text{sign}(r_1 r_4 + r_2 r_5)(1 - r_4^2 - r_5^2)^{1/2}$$
$$[r_7 \ r_8 \ r_9]^T = [r_1 \ r_2 \ r_3]^T \times [r_4 \ r_5 \ r_6]^T$$

where "x" denotes a vector cross product. If this results in a positive f_x, to be computed in Stage 2 using the solved rotation matrix, we retain this solution. Otherwise we reverse the sign of f_x, and modify the solution to complete the rotation matrix computation as follows:

$$r_3 = - (1 - r_1^2 - r_2^2)^{1/2}$$
$$r_6 = \text{sign}(r_1 r_4 + r_2 r_5)(1 - r_4^2 - r_5^2)^{1/2}$$

$$[r_7 \ r_8 \ r_9]^T = [r_1 \ r_2 \ r_3]^T \times [r_4 \ r_5 \ r_6]^T.$$

Remark: The resulting computed matrix R may not be orthonormal. It is then necessary to apply an orthonormalization procedure to R. References to available orthonormalization procedures are provided at the end of this chapter.

B. STAGE 2: COMPUTATION OF t_z, k, f_x AND f_y

Taking R, t_x and t_y to be known, one is able to estimate the remaining parameters, t_z, k, f_x and f_y. From (2.2.12), for the ith calibration point,

$$\begin{bmatrix} -X_i & x_i & -x_i r_i^2 \end{bmatrix} \begin{bmatrix} t_z \\ f_x \\ k f_x \end{bmatrix} = X_i w_i \tag{2.3.2}$$

where

$$x_i \equiv r_1 x_{w,i} + r_2 y_{w,i} + t_x$$
$$w_i \equiv r_7 x_{w,i} + r_8 y_{w,i}$$

and r^2 is given in (2.2.18). Whenever the number of object calibration points N is greater than three, an overdetermined system of linear equations is established that can then be solved by a linear least-squares algorithm for the unknowns k, t_z and f_x. After f_x is obtained, we compute the other intrinsic parameters,

$$f_y = f_x \, \mu$$
$$k = (k f_x) f_x^{-1}$$

IV. A FAST RAC-BASED ALGORITHM

The idea behind the simplification of Tsai's RAC algorithm is that during the first stage of camera calibration, if one selects only calibration points which lie along the x and y axes of the world coordinate system, the camera radial alignment constraint equation becomes greatly simplified.   More specifically, the middle row and middle column of the calibration points array of the camera calibration board define the world x_w and y_w axes. Tsai's original algorithm uses a five-variable linear least-squares estimation algorithm in stage one, and a three-variable linear least squares in stage two.   The simplified method described in this section computes three two-variable linear least-squares. Because a two variables linear least squares problem has a simple closed-form solution, a significant reduction of the camera calibration computation time results. As in Tsai's   method, for this method to be applicable, it is required that the ratio of camera scale factors μ and the image center be given *a priori*.

Similar to the original RAC-based calibration algorithm, the simplified algorithm also consists of two stages. In the first stage, only calibration points along the world x_w and y_w axes (refer to Figure 2.4.1) are used for camera parameter estimation. This allows a quick estimation of the rotation matrix R and two of the translation parameters t_x and t_y, due to the highly simplified structure of the RAC equation. In the second stage, the remaining parameters are computed, using the entire set of visible calibration points.

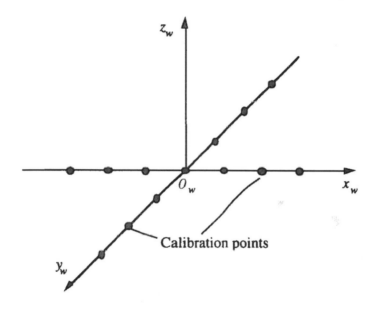

Figure 2.4.1 Calibration points for the first stage
of the fast RAC-based camera calibration method

A. STAGE 1: COMPUTATION OF THE ROTATION MATRIX R AND THE TRANSLATION PARAMETERS t_x AND t_y

The algorithm is valid only if t_x and t_y do not vanish simultaneously. Assuming that $t_y > 0$, we will still use the vector v defined in the last section. Utilizing the x_w axis points ($z_w = y_w = 0$) yields the RAC equation

$$Y(r_1 x_w + t_x) = \mu X(r_4 x_w + t_y)$$

and consequently the following equations in v_1 and v_4 can be obtained:

$$\begin{bmatrix} x_{w,i} Y_i & -\mu x_{w,i} X_i \end{bmatrix} \begin{bmatrix} v_1 \\ v_4 \end{bmatrix} = \mu (X_i - Y_i \frac{X_0}{Y_0};$$

which can be solved using a two-variable linear least squares procedure. Similarly, using the y_w axis points ($z_w = x_w = 0$) yields the RAC equation in v_2 and v_5,

$$\begin{bmatrix} y_{w,i} Y_i & -\mu y_{w,i} X_i \end{bmatrix} \begin{bmatrix} v_2 \\ v_5 \end{bmatrix} = \mu (X_i - Y_i \frac{X_0}{Y_0})$$

The variable v_3 is obtained directly from substituting $z_w = y_w = x_w = 0$ into the RAC equation,

$$v_3 = \mu \frac{X_0}{Y_0}$$

Note that the origin (X_0, Y_0) of the coordinate frame, rather than being obtained from measuring a single calibration point, may be obtained from intersecting the entire x_w and y_w axes to improve on the accuracy.

After v_1, v_2, ..., v_5 are obtained, the rotation matrix R and the translations t_x and t_y can be obtained using Tsai's procedure described earlier (Section III.A.)

B . STAGE 2:   COMPUTATION OF t_z, k, f_x AND f_y

For the estimation of the remaining parameters, t_z, k, f_x and f_y, calibration points from the entire calibration board can be used (refer to Figure 2.4.2). Interestingly, this phase of the camera calibration involves also a two-variable linear least squares solution.

By (2.2.16), the origin of the world coordinate system yields the relationship:

$$\frac{X_0}{1 - kr_0^2} = f_x \frac{t_x}{t_z} \qquad (2.4.1)$$

Substituting t_z of (2.4.1) and $z_w = 0$ into (2.2.21) yields, for each calibration point, the equation

$$X_i w_i \quad -X_i X_0^1 t_x \, r_0^2 + x_i \, r_i^2 \, \Big] \begin{bmatrix} f_x^1 \\ k \end{bmatrix} = x_i - X_i X_0^1 t_x$$

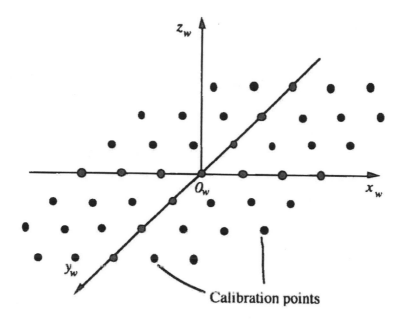

Calibration points

Figure 2.4.2 In the second stage,
calibration points are not restricted to lie on x_w and y_w

where

$$x_i = r_1 x_{w,i} + r_2 y_{w,i} + t_x$$
$$w_i = r_7 x_{w,i} + r_8 y_{w,i}$$

The intrinsic parameters k and f_x can be solved by a linear least-squares algorithm or by a closed form solution. After f_x is found, t_z can be solved from equation (2.4.1).

V. OPTICAL AXIS PERPENDICULAR TO THE CALIBRATION BOARD

The situation in which the camera optical axis is perpendicular (or almost perpendicular) to the calibration board is sometimes inevitable. This can happen, for instance, whenever a camera is mounted on the roll axis of a SCARA arm. In such a case, Tsai's calibration algorithm fails due to singularities, as is evident from the algorithm presented in Section III. Whenever the calibration board is parallel to the image plane, the intermediate parameter w_i in (2.3.2) becomes zero, resulting in a homogeneous equation. This results in inability to uniquely solve for the remaining unknown parameters. The same also happens in the fast RAC-based algorithm. In a practical setup, the camera axis may be nearly perpendicular to the calibration board, in which case Tsai's algorithm may become ill-conditioned. In this section, we present a calibration algorithm that deals with such near-singular camera configuration.

A. MODIFICATION OF THE CAMERA MODEL

In the case of the near singular camera configuration, the rotation matrix R may be represented by Euler angles in the following form,

$$R = Rot(z, \alpha)Rot(y, \beta)Rot(x, \gamma)$$

That is

$$R \equiv \begin{bmatrix} r_1 & r_2 & r_3 \\ r_4 & r_5 & r_6 \\ r_7 & r_8 & r_9 \end{bmatrix} = \begin{bmatrix} c\alpha c\beta & c\alpha s\beta s\gamma - s\alpha c\gamma & c\alpha s\beta c\gamma + s\alpha s\gamma \\ s\alpha c\beta & s\alpha s\beta s\gamma + c\alpha c\gamma & s\alpha s\beta c\gamma - c\alpha s\gamma \\ -s\beta & c\beta s\gamma & c\beta c\gamma \end{bmatrix} \tag{2.5.1}$$

where $c\beta \equiv cos\beta$, and similar short-hand-notations are used for other trigonometric functions. Since the sensor plane is nearly parallel to the calibration board, β and γ are very small. Thus, $cos\beta \cong 1$, $sin\beta \cong \beta$, $cos\gamma \cong 1$ and $sin\gamma \cong \gamma$. By substituting these approximations into (2.5.1) and neglecting the product term $\beta\gamma$, one obtains

$$\begin{bmatrix} r_1 & r_2 & r_3 \\ r_4 & r_5 & r_6 \\ r_7 & r_8 & r_9 \end{bmatrix} \cong \begin{bmatrix} c\alpha & -s\alpha & c\alpha\beta + s\alpha\gamma \\ s\alpha & c\alpha & s\alpha\beta - c\alpha\gamma \\ -\beta & \gamma & 1 \end{bmatrix} \tag{2.5.2}$$

Consider now the case that all calibration data are obtained from a single planar calibration board, where z_w is a constant. Without any loss of generality, one can assume that $z_w = 0$. Substituting the elements of R, given in (2.5.2), into (2.2.21) and (2.2.22), under the condition $z_w = 0$, yields

$$\frac{X}{1 - kr^2} = f_x \frac{c\alpha x_w - s\alpha y_w + t_x}{-\beta x_w + \gamma y_w + t_z} \tag{2.5.3}$$

$$\frac{Y}{1 - kr^2} = f_y \frac{s\alpha x_w + c\alpha y_w + t_y}{-\beta x_w + \gamma y_w + t_z} \tag{2.5.4}$$

These equations define a camera model that is valid under the condition that the camera plane is nearly parallel to the calibration board. It is assumed that the intrinsic parameters μ and f_x (and f_y) are pre-calibrated. The parameters to be calibrated are the six independent extrinsic parameters of R and t as well as the distortion coefficient k.

Whenever the planes are exactly parallel, that is $\beta = \gamma = 0$, the above camera model can be further simplified. It should be stressed that in such a case, f_x (or f_y) can never be recovered together with t_z since among these three parameters, only two are independent.

Note that the matrix on the right-side of (2.5.2) is not necessarily an orthonormal matrix. One way to remedy this is to use (2.5.1) instead of (2.5.2) to compute the rotation matrix R once α, β, and γ are calibrated.

B. A CALIBRATION ALGORITHM

This linear algorithm, like the original Tsai's algorithm, consists of two stages. In the first stage, the parameters α, t_x and t_y are determined. The remaining unknown parameters are then computed in the second stage.

Stage 1. Computation of α, t_x and t_y

The algorithm can be used only if t_x and t_y do not vanish simultaneously. Assume that $t_y \neq 0$. By dividing (2.5.3) by (2.5.4) and with some additional manipulation, one obtains

$$x_w Y c \alpha t_y^{-1} - y_w Y s \alpha t_y^{-1} + Y t_x t_y^{-1} - x_w X \mu s \alpha t_y^{-1} - y_w X \mu c \alpha t_y^{-1} - X \mu = 0$$
(2.5.5)

By defining the vector v as

$$v \equiv [v_1 \quad v_2 \quad v_3]^T \equiv [c \alpha t_y^{-1} \quad s \alpha t_y^{-1} \quad t_x t_y^{-1}]^T$$

Equation (2.5.5) can be rewritten as follows

$$(x_w Y - y_w X \mu)v_1 - (y_w Y + x_w X \mu)v_2 + Y v_3 = X \mu \qquad (2.5.6)$$

By taking a sufficient number of calibration points, one can solve for v using a three-variable linear least squares procedure. Note that a system of equations of the form (2.5.6) has a unique solution, as long as these calibration points are non-colinear.

After v is computed, α is obtained

$$\alpha = A tan2(v_1, v_2).$$

Because v_1 and v_2 cannot simultaneously vanish, one can determine t_y in the following way. If v_2 is nonzero,

$$t_y = sin\alpha/v_2,$$

otherwise,

$$t_y = cos\alpha/v_1.$$

Finally,

$$t_x = v_3 t_y.$$

Stage 2.   Computation of β, γ, and t_z

From (2.5.3) and (2.5.4),

$$- Xx_w\beta + Xy_w\gamma + Xt_z = f_x(1 - kr^2)(c\alpha x_w - s\alpha y_w + t_x)$$
$$- Yx_w\beta + Yy_w\gamma + Yt_z = f_y(1 - kr^2)(s\alpha x_w + c\alpha y_w + t_y).$$

By rearranging the above equations,

$$- Xx_w\beta + Xy_w\gamma + Xt_z + r^2pf_xk = p \qquad (2.5.7)$$

$$- Yx_w\beta + Yy_w\gamma + Yt_z + r^2qf_yk = q \qquad (2.5.8)$$

where

$$p = c\alpha x_w - s\alpha y_w + t_x$$
$$q = s\alpha x_w + c\alpha y_w + t_y.$$

The unknown parameters β, γ, t_z and k can be solved by a linear least-squares algorithm if a sufficient number of calibration points is available. Again a system of equations of the form (2.5.7) and (2.5.8) has a unique solution as long as these calibration points are non-colinear.

VI. NONLINEAR LEAST-SQUARES APPROACH

The cost function associated with a camera calibration process is in general non-convex. A way of enhancing convergence to a global optimum is to employ a quick and simple suboptimal algorithm, just for the sake of creating initial conditions in the neighborhood of the optimal solution. This is a key idea behind the nonlinear algorithm discussed in this section that was originally proposed by Weng et. al.

According to this method, the parameters of the camera model are identified in two phases. In the first preparatory phase, a linear procedure is devised to find the nominal values of the unknown parameters for the second phase. These nominal parameters are then used as initial conditions for the nonlinear least squares algorithm that finds the globally optimal set of parameters of the camera model.

A. A LINEAR ESTIMATION PROCEDURE

The linear solution presented next determines the rotation matrix R, the translation vector t, the image center (C_x, C_y), and the scale factors f_x and f_y. Note that the image center is indeed solved for in this procedure. However, the radial distortion parameter k is assumed zero.

For each visible calibration point, two equations are established from the camera model,

$$(r_7 x_w + r_8 y_w + r_9 z_w)(X_f - C_x) - f_x\,(r_1 x_w + r_2 y_w + r_3 z_w) = 0 \quad (2.6.1)$$
$$(r_7 x_w + r_8 y_w + r_9 z_w)(Y_f - C_y) - f_y\,(r_4 x_w + r_5 y_w + r_6 z_w) = 0 \quad (2.6.2)$$

For the ten unknown parameters, at least five calibration points are required to provide ten nonlinear equations of the form (2.6.1) and (2.6.2).

In order to derive a closed-form solution for the camera parameters, the following intermediate parameters are defined:

$$\eta_1 = f_x R_1 + C_x R_3$$
$$\eta_2 = f_y R_2 + C_y R_3$$
$$\eta_3 = R_3$$
$$\eta_4 = f_x t_x + C_x t_z$$
$$\eta_5 = f_y t_y + C_y t_z$$
$$\eta_6 = t_z$$

where column vectors R_1, R_2 and R_3 are respectively the transposed first, second, and third rows of the rotation matrix R. Thus η_1, η_2 and η_3 are 3X1 vectors and η_4, η_5 and η_6 are scalars. The set of equations in the form of (2.6.1) and (2.6.2) provided by n calibration points can then be expressed in a matrix form

$$A\eta = 0 \qquad\qquad (2.6.3)$$

where A is a $(2n)$X12 matrix shown below,

$$A = \begin{bmatrix} -x_{w,1} & -y_{w,1} & -z_{w,1} & 0 & 0 & 0 & X_{f,1}x_{w,1} & X_{f,1}y_{w,1} & X_{f,1}z_{w,1} & -1 & 0 & X_{f,1} \\ 0 & 0 & 0 & -x_{w,1} & -y_{w,1} & -z_{w,1} & Y_{f,1}x_{w,1} & Y_{f,1}y_{w,1} & Y_{f,1}z_{w,1} & 0 & -1 & Y_{f,1} \\ \vdots & \vdots & \vdots & \vdots & \vdots & \vdots & \vdots & \vdots & \vdots & \vdots & \vdots & \vdots \\ \vdots & \vdots & \vdots & \vdots & \vdots & \vdots & \vdots & \vdots & \vdots & \vdots & \vdots & \vdots \\ -x_{w,n} & -y_{w,n} & -z_{w,n} & 0 & 0 & 0 & X_{f,n}x_{w,n} & X_{f,n}y_{w,n} & X_{f,n}z_{w,n} & -1 & 0 & X_{f,n} \\ 0 & 0 & 0 & -x_{w,n} & -y_{w,n} & -z_{w,n} & Y_{f,n}x_{w,n} & Y_{f,n}y_{w,n} & Y_{f,n}z_{w,n} & 0 & -1 & Y_{f,n} \end{bmatrix}$$

and η is the 12X1 vector of all unknown intermediate parameters:

$$\eta = \begin{bmatrix} \eta_1 \\ \eta_2 \\ \eta_3 \\ \eta_4 \\ \eta_5 \\ \eta_6 \end{bmatrix}.$$

As mentioned earlier, in conjunction with the Perspective Transformation Matrix method, this system of linear homogeneous equations (2.6.3) does not have a unique solution. Among the multiple solutions, the one that corresponds to the "true" camera parameters must satisfy the following two conditions:

a. The norm of vector η_3 must be unity since the vector is the transpose of the last row of the rotation matrix R.
b. The sign of η_6 must be compatible with the position of the camera in the world coordinate system. Namely, η_6 must be positive if the camera is in front of the (x,y) plane of the world coordinate system, and should be negative if the camera is behind the plane.

Because the world coordinate system can be defined such that t_z is either positive or negative, one way to solve for η is to impose a temporary constraint

$$\eta_6 = \text{sign}(t_z) = 1.$$

and transform (2.6.3) into a nonhomogeneous linear system,

$$A^1 \eta^1 + a^1 = 0$$

where A^1 denotes the matrix consisting of the first 11 columns of A, a^1 denotes the last column of A, and η^1 represents the corresponding reduced-order unknown vector. In the presence of noise, the system of linear equations can be solved using the least squares algorithm.

After η^1 is obtained, a scaled version of the intermediate parameter vector, which satisfies both conditions (a) and (b), is determined as follows:

$$
s = \begin{bmatrix} s_1 \\ s_2 \\ s_3 \\ s_4 \\ s_5 \\ s_6 \end{bmatrix} = \pm \frac{1}{\|\eta_3\|} \begin{bmatrix} \eta_1 \\ \eta_2 \\ \eta_3 \\ \eta_4 \\ \eta_5 \\ 1 \end{bmatrix}
$$

where the sign is chosen to satisfy condition (b) mentioned above.

After s is computed, the initial estimate of the camera parameters can be solved from the following equations,

$$
C_x = s_1{}^T s_3
$$
$$
C_y = s_2{}^T s_3
$$
$$
f_x = - \|s_1 - C_x s_3\|
$$
$$
f_y = \|s_2 - C_x s_3\|
$$
$$
t_x = (s_4 - C_x s_6)/f_x
$$
$$
t_y = (s_5 - C_y s_6)/f_y
$$
$$
t_z = s_3
$$
$$
R_1 = (s_1 - C_x s_3)/f_x
$$
$$
R_2 = (s_2 - C_y s_3)/f_v
$$
$$
R_3 = \eta_3.
$$

The rotation matrix solution R may not be orthonormal. Therefore an orthonormalization procedure may be applied to improve on the accuracy of the estimated parameters. After the new rotation matrix is computed, other parameters can be recomputed using the above equations, with s_3 replaced by R_3. However, it should be noted that the orthonormalization process, in this approach, is not as critical as in the RAC-based camera calibration approach, since the linear solution given in this section only provides initial conditions for the nonlinear least squares solution outlined in the next section.

B. A NONLINEAR ESTIMATION PROCEDURE

To be able to apply any nonlinear least squares algorithm to the problem of camera calibration, a cost function has to be formed. Using the model that includes the image center coordinates and the radial lens distortion coefficient, we define the residuals:

$$F_{x,i} \equiv \frac{X_{f,i} - C_x}{1 - kr_i^2} - f_x \frac{r_1 x_{w,i} + r_2 y_{w,i} + r_3 z_{w,i} + t_x}{r_7 x_{w,i} + r_8 y_{w,i} + r_9 z_{w,i} + t_z}$$

$$F_{y,i} \equiv \frac{Y_{f,i} - C_y}{1 - kr_i^2} - f_y \frac{r_4 x_{w,i} + r_5 y_{w,i} + r_6 z_{w,i} + t_y}{r_7 x_{w,i} + r_8 y_{w,i} + r_9 z_{w,i} + t_z}$$

Ideally, $F_{x,i}$ and $Fy_{x,i}$ are both zero. Due to uncertainties or errors in the system modeling and calibration measurements, these entities are in general nonzero. The objective is then to minimize the least squares error

$$F(R, t, C_x, C_y, f_x, f_y, k) = \sum_{i=1}^{m} \left(F_{x,i}^2 + F_{y,i}^2 \right) \qquad (2.6.4)$$

by choosing both the intrinsic and extrinsic parameters that are listed as variables in F, where m is the number of image points used for calibration. If some of the parameters need not be estimated, these can then be set to their initial values. Note that proper weights may be incorporated into the cost function (2.6.4).

Whenever any nonlinear algorithm is employed, convergence and stability of the algorithm are always issues. An algorithm suggested by Weng that solves for the distortion parameters and other camera parameters alternately in each iteration can be implemented as follows: In the first stage, all the distortion parameters (in the model presented here, there is only one such parameter k) are kept fixed, and other parameters are updated using a linearized error model. In the second stage, a system of linear equations is used to solve for the distortion parameters, assuming that all the other parameters are constant. This process terminates whenever no improvement is made on the estimates.

Some of the implementation details of this strategy can be modified. For instance, the image center parameters can form one group and the remaining parameters can form the other group. These two groups of parameters can be solved for alternately, in which case the image center can be determined linearly in each iteration. It should be pointed out that if image center parameters are assumed known, the problem is reduced to a standard nonlinear least squares and any appropriate nonlinear least squares procedure can be applied to solve it, using the initial conditions provided by the linear procedure presented in Section VI.A.

Another remark is that the calibration board must be placed in such a way that the image plane is not parallel to the calibration board; otherwise the algorithm becomes unstable.

VII. ESTIMATION OF
THE RATIO OF SCALE FACTORS

The scale factors f_x and f_y are fixed for a particular camera/vision system. Therefore μ, the ratio of f_y and f_x, needs to be identified only once as long as the same camera/vision system is used. To estimate μ, one may need to use calibration points taken from different planes. The method presented in this section requires that the calibration board be moved parallel to itself along a rail, but not necessarily in the direction of the board plane normal.

By moving the calibration board to two parallel locations, a pair of constraint equations in terms of the ratio of scale factors μ, a translational parameter and two surface parameters of the calibration board can be derived from the RAC equation. The parameters μ cannot be uniquely determined from this pair of equations. However by utilizing the property that μ does not change as the camera moves to more locations, more equations can be generated. More specifically, if the camera is moved to three random locations while keeping the calibration board fully visible, six constraint equations are derived in terms of six unknowns, one of which is the ratio of scale factors. A nonlinear least squares algorithm is then applied to estimate these unknown parameters. Two alternative schemes are discussed next.

A. SINGLE CAMERA METHOD
The scheme requires that the camera be moved to three random locations while keeping the calibration points within its field of view, and that for each camera position, the calibration board be moved known distances along a rail to two parallel locations. This measurement procedure produces six constraint equations in terms of six unknowns, among which are the ratio of scale factors, the direction vector of the board motion (two variables) and three values of t_y, which is a camera translation parameter. Interestingly, among all camera parameters, only t_y needs to be explicitly solved for. The estimation procedure is similar to the ones described earlier with the exception that μ is treated here as an unknown.

Suppose that the camera is at the first position and the calibration board is also at its first position. Let

$$\{v_1, v_2, v_3, v_4, v_5\} \equiv \{r_1 \mu^{-1} t_y^{-1}, r_2 \mu^{-1} t_y^{-1}, t_x \mu^{-1} t_y^{-1}, r_4 t_y^{-1}, r_5 t_y^{-1}\}$$

where t_y corresponds to the first camera position and the first board position. The projection of the origin of the world coordinate system should be set away from the image center, to assure that $t_y \neq 0$. The intermediate variables v_i, for $i = 1, 2, ..., 5$, can be computed by the technique of Section III.A.

The calibration board is now moved by a known distance s to the second location along the rail, while the camera is kept at its original place. For this camera and board location, define

$$\{v_1', v_2', v_3', v_4', v_5'\} \equiv \{r_1 \mu^{-1} t_y'^{-1}, r_2 \mu^{-1} t_y'^{-1}, t_x' \mu^{-1} t_y'^{-1}, r_4 t_y'^{-1}, r_5 t_y'^{-1}\}$$

Note that only the values of t_x and t_y change. The following derivation is for the solution of $\{v_1', v_2', v_3', v_4', v_5'\}$.

Denote by $n \equiv [n_x, n_y, n_z]^T$ the unit direction vector of the rail. Let $\{x_{w0}, y_{w0}, 0\}$ be the location of a calibration point in $\{x_w, y_w, z_w\}$ when the calibration board is at its first position, and $\{x_{w1}, y_{w1}, z_{w1}\}$ be the coordinates of that point after the calibration board is shifted to its second location. Then

$$x_{w1} = x_{w0} + sn_x$$
$$y_{w1} = y_{w0} + sn_y$$
$$z_{w1} = sn_z$$

By the camera model given in (2.2.16) and (2.2.17),

$$\mu \frac{X}{Y} = \frac{r_1 x_{w1} + r_2 y_{w1} + r_3 z_{w1} + t_x}{r_4 x_{w1} + r_5 y_{w1} + r_6 z_{w1} + t_y}$$

$$= \frac{r_1 x_{w0} + r_2 y_{w0} + s(r_1 n_x + r_2 n_y + r_3 n_z) + t_x}{r_4 x_{w0} + r_5 y_{w0} + s(r_4 n_x + r_5 n_y + r_6 n_z) + t_y} \tag{2.6.1}$$

Define

$$t_x' \equiv s(r_1 n_x + r_2 n_y + r_3 n_z) + t_x \tag{2.6.2}$$
$$t_y' \equiv s(r_4 n_x + r_5 n_y + r_6 n_z) + t_y \tag{2.6.3}$$

Equation (2.6.1) can be arranged in a similar form to that of (2.3.1). The vector $\{v_1', v_2', v_3', v_4', v_5'\}$ is then computed using the algorithm given in Section III.A.

The next step is to determine μ and the unity normal components to the calibration board. Define

$$\{v_6, v_7\} \equiv \{r_3\, \mu^{-1} t_y^{-1}, r_6 t_y^{-1}\}$$
$$\{v_6', v_7'\} \equiv \{r_3\, \mu^{-1} t_y'^{-1}, r_6 t_y'^{-1}\}$$

Dividing both sides of (2.6.2) by μt_y and both sides of (2.6.3) by t_y yields

$$v_3'\, v_1/v_1' = s(v_1 n_x + v_2 n_y + v_6 n_z) + v_3$$
$$v_1/v_1' = s(v_4 n_x + v_5 n_y + v_7 n_z) + 1$$

By rearranging the above equations, squaring both sides, and using the relationships $v_6^2 = \mu^2 t_y^{-2} - v_1^2 - v_2^2$ and $v_7^2 = t_y^{-2} - v_4^2 - v_5^2$, one obtains

$$\left[1 - \left(v_1^2 + v_2^2\right)(\mu t_y)^2\right]\left(1 - n_x^2 - n_y^2\right) = \left(\frac{v_3' v_1}{s v_1'} - v_1 n_x - v_2 n_y - \frac{v_3}{s}\right)^2 (\mu t_y)^2$$

(2.6.5)

$$\left[1 - \left(v_4^2 + v_5^2\right)t_y^2\right]\left(1 - n_x^2 - n_y^2\right) = \left(\frac{v_1}{s v_1'} - v_4 n_x - v_5 n_y - \frac{1}{s}\right)^2 t_y^2$$

(2.6.6)

Equations (2.6.5) and (2.6.6) are the constraint equations used to solve for the ratio of scale factors and the unit normal components. There are two equations with four unknowns μ, t_y, n_x and n_y. If the camera remains at one position, these unknowns cannot be solved uniquely. However since among μ, t_y, n_x and n_y, only t_y changes as the camera moves, more constraint equations than new unknowns, for solution of μ, n_x and n_y can be obtained by adding multiple camera positions. By placing the camera at three positions 6 constraint equations are obtained in terms of 6 unknowns μ, n_x, n_y, $t_{y,1}$, $t_{y,2}$ and $t_{y,3}$. These unknowns are then determined using a nonlinear least squares algorithm such as the Gauss-Newton algorithm. The solutions are not unique when system errors (that is, uncertainties of calibration point locations in both world and image coordinate systems) exist. The algorithm can tolerate a crude determination of the initial value n_0. Other initial values can be derived using n_0. If the initial values of the other unknowns are sufficiently close to the true values, the algorithm converges to the optimal point very rapidly. The following example demonstrates the selection of initial values for the unknowns.

Example: Whenever the normal vector is almost parallel to the z axis of $\{x_w, y_w, z_w\}$, n_0 can be chosen as $[0, 0, 1]^T$. Substituting n_0 into (2.6.5) and (2.6.6) yields

$$\frac{1}{(\mu t_y)^2} - v_1^2 - v_2^2 = \left(\frac{v_3' v_1}{s v_1'} - \frac{v_3}{s}\right)^2$$

$$t_y^{-2} - v_4^2 - \frac{2}{3} = \left(\frac{v_1}{s v_1'} - \frac{1}{s}\right)^2$$

Solving for μ from the above equations yields

$$\mu = \sqrt{\frac{s^{-2}\left(\frac{v_1}{v_1'} - 1\right)^2 + v_4^2 + v_5^2}{s^{-2}\left(v_1 \cdot \frac{v_1}{v_1'} - 1\right)^2 + v_1^2 + v_2^2}} \tag{2.6.6}$$

and

$$t_y = \frac{1}{\sqrt{s^{-2}\left(\frac{v_1}{v_1'} - 1\right)^2 + v_4^2 + v_3^2}} \tag{2.6.7}$$

The sign of t_y should be determined based on the particular physical setup of the system.   Assuming that the camera is moved to three locations, three values of μ can be calculated from (2.6.6).   Then parameter μ_0, the initial value of μ, can be chosen as the average of these values.   Parameters $t_{y,10}$, $t_{y,20}$ and $t_{y,30}$ the initial values of $t_{y,1}$, $t_{y,2}$ and $t_{y,3}$, respectively, are solved from (2.6.7).

B. STEREO CAMERAS METHOD
Whenever the camera/vision system consists of two cameras, it is possible to determine the two ratios of scale factors along with the unit normal of the rail.   This time the calibration board is moved a distance s along the direction n.   Each camera needs to be placed at only two different locations.   Similar to the discussion given in the last section, 8 constraints equations are derived with 8 unknowns which are n_x, n_y, and two sets of $\{\mu,$ $t_{y,1}, t_{y,2}\}$.   Each set of unknowns μ, $t_{y,1}$ and $t_{y,2}$ is related to a single camera.   These 8 unknowns are to be solved from the constraints equations using a nonlinear least squares algorithm.

This method requires fewer motions along the rails, compared to the monocular method.   This reduces errors due to uncertainties in the displacements along the rail.

VIII. ESTIMATION OF THE IMAGE CENTER

The Radial Alignment Constraint equation is affected by the values of the image center coordinates C_x and C_y. Under the assumption that only the radial lens distortion exists, the RAC is satisfied when the image center is accurate. Otherwise a residual exists that may be used for least squares identification of the image center. It may appear natural to minimize these residuals by selection of proper image center coordinates through a gradient-based search algorithm. Unfortunately the RAC is highly nonlinear in terms of the image center coordinates, and as such there may exist multiple local minima. Any suitable direct search method will do better in this case. One such method known as *the Polytope method* (also termed *the Simplex method*) is developed in detail in this section. The algorithm does not use any gradient information.

Let a cost function F be the sum of squares of the RAC residuals for all calibration points described by equation (2.6.4). The estimation task is again divided into two stages. In the first stage, an initial value of the image center is assumed given. The camera model is then computed using the method described in Section III. In the second stage, the solved parameters in the camera model are substituted into the cost function. The Polytope algorithm is then applied to obtain a better image center estimate. The above two stages are iteratively applied until the difference of two consecutively estimated image center values is below a prescribed threshold. The polytope algorithm specialized to the estimation of the image center is explained next.

Let $x \equiv [C_x, C_y]^T$. At each stage of the algorithm, three points x_h, x_g and x_l are retained, together with the cost function values at these points, which are ordered so that $F_h > F_g > F_l$. The three points, referred to as a "polytope", are chosen to be non-colinear (refer to Figure 2.8.1). At each iteration, a new polytope is generated by introducing a new point to replace the "worst" point x_h.

The algorithm starts with a random selection of 3 non-colinear points and it thereafter consists of the following steps:

1. **Calculation of cost function values**
   In this step, the best point x_l and worst point x_h are established.

2. **Determination of a reflection point**
   The reflection point x_r is calculated by

$$x_r = x_l + x_g - x_h$$

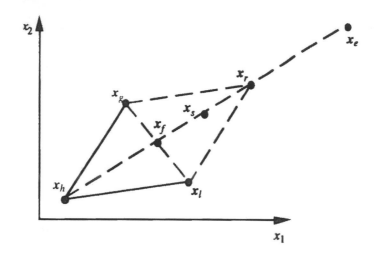

Figure 2.8.1 Illustration of the Polytope algorithm

The cost function F_r is evaluated. If $F_r \geq F_h$, the algorithm goes to Step 3, otherwise to Step 4.

3. **Step contraction**
    Let

$$x_s = (1 - \lambda)x_h + \lambda x_r$$

where $0 < \lambda < 1$, and $\lambda \neq 0.5$. The cost F_s is calculated. If $F_s < F_g$, we let $x_h = x_s$ and $F_h = F_s$. The algorithm returns to Step 1. If $F_s \geq F_g$, it goes to Step 5.

4. **Step expansion**
    Let

$$x_e = (1 - \rho)x_h + \rho x_r$$

where $\rho > 1$. F_e is calculated. If $F_e < F_r$ we let $x_h = x_e$ and $F_h = F_e$, otherwise let $x_h = x_r$ and $F_h = F_r$. The algorithm returns to Step 1.

5. **Polytope contraction**

The algorithm substitutes $(x_i + x_l)/2$ for x_i $i = g$, h; and then returns to Step 2.

The above procedure is repeated until $(F_h - F_l)^2 < \epsilon$, where ϵ is a sufficiently small number.

Remarks:
1. Let x_0 be an initial point. The other two initial points can be chosen by

$$x_1 = x_0 + he_x, \quad x_2 = x_0 + he_y$$

where $e_x = [1, 0]^T$ and $e_y = [0, 1]^T$, and h is the initial step length. The size of h has a significant impact on the estimation results. For the image center problem, we empirically found that a good choice for h is 5-10 pixels.
2. The contraction coefficient λ can be taken to be about 0.25.
3. The expansion coefficient ρ can be taken to be about 2. The estimation results are not very sensitive to the choices of λ and ρ. Simulation results of image center estimation are presented in Section X.A.

IX. PERSPECTIVE PROJECTION DISTORTION OF CIRCULAR CALIBRATION POINTS

Camera calibration requires the computation of the projection of the center of each calibration point onto the image plane. This center projection is not directly measurable. A common practice is to compute the centroid of the image pattern as an estimate of the center projection. Many of the calibration methods, discussed earlier, require that the calibration board be non-perpendicular to the camera optical axis. That is, that the camera must be tilted with respect to the calibration board. Consequently all circular calibration points appear as ellipses in the image. This section analyzes the distortion effect due to this phenomenon.

A. A SPECIAL CASE – ONE DIMENSIONAL DISTORTION
As is often true in practice, the uncertainties in the x and y image coordinates, corresponding to the center of a circular object point, are not the same. By careful arrangement of the camera measurement system, distortion effects in one of the directions can be made practically negligible.

It is assumed in this section that the tilt angle between the sensor plane and the calibration board can be described in terms of a single angle of rotation about the y axis of the camera coordinate system. It is further assumed that the rotations about the x and z axes of the camera coordinate system are negligible. Later in this section a quantitative analysis reveals to what extent these assumptions are valid. The set up of most single and stereo camera systems can in practice be modeled in this way as a good approximation.

By the assumptions made, there is no image distortion in the y direction. The binary image pattern of an individual object point is thus symmetrical about a line parallel to the x axis of the image coordinate system.

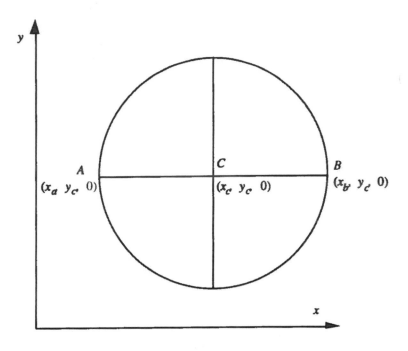

Figure 2.9.1 Definition of world coordinates for the special case of a one-dimensional perspective projection distortion

Let the calibration board be placed on the plane $z_w = 0$. Consider now a particular circular object point on this board. The world coordinates of its center are $(x_{w,c}, y_{w,c}, 0)$. Denote the projected image coordinates of this point by (X_c, Y_c). A line is defined so that it is parallel to the x_w axis and passes through the center of the calibration point. The line intersects with the circle at Points A and B. The world coordinates of these two points are denoted as $(x_{w,a}, y_{w,c}, 0)$ and $(x_{w,b}, y_{w,c}, 0)$. Let the image coordinates of these points be (X_a, Y_a) and (X_b, Y_b) such that $X_a < X_c < X_b$ (refer to Figure 2.9.1).

Let θ denote the camera rotation angle about the y axis of the camera coordinate system. Thus the camera rotation matrix R of Equation (2.2.1) becomes

$$R = \begin{bmatrix} c\theta & 0 & s\theta \\ 0 & 1 & 0 \\ -s\theta & 0 & c\theta \end{bmatrix}$$

where $s\theta \equiv sin\theta$, and $c\theta \equiv cos\theta$. With this rotation matrix being substituted into the camera model (2.2.16) and (2.2.17), one obtains,

$$\frac{X_a}{1 - k\left(\mu^2 X_a^2 + Y_c^2\right)} = f_x \frac{c\theta x_{w,a} + t_x}{-s\theta x_{w,a} + t_z} \tag{2.9.1}$$

$$\frac{X_b}{1 - k\left(\mu^2 X_b^2 + Y_c^2\right)} = f_x \frac{c\theta x_{w,b} + t_x}{-s\theta x_{w,b} + t_z} \tag{2.9.2}$$

$$\frac{X_c}{1 - k\left(\mu^2 X_c^2 + Y_c^2\right)} = f_x \frac{c\theta \dfrac{x_{w,a} + x_{w,b}}{2} + t_x}{-s\theta \dfrac{x_{w,a} + x_{w,b}}{2} + t_z} \tag{2.9.3}$$

Define the *eccentricity coefficient* of the image pattern of a calibration point, along the x direction of the world coordinate system, as

$$\gamma_x \equiv \frac{X_c - X_a}{X_b - X_c} \tag{2.9.4}$$

Later in this section, whenever no confusion arises, the subscript x of the eccentricity coefficient γ will be omitted. Substitution of (2.9.1)-(2.9.3) into (2.9.4) yields

$$\gamma = \frac{1 - s\theta \dfrac{x_{w,b}}{t_z}}{1 - s\theta \dfrac{x_{w,a}}{t_z}} \tag{2.9.5}$$

where use has been made of the approximation

$$1 - k\left(\mu^2 X_a^2 + Y_c^2\right) \cong 1 - k\left(\mu^2 X_b^2 + Y_c^2\right) \cong 1 - k\left(\mu^2 X_c^2 + Y_c^2\right)$$

which is justifiable in practice since

$$1 \gg 2 k \mu^2 X_a (X_b - X_a)$$

By doing this, we have ignored the effect of lens distortion on the image coordinate computation of the object point.

The x component of the image coordinates, corresponding to the object point center, is

$$X_c = X_a + \frac{\gamma}{1+\gamma}(X_b - X_a) \qquad (2.9.6)$$

The estimation error E_I of the calibration point image coordinates, due to this type of distortion, is therefore

$$E_I = (X_b - X_a)\left|\frac{\gamma}{1+\gamma} - 0.5\right|$$

$$= 0.5(X_b - X_a)\left|\frac{\gamma-1}{\gamma+1}\right|$$

The equivalent 3D error E_w for the center of the object point in the world coordinate system is then

$$E_w = r\left|\frac{\gamma-1}{\gamma+1}\right| \qquad (2.9.7)$$

where r is the radius of the calibration point.

From (2.9.5) and (2.9.7), errors due to this type of distortions depend on the distance between the camera sensor plane and the object point, the angle between the sensor plane and the calibration board, the position of the calibration points, and the size of the object point. This kind of distortion can be ignored if the tilt angle θ is very small, the distance t_z is very large, or the size of the calibration point is very small.

Unfortunately, very small values of θ introduce relatively large numerical errors in the RAC camera calibration algorithm. An increase of t_z expands the camera field of view and in turn reduces the resolution. This affects the absolute accuracy of the camera measurement system. Finally the shrinking of the object point increases the effect of quantization noise for the image coordinate estimation of the object point.

As also seen from (2.9.5) and (2.9.7), the uncertainty in the image coordinate estimation of object points due to perspective projection is independent of the camera focal length and the scale factors. Its dependence on the lens distortion coefficient is negligible. Therefore distortions due to the perspective projection can be corrected separately. The following examples illustrate the magnitude of the perspective distortion effect in several situations.

Example 1: Let $t_z = 150$ mm, $r = 1$ mm, and $\theta = -30$ degrees. Also let $x_{w,a} = -x_{w,b} = -1$ mm. By (2.9.5), $\gamma = 1.0067$, and by (2.9.7), $E_w = 0.0066$ mm.

Example 2: Let $x_{w,a} = 39$ mm, $x_{w,b} = 41$ mm, and $r = 1$ mm. Other parameters are the same as those in Example 1. In this case, $\gamma = 1.0059$, and $E_w = 0.0058$ mm.

The given parameters in the first two examples are typical of commercial camera systems. Thus the eccentricity of the projected patterns of calibration points, if not corrected, may introduce errors in image coordinate estimation.

Example 3: Let $t_z = 500$ mm, $r = 25$ mm, and $\theta = -30$ degrees. Also let $x_{w,a} = -x_{w,b} = -50$ mm. Then $\gamma = 1.0513$, and $E_w = 1.25$ mm.

This example illustrates that if the image pattern of the calibration point is large, proper measures have to be taken to compensate for the errors.

Whenever (2.9.5) is applied to camera calibration, a remaining problem is that prior to calibration, the parameters θ and t_z are unknown. To overcome the problem, let us consider the eccentricity coefficient error, dy due to $d\theta$ and dt_z, where $d\theta$ and dt_z are respectively the errors of θ and t_z. By (2.9.5), the total derivative of γ with respect to $d\theta$ and dt_z is

$$d\gamma = \frac{x_{w,a} - x_{w,b}}{t_z\left(1 - s\theta \frac{x_{w,a}}{t_z}\right)^2}\left(c\theta\, d\theta + \frac{s\theta}{t_z}dt_z\right)$$

An upper bound for $|dy|$ is therefore

$$|d\gamma| = \frac{x_{w,a} - x_{w,b}}{\left|t_z\left(1 - s\theta \frac{x_{w,a}}{t_z}\right)^2\right|}\left(\left|c\theta\, d\theta\right| + \left|\frac{s\theta}{t_z}dt_z\right|\right) \tag{2.9.8}$$

By (2.9.7), the bound on the error variation $|dE_w|$ due to dy is obtained:

$$|dE_w| \le \frac{2r|d\gamma|}{(\gamma + 1)^2} \tag{2.9.9}$$

Example 4: Assume that $d\theta = 3$ degrees, and $dt_z = 5$ mm. Other parameters are kept as in Example 1. By (2.9.8), $dy \le 0.0008$, and by (2.9.9), $dE_w \le 0.0008$ mm.

Example 4 suggests that small uncertainties in camera positions do not contribute significant errors if the image pattern if the point is small.

Example 5: Assume that $d\theta = 3$ degrees, and $dt_z = 5$ mm. Other parameters are given in Example 3. In this case, $dy \le 0.01$ and $dE_w \le 0.125$ mm.

Example 5 suggests that small uncertainties in camera positions do contribute notable errors if the image pattern of the point is large.

A method to obtain t_z and θ for distortion compensation is to use measurement instruments to determine roughly the values of t_z and θ, since by the examples, errors due to $d\theta$ and dt_z are secondary comparing with errors due to large camera tilt angles.

Let (X_c, Y_c) be the image coordinates of the center of the kth calibration point to be estimated. The distortion correction procedure for each calibration point is as follows:

1. Rough measurements of t_z and θ.
2. Computation of γ for the kth object point.
3. Circle fitting of the binary image pattern of the kth calibration point. Y_c is then taken to be the y-axis component of the circle center.
4. Drawing of a line, which passes through the center of the circle and is parallel to the x axis of the image coordinate system. The line intersects the circle at points a and b. Let the x coordinates of the two intersection points be X_a and X_b, where $X_a < X_b$. Equation (2.9.6) is then used to compute X_c.

If errors caused by the uncertainties in t_z and θ are intolerable, one needs to use a camera calibration process to compute an initial camera model. From the computed camera model, t_z and θ are then determined. This method can provide more precise values of t_z and θ comparing with direct measurement. This scheme shall be used to replace Step 1 of the above procedure.

B. THE GENERAL CASE

For certain camera measurement systems, it is often inconvenient, or even impossible, to restrict multiple camera tilt angles to be all about a common axis. For example, if three cameras are used for 3D measurement, it is undesirable to let all three cameras be rotated only about the y axis of the world coordinate system. Thus, it is necessary to extend the method given in the previous section to the general case.

Elements of the rotation matrix R and the translation vector t can be used to specify the degree of distortion due to perspective projection, in the same way as the distance t_z and the angle θ were used to describe this type of distortion in Section IX.A. However these quantities can only be obtained after an initial camera calibration.

It is assumed that an initial calibration has been done, ignoring perspective projection of the calibration points so that initial values of R and

t are available. Let the calibration board be placed at $z_w = 0$. A particular calibration point is now considered. The world coordinates of its center is $(x_{w,c}, y_{w,c}, 0)$. Denote the projected image coordinates by (X_c, Y_c). A line is drawn which is parallel to the x_w axis and passes through the center of the calibration point. The line intersects with the circle boundary at Points A and B. Thus the world coordinates of the two points are $(x_{w,a}, y_{w,c}, 0)$ and $(x_{w,b}, y_{w,c}, 0)$. Let the corresponding image coordinates be (X_a, Y_a) and (X_b, Y_b). Similarly another line is drawn that is parallel to the y_w axis and passes through the center of the calibration point. The line intersects with the boundary at Points D and E. The world coordinates of the two points are $(x_{w,c}, y_{w,d}, 0)$ and $(x_{w,c}, y_{w,e}, 0)$. Let the corresponding image coordinates be (X_d, Y_d) and (X_e, Y_e). It is further assumed that $Y_d < Y_c < Y_e$ (refer to Figure 2.9.2).

Similar to (2.9.4), the *eccentricity coefficient* of the image pattern of a calibration point along the y direction of the world coordinate system is defined as

$$\gamma_y \equiv \frac{Y_c - Y_d}{Y_e - Y_c} \tag{2.9.10}$$

A general rotation matrix can be modeled by three basic rotations

$$R = Rot(z, \psi) \, Rot(y, \theta) \, Rot(x, \phi)$$

If the scale factors along the x and y axes of the image plane are assumed to be the same, the rotation about the z axis of the camera coordinate system does not change the shape of the image pattern of the camera calibration point; it only changes the orientation of the image pattern. In order to analyze the distortion effect due to the perspective projection, it is assumed that $\psi = 0$. Thus

$$R = \begin{bmatrix} c\theta & s\phi s\theta & s\phi c\theta \\ 0 & c\phi & -s\phi \\ -s\theta & s\phi c\theta & c\phi c\theta \end{bmatrix}$$

where $s\phi \equiv \sin\phi$ and $c\phi \equiv \cos\phi$.

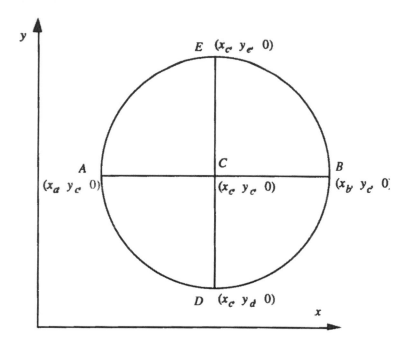

Figure 2.9.2 Definition of world coordinates for the general case of
perspective projection

Employing a technique similar to the one used in deriving (2.9.4) results
in the following expressions,

$$\gamma_x = \frac{1 - \dfrac{s\theta x_{w,b} - c\theta s\phi y_{w,c}}{t_z}}{1 - \dfrac{s\theta x_{w,a} - c\theta s\phi y_{w,c}}{t_z}} \qquad (2.9.11)$$

$$\gamma_y = \frac{1 - \dfrac{s\theta x_{w,c} - c\theta s\phi y_{w,e}}{t_z}}{1 - \dfrac{s\theta x_{w,c} - c\theta s\phi y_{w,d}}{t_z}} \qquad (2.9.12)$$

Using the Euclidian distance, the error E_I on the estimation of the image
coordinates of the calibration point due to this type of distortion is

$$E_I = \sqrt{\left[(X_b - X_a)\left(\frac{\gamma_x}{1 + \gamma_x} - 0.5\right)\right]^2 + \left[(X_e - X_d)\left(\frac{\gamma_y}{1 + \gamma_y} - 0.5\right)\right]^2}$$

$$= \sqrt{\left[\frac{(X_b - X_a)}{2}\left(\frac{1 - \gamma_x}{1 + \gamma_x}\right)\right]^2 + \left[\frac{(X_e - X_d)}{2}\left(\frac{1 - \gamma_y}{1 + \gamma_y}\right)\right]^2}$$

The equivalent 3D error E_w for the center estimation of the calibration point in the world coordinate system is

$$E_w = r\sqrt{\left(\frac{1 - \gamma_x}{1 + \gamma_x}\right)^2 + \left(\frac{1 - \gamma_y}{1 + \gamma_y}\right)^2} \qquad (2.9.13)$$

Remarks:
1. Errors caused by a small tilt angle in the y direction can be analyzed using (2.9.11), which allows the assessment as to what extent errors due to small tilt angles can be ignored.
2. Equation (2.9.4) is obtained from (2.9.11) by setting $\phi = 0$. Equation (2.9.7) is obtained from (2.9.13) by setting $\gamma_y = 1$. Due to the striking similarities between the formulas presented in Section IX.A and this section, many relevant remarks given before apply here as well.

In practice, the following method can be used to estimate γ_x and γ_y. The initial camera model (R and t) is used to obtain the projections of $(x_{w,i}, y_{w,i}, 0)$ for $i = a, b, c, d, e$. The projected points are denoted by (X_i', Y_i') for $i = a, b, c, d, e$, respectively. Both γ_x and γ_y are then computed by replacing (X_i, Y_i) with (X_i', Y_i') in (2.9.4) and (2.9.10). E_w in turn is computed from (2.9.13).

There are several ways to estimate the image coordinates of the calibration point center, after (X_i', Y_i') ($i = a, b, c, d, e$) become available using an initial camera model. A simple and effective scheme is given next.

A circle, say Circle A, is fitted with (X_i', Y_i'), $i = a, b, d, e$. Denote by (X', Y') the center of Circle A. Another circle, say Circle B, is fitted with the image pattern of the calibration point. Denote by (X, Y) the center of Circle B. The centers of both circles are known after performing circle fitting. Ideally the two centers coincide. Due to the uncertainties in the initial camera model, these centers may not be coincident. Since (X_c', Y_c') can also be computed using the initial camera model, the offset between (X_c', Y_c') and (X', Y') is known. This offset can be used as the estimate of the offset between (X_c, Y_c) and (X, Y). Mathematically, this can be written as

$$X_c = X + X_c' - X' \qquad (2.9.14)$$
$$Y_c = Y + Y_c' - Y' \qquad (2.9.15)$$

The distortion correction procedure for each calibration point is as follows:

1. Computation of (X_i, Y_i) for $i = a, b, d, e$ and (X_c', Y_c') for the kth calibration point using an initial camera model.
2. Fitting of (X_i, Y_i), $i = a, b, d, e$, with a circle to obtain (X', Y'), the center of Circle A.
3. Fitting of the binary image pattern of the kth calibration point with a circle to obtain (X, Y), the center of Circle B.
4. The image coordinates (X_c, Y_c) of the kth calibration point are then approximated by applying (2.9.14) and (2.9.15).

If the diameter of Circle A is not very close to that of Circle B, scaling may be necessary to compute the offset. Let the radii of Circles A and B be R_A and R_B, respectively. Then

$$X_c = X + \rho(X_c' - X')$$
$$Y_c = Y + \rho(Y_c' - Y')$$

where $\rho \equiv R_B/R_A$.

After the perspective projection distortion for all calibration points are corrected, these can then be used for a recomputation of camera model parameters.

X. SIMULATION AND EXPERIMENTAL RESULTS

A. SIMULATION STUDY OF IMAGE CENTER ESTIMATION

To test the effectiveness of the image center estimation algorithm given in Section VIII, an artificial stereo-camera measurement system was simulated. There were presumably 30 error-free calibration points on the "calibration board". The two "cameras" were assumed to be 150 mm away from the board and tilted at certain angles with respect to it. Proper rotation and translation parameters of the camera model for each "camera", corresponding to the setup, had been given *a priori*. Among the intrinsic parameters, the focal lens was assumed known and unchanged in the simulation. The image coordinates of the calibration points were computed while varying the lens distortion and the image center. A uniformly distributed noise of varying intensity was added to the "true" image coordinates. The errors introduced to the image center computation were the lens distortion and the image coordinate uncertainties of the calibration points.

After the world coordinates of calibration points and the corresponding image coordinates became available, the camera models were estimated using

20 calibration points. These camera models are functions of the image center, the lens distortion coefficient, and the image coordinate uncertainties of the calibration points. For a given lens distortion, the cost function in (2.6.4) is a function of the image center.

The algorithm given in Section VIII was then used to minimize (2.6.4) by searching for C_x and C_y. After the initial camera model (obtained from an initial image center) and final camera model (obtained from the calibrated image center) of the two "cameras" became available, these estimated models were verified by 10 test points from the same plane. The computed world coordinates of the test points were then compared with the "true" world coordinates of these test points. Let C_x and C_y be the coordinates of the true image center and $C_{x,0}$ and $C_{y,0}$ be those of the initial image center. Let $\Delta C_x \equiv C_{x,0} - C_x$ and $\Delta C_y \equiv C_{y,0} - C_y$, and $\Delta C \equiv max\{\Delta C_x, \Delta C_y\}$. A typical simulation result is listed in Tables 2.10.1 and 2.10.2.

Table 2.10.1 3D error of the verification points vs. image center offset under various radial lens distortion coefficients with no image noise

ΔC (pixel)	3D error (mm)		
	$k = 0$	$k = 10^8$	$k = 10^{-4}$
0	4.95×10^{-11}	4.95×10^{-11}	4.95×10^{-11}
5.6	3.13×10^{-3}	3.13×10^{-3}	2.68×10^{-3}
11.3	6.07×10^{-3}	6.07×10^{-3}	5.24×10^{-3}
22.6	1.14×10^{-2}	1.14×10^{-2}	9.99×10^{-3}

Table 2.10.2 3D error of the verification points vs. image center offset under various radial lens distortion coefficients with lens distortion $k = 10^{-4}$

ΔC (pixel)	3D error (mm)	
	(image error = 0.01 pixel)	(image error = 0.1 pixel)
0	1.29×10^{-3}	1.30×10^{-11}
5.6	3.22×10^{-3}	1.35×10^{-2}
11.3	5.59×10^{-3}	1.41×10^{-2}
22.6	1.01×10^{-2}	1.66×10^{-2}

Based on the simulation study, the following comments can be made:

1. The image center estimation algorithm consistently converged to the "true" image center for nonzero lens distortions and for negligible image coordinate uncertainties of the calibration points, even when $|\Delta C_x|$ and

    $|\Delta C_y|$ were as large as 50 pixels. In such a case, the 3D errors were zero, and are therefore not listed.
2. Image center offsets contributed moderately to 3D errors regardless of lens distortion effects (Table 2.10.1).
3. Whenever there is no lens distortion, the algorithm fails because in this case the measurement residuals are zero. One is unable to estimate image center coordinates if the relevant measurement information is unavailable.
4. Image coordinate uncertainties (image errors) of the calibration points contributed significant 3D errors even when the image center was perfectly known (Table 2.10.2).
5. The algorithm did not converge to the "true" image center whenever the image coordinate uncertainties of calibration points were significant (about 0.3 pixel or more). Further studies are needed to improve the robustness of the image center estimation algorithm against this type of uncertainties.

B. SIMULATION STUDY FOR THE ESTIMATION OF THE RATIO OF SCALE FACTORS

To test the effectiveness of the scale factor estimation algorithm given in Section 4, an artificial stereo-camera measurement system was simulated. There were 30, presumably error-free, calibration points on the "calibration board". The two "cameras" were assumed to be 150 mm away from the board and tilted at certain angles with respect to it. Proper rotation and translation parameters of the camera model for each "camera", corresponding to the setup, had been given *a priori*. Among the intrinsic parameters, the focal length was assumed known and unchanged in the simulation. A uniformly distributed noise of varying intensity was added to the "true" image coordinates.

It was found that the ratio of scale factors μ can be estimated consistently up to the first three decimals even for image center offsets that were made as large as 20 pixels and for uncertainties of image coordinates of about 0.1 of a pixel (Table 2.10.3). In conclusion, μ was found to be insensitive to image center offsets, lens distortions and uncertainties of image coordinates. In practice, μ needs to be calibrated only once for a particular camera-vision system.

Table 2.10.3 Estimates of the ratio of scale factors
vs. image center offsets with radial lens distortion coefficient $\alpha = 10^{-4}$

Image center offset	μ (image error = 0.01 pixel)	(image error = 0.1 pixel)
0. 0	1.286300	1.286093
5.6	1.286282	1.286074
11.3	1.286261	1.286051
22.6	1.286210	1.285999

C. EXPERIMENTAL STUDY OF THE RAC-BASED CAMERA CALIBRATION

The experimental system consisted of two CCD cameras (Electrophysics, model number CCD1200), two 25 mm TV lenses, an IBM-AT personal computer, an ITEX[R] video imaging board with its driver software, a camera calibration board, and a coordinate measuring machine (CMM). The CMM was replacing the rail to move the calibration board to different locations along a *single* direction. The relative position of the cameras and the calibration board is shown in Figure 2.10.1.

Each CCD camera has 510H by 492V picture elements with a 8.8x6.6 mm^2 sensing area. The camera calibration board is a custom-ordered glass plate painted with vapor deposited metallic chromium. It has 10x10 dot array points with center to center distance of 10 mm, and diameter (of each point) of 2 mm. The rated flatness of the calibration board is ± 0.003 mm and the rated center to center accuracy of the calibration points is ± 0.002 mm (Figure 2.10.2).

The vision algorithms for camera calibration were written in Microsoft-C^R. To estimate the image coordinates of a calibration point, an adaptive thresholding algorithm was devised. The algorithm first smoothed an image of the camera calibration board by applying a 3x3 low-pass mask three times. It then consecutively detected each calibration point and computed the histogram within a window surrounding each point. The intersection of two Gaussian curves which fit the histogram was chosen as the threshold value for binalizing the calibration point. The centroid of the image of a calibration point was selected as the estimate of its image coordinates. This procedure could yield image coordinates accuracy to within 1/5 of a pixel. As the image pattern of each calibration point was small, distortion due to perspective projection was negligible. No effort was spent to compensate for such errors.

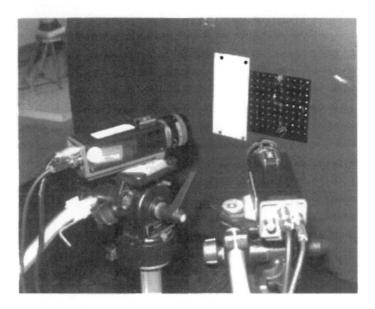

Figure 2.10.1 Experimental setup for camera calibration

Figure 2.10.2 Camera calibration board

To estimate the ratios of the scale factors μ of the two cameras, each camera was placed at three locations. The angle between each camera's sensing plane and the calibration board was about 30 degrees at all locations. The calibration board was moved to two positions along a straight line. The distance from each camera to the calibration board at its center position was about 150 mm. The distance between two locations of the calibration board along the rail was 20 mm. The field of view was about 40x50 mm^2. The algorithm described in Section VII was applied to compute μ. Since the cameras and lenses of the same model were used, identical values of μ for two cameras were obtained. The initial value of μ was about 1.534, and the estimated value was 1.2402.

The algorithms given in Sections III and IV were employed to compute camera models utilizing the estimated μ. For Tsai's RAC-based algorithm, 16 points were used for the identification of the camera parameters and 8 points were used to estimate the 3D accuracy of these models. Tables 2.10.4 and 2.10.5 list the estimated 3D accuracy of the identified camera models for a single camera with two views. The results obtained by using two cameras with single view were compatible with those listed in Tables 2.10.4 and 2.10.5.

Table 2.10.4 3D error with a single camera (two views)
for a field-of-view = 40x50 mm^2
Test points are on the same plane as the calibration points

	Average Error (mm)			Maximum Error (mm)		
	x	y	z	x	y	z
Tsai's RAC	0.0045	0.0043	0.0094	0.0109	0.0094	0.0225
Fast RAC	0.0061	0.0117	0.0129	0.0162	0.0252	0.0347

Table 2.10.5 3D error with a single camera (two views)
for a field-of-view = 40x50 mm^2
Test points are on a plane 20 mm away from the calibration board plane

	Average Error (mm)			Maximum Error (mm)		
	x	y	z	x	y	z
Tsai' RAC	0.0077	0.0045	0.0235	0.0290	0.0102	0.0459
Fast RAC	0.0520	0.0245	0.0530	0.0661	0.0504	0.1019

Table 2.10.6 Comparison of CPU times
for Tsai's and fast RAC-based calibration algorithms

# of measurements	Tsai's RAC (Seconds)	Fast RAC (Seconds)
10	0.88	0.16
20	1.26	0.22
30	1.48	0.33

Table 2.10.6 lists the average CPU time required for the computation of Tsai's original algorithm and the fast RAC-based algorithms using different number of measurements. Note that only the time needed to compute the set of camera model parameters is listed, given that the world and image coordinates of the calibration points have been computed.

The following observations can be made:

1. For the original RAC-based algorithm, whenever the test points are on the same plane as the calibration points, the average accuracy is 1 part in 10000 at the x and y directions.
2. For the original RAC-based algorithm, when the test points are 20 mm away from the plane of the calibration points, the average accuracy is 1 part in 5000 at the x and y directions.
3. As a well-known shortcoming of the stereo vision technique for coordinate measuring, the accuracy at the z direction is consistently lower than that in the x and y directions.
4. The fast RAC-based algorithm, as expected, demonstrates a tradeoff between computational efficiency and accuracy. The original RAC-based Algorithm is superior in terms of accuracy. One reason is that only a limited number of calibration points (7 points) were used in the first stage computation of the fast algorithm. Improvements can be made for the simplified RAC-based algorithm by increasing the number of calibration points contained within the field of view. The fast algorithm justifies its name by being faster.
5. Although the uncalibrated image center was 25 pixels away from the estimated image center, the impact on the 3D measurement accuracy is very little.   This is probably due to the small radial distortions of the lens used in the experiments and the relatively large image coordinate uncertainties of the calibration points.

D. EXPERIMENTAL RESULTS FROM A NEAR SINGULAR CAMERA CONFIGURATION

To estimate the intrinsic parameters of the camera, the angle between the sensing plane of the camera and the calibration board was set to about 30 degrees at all locations. The distance from each camera to the calibration board at its center position was about 110 mm. The field of view was about 40x30 mm^2. The algorithm given in Section III was applied to compute the intrinsic parameters. The estimated μ, f_x, and f_y were 1.2401506, 1.5967550 and 1.9802167.

The camera was then placed to be nearly-parallel to the calibration board. The algorithm given in Section V was employed to compute the unknown parameters of the camera models with the estimated intrinsic parameters μ, f_x, and f_y. Eight points were used for the identification of the camera parameters and another four points for the estimation of the 3D accuracy of the model. Table 2.10.7 lists the identified camera parameters when the camera was moved to various positions by a robot arm. Table 2.10.8 presents the verification results that correspond to the camera models given in Table 2.10.7. Note that the accuracy may be further improved by using a better synchronization circuit to lock to the incoming video signal. Currently, calibration errors due to video synchronization errors can be as large as half of a pixel.

Table 2.10.7 Identified camera parameters in different tests
for different camera positions

	α (rad.)	β (rad.)	γ (rad.)	t_x (mm)	t_y (mm)	t_z (mm)	k (10^{-8})
Test 1	0.1175	0.0459	0.0092	-15.4438	-12.1306	110.6165	6.7998
Test 2	0.0643	0.0319	0.0153	-14.8638	- 8.9083	111.3940	8.6977
Test 3	0.0145	0.0483	0.0164	-16.9317	-10.5072	111.1975	7.9975
Test 4	0.0506	0.0469	0.0170	-15.9187	-11.9978	110.6429	8.8992
Test 5	0.0207	0.0444	0.0215	-14.4890	- 9.6100	111.0682	5.2833

Table 2.10.8   Verification results
corresponding to the camera models given in Table 2.10.7

	Mean Error		Maximum Error	
	x (pixel)*	y (pixel)	x (pixel)	y (pixel)
Test 1	0.2315	0.1317	0.3465	0.4080
Test 2	0.3076	0.2188	0.4685	0.3640
Test 3	0.2913	0.1347	0.5568	0.2597
Test 4	0.2859	0.1243	0.6068	0.3034
Test 5	0.3251	0.1686	0.5070	0.3832

*   In the x-direction, each pixel size is roughly 0.08 mm; and

in the y-direction, each pixel size is roughly 0.0645 mm.

XI.   SUMMARY AND REFERENCES

There is a rich literature on the subject of camera calibration. Existing techniques for camera calibration can be classified into two main categories: Nonlinear and linear methods.

In the category of nonlinear methods, equations that relate the camera parameters to be estimated with the 3D coordinates of the calibration points and their image projections are established.   A nonlinear optimization procedure is then applied to estimate the unknown parameters by minimizing some sort of error residuals. Most of the classical calibration techniques in photo-grammetry belong to this category.   One advantage of this type of technique is that the camera model can have sufficient number of parameters to characterize various types of errors. Another advantage is that this type of algorithm is potentially ultra-accurate, providing that the error sources are properly modeled, and the algorithm converges correctly. One more advantage is its simplicity of the algorithm.   There is an abundance of optimization software that can be used to estimate the camera parameters once the problem is properly formulated.

There are some drawbacks associated with the nonlinear approaches. Stability of the algorithm is an important issue. If the intrinsic parameters used to represent different types of distortion are not independent of one another, singularities may occur in the optimization process. Further, when gradient-based iterative procedures are applied to solve the problem, the selection of initial conditions is crucial for a proper convergence of the algorithm. Finally, a nonlinear optimization algorithm normally needs more computational time comparing with a linear algorithm.

In contrast to the nonlinear method, a linear approach enables the camera parameters to be solved noniteratively directly from the equations that relate the 3D coordinates of calibration points to their image projections. A multi-stage procedure is often needed in this category. A set of intermediate parameters is defined in terms of the original parameters. The intermediate parameters can be computed by solving a system of linear equations, and the unknown camera parameters are then determined from the solved intermediate parameters. Speed and stability are two major advantages of this type of algorithms since no iterations are involved. A major limitation of the linear approaches is their accuracy.

To automate the robot calibration process, the speed of camera calibration has to be enhanced. An automated robot calibration process, which utilizes *monocular* "eye-on-hand" configuration and repeated camera calibrations, is a good example of applications that require the performing of a camera calibration in almost real time. Ideally, a single camera calibration task should be completed while the robot changes from one measurement configuration to another. A typical error-model based robot kinematic identification procedure requires 12-15 end-effector pose measurements in well-selected robot configurations. With present technology, it is therefore not unreasonable to expect a full robot calibration cycle (utilizing off-the-shelf cameras and lenses and a 486 personal computer) to take only a few minutes.

Among the linear algorithms, a method known as Tsai's (1987) Radial Alignment Constraint (RAC) camera calibration, is viewed by many researchers as a compromise between accuracy, on one hand, and speed, on the other hand. The identification of all extrinsic parameters and intrinsic parameters from given sets of data points, is a nonlinear optimization problem. Tsai's method is based on his observation that with suitable change of variables, the RAC equations suggest a camera calibration solution in *two* linear least squares steps. The solution is generally suboptimal, in comparison to methods that seek global optimum to the nonlinear least squares problem, but is relatively fast.

It was suggested in Zhuang, Roth, Xu and Wang (1993) that with some modifications, the RAC-based camera calibration, embedded within the robot eye-on-hand calibration process, is a viable candidate for on-line robot calibration to be done on the manufacturing floor. The modification involves a simplified RAC-based camera calibration algorithm to allow about a 5:1

computation time reduction with respect to Tsai's original algorithm.

When *stereo* cameras are mounted on the hand of a robot manipulator, the cameras can be calibrated in advance, in which case the speed of camera calibration is no longer important. Nonlinear optimization methods can thus be applied to acquire poses of the manipulator end-effector at different manipulator configurations. Weng, Cohen, and Herniou (1992) proposed a two-phase approach to perform camera calibration. In the first phase, a linear algorithm is applied to determine the initial conditions of the unknown parameters. In the second phase, the camera parameters are iteratively improved by using the initial conditions. Since the initial conditions, obtained in the first phase, are close to the optimal ones, only a few iterations are needed in the second phase.

For a proper implementation of a calibration algorithm, some additional issues have to be considered. In Tsai's original paper (1987), the ratio of the scale factor was estimated using non-coplanar calibration points. Non-coplanar points are normally obtained by moving the calibration board along an axis that is parallel to the normal of the plane defined by the board. This chapter presented a low-cost method of estimating the ratio of scale factors. It should be emphasized that this operation needs to be done only once for a given camera-vision system set-up.

It has been a common practice, in the computer vision and robotics areas, to choose the center of the image frame buffer as the origin of the image coordinate system. It was reported by Tsai that altering the image center by as much as ten pixels does not significantly influence the accuracy of 3D measurement using the calibrated camera. Experiments on the effects of image center uncertainties on 3D measurements were also conducted in the FAU Robotics Lab, and it was observed that, for the purpose of calibrating the PUMA 560 robot, even when the image center coordinates are 20 pixels away from the nominal ones, the impact on the 3D error is negligible. If a very large image center offset is experienced in implementing camera-assisted robot calibration procedures, the image center needs to be pre-calibrated. Fortunately, the image center needs only to be calibrated once for a given camera system. A couple of techniques were proposed by Lenz and Tsai (1987) for the estimation of image centers. One of the methods requires using a laser beam and a four degrees of freedom mechanism to adjust the position and orientation of the laser beam. Another method formulates the problem as that of minimization, employing subsequently a gradient-based solution algorithm.

A camera calibration board normally consists of an array of circular points of relatively small diameter. The joint action of camera calibration and robot pose measurements requires the acquisition of a large enough subset of this calibration point array. If a small number of calibration points is measured in each image frame, the image size of each calibration point in such cases is relatively large. As a result, a large distortion in the centroid location occurs due to perspective projection. The impact of such distortions on 3D accuracy may be significant. An error analysis of such distortion effects and simple schemes for their compensation was originally provided by Zhuang, Roth, Xu and Wang (1993).

Tsai's Radial Alignment Constraint (RAC) method will fail if the camera sensor plane is parallel to the plane defined by the calibration board. This problem exists in many other calibration techniques. Zhuang and Wu (1996) suggested that, by a slight modification of the RAC-based camera calibration method, the unknown parameters of the camera can still be calibrated even when the sensor plane is parallel or nearly-parallel to the camera calibration board. An application example of this approach is the calibration of a SCARA arm with a hand-mounted camera, where the extrinsic parameters need to be repeatedly calibrated to reconstruct robot end-effector poses at various robot measurement configurations.

There are basically two ways to assess the accuracy performance of a camera system after calibration. The first is a 2D approach, and the second is a 3D approach. In the 2D method, the calibrated camera model is used to predict the image projections of object points, and compare these with the measured image points. In the 3D method, the world coordinates of object points are computed by a pair of calibrated camera models using the measured image projections. The computed world coordinates are then compared with the true world coordinates of these object points.

A procedure that does orthonormalization of a rotation matrix is outlined in Section IV.D, Chapter 5. More techniques can be found in Zhuang, Roth and Sudhakar (1992).

Finally, self-calibration of cameras is currently a very active research area. Readers are referred to Maybank and Faugeras (1990), Trivedi (1991) and Zhuang (1995).

Chapter 3

KINEMATIC MODELING FOR ROBOT CALIBRATION

I. INTRODUCTION

Kinematic modeling is the first step in a robot calibration process. A robot kinematic model relates the outputs of the robot joint sensors to the pose of the robot end-effector. One of the goals of robot calibration is to improve the accuracy of the robot by modifying its kinematic model parameters residing in the robot controller.

Kinematic modeling can be extended to include an entire robotic workcell. By such kinematic modeling, the relative positions of various system components are established in a unique way. A convenient mathematical approach for describing geometric relationships among various system components is through homogeneous transformations. The modeling starts by establishing reference coordinate systems for the different subsystems, as shown in Figure 3.1.1. Next, local coordinate frames are assigned to each moving part. For robot manipulator, a link coordinate system is assigned at each joint of the robot. A set of 4x4 homogeneous transformation matrices is then introduced to relate each link coordinate system to its neighboring coordinate system.

As argued by robot calibration researchers, a kinematic model suitable for robot calibration should be complete and proportional. A *complete* kinematic model is one that has the capability of relating the joint displacements to the tool pose for any manipulator geometry, while allowing for the arbitrary placement of the world frame and arbitrary assignment of the manipulator zero position. Another way of defining completeness is to say that a complete model has enough parameters to express any variation of the actual robot structure away from the nominal design. To be complete, the model must contain the required number of independent kinematic parameters. This number, for a serial manipulator which consists of rigid links connected by lower pair joints, is $4R + 2P + 6$, where R and P are respectively the number of revolute and prismatic joints.

Proportionality (or *Parametric continuity*, as coined by some researchers) implies no model singularities. Formally, a kinematic model is parametrically continuous if continuous changes of position and orientation of any robot joint axis result in continuous changes of the model link parameters. Lack of parametric continuity in the model may invalidate linearized accuracy error models and may also cause numerical instabilities in the kinematic identification process. The concepts of completeness and proportionality will be revisited in much more detail later in this chapter.

The chapter organization is as follows: Section II is a review of basic concepts in kinematics that are relevant for manipulation in general and robot kinematic modeling in particular. These include Rodrigues' equation, homogeneous transformations, the general rotation transformation matrix and more.

In Section III the classical Denavit-Hartenberg (D-H) modeling convention is reviewed. The shortcomings of this method for the robot calibration application are explained, and alternative models such as DH-modifications and the Zero Reference Model are briefly introduced.

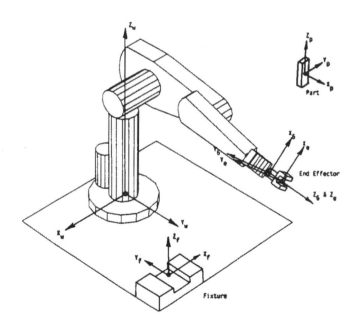

Figure 3.1.1 A schematic robotic workcell
(From Mooring, B. W., Z. S. Roth, and M. Driels (1991). *Fundamentals of Manipulator Calibration*, John Wiley & Sons Inc. With permission.)

The Complete and Parametrically Continuous (CPC) model and its variations, presented in Section IV, is a natural follow-up of earlier robot models. It is the central part of the chapter. The model is shown to be complete, minimal and parametrically continuous. Section IV also includes examples for constructing the CPC model for various robot geometries. The relationship between this model and its modified variation (MCPC) and to other models are analyzed in detail in Sections IV and V. Section VI is a

general treatment of parameter continuity. Singularities of the MCPC model
are discussed in Section VII.   The chapter concludes with discussions and
references.

II.   BASIC CONCEPTS IN KINEMATICS

The most general spatial displacement of a rigid body can be characterized
as a screw displacement, that is a translation along and a rotation about a
screw axis.   If we let k denote the direction of the screw axis, then the
displacement of a rigid body comprises of a rotation of angle ϕ about k and a
translation of distance d parallel to k. (d, ϕ) are the screw parameters for the
displacement.

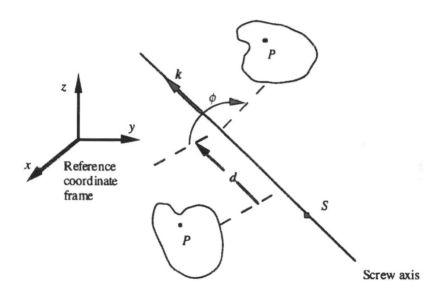

Figure 3.2.1 The screw axis and screw parameters

Let P denote an arbitrary point on a rigid body, and S denote an arbitrary
point on the screw axis.   Let r_s be the position of S in a reference frame.
Also let $r_p^{(1)}$ and $r_p^{(2)}$ denote the positions of point P in the same reference,
where the superscripts (1) and (2) denote the first and second positions of the
point in the reference frame, respectively.   It can then be shown that $r_p^{(2)}$ is
related to $r_p^{(1)}$ through the parameters of r_s, k, d, and ϕ, as follows:

$$r_p^{(2)} = (r_p^{(1)} - r_s)\cos\phi + k \times (r_p^{(1)} - r_s)\sin\phi$$
$$+ [(r_p^{(1)} - r_s)^{\cdot} k]k(1 - \cos\phi) + r_s + dk \qquad (3.2.1)$$

This is a version of the well known Rodrigues' formula. Equation (3.2.1) in matrix form is:

$$\begin{bmatrix} r_{p,x}^{(2)} \\ r_{p,y}^{(2)} \\ r_{p,z}^{(2)} \end{bmatrix} = R \begin{bmatrix} r_{p,x}^{(1)} \\ r_{p,y}^{(1)} \\ r_{p,z}^{(1)} \end{bmatrix} + \begin{bmatrix} p_x \\ p_y \\ p_z \end{bmatrix}$$

where the 3×3 rotation matrix R is

$$R = \begin{bmatrix} n_x & o_x & a_x \\ n_y & o_y & a_y \\ n_z & o_z & a_z \end{bmatrix} = \begin{bmatrix} k_x^2 v\phi + c\phi & k_x k_y v\phi - k_z s\phi & k_x k_z v\phi + k_y s\phi \\ k_x k_y v\phi + k_z s\phi & k_y^2 v\phi + c\phi & k_y k_z v\phi - k_x s\phi \\ k_x k_z v\phi - k_y s\phi & k_y k_z v\phi + k_x s\phi & k_z^2 v\phi + c\phi \end{bmatrix}$$
$$(3.2.3)$$

where $v\phi \equiv (1 - \cos\phi)$, $s\phi \equiv \sin\phi$ and $c\phi \equiv \cos\phi$. Furthermore,

$$\begin{bmatrix} p_x \\ p_y \\ p_z \end{bmatrix} = \begin{bmatrix} dk_x - s_x(n_x - 1) - s_y o_x - s_z a_x \\ dk_y - s_x n_y - s_y(o_y - 1) - s_z a_y \\ dk_z - s_x n_z - s_y o_z - s_z(a_z - 1) \end{bmatrix} \qquad (3.2.4)$$

The matrix R in (3.2.3) is the general rotation matrix by an angle ϕ about an axis k. This matrix may be denoted as $Rot(k, \phi)$. The three unit vectors a, o and n are known as the "approach", "orientation" and "normal" vectors, respectively.

The converse problem, of finding the axis of rotation k and angle of rotation ϕ from a given numerical value of the matrix $Rot(k, \phi)$, can be solved from (3.2.3) in a straightforward manner. Readers are referred to Paul's textbook (1981) for details.

Equations (3.2.2)-(3.2.4) can be written more compactly using the 4×4 homogeneous transformation matrix T,

$$\begin{bmatrix} r^{(2)} \\ 1 \end{bmatrix} = T \begin{bmatrix} r^{(1)} \\ 1 \end{bmatrix}$$

where

$$T = \begin{bmatrix} R & p \\ 0_{1 \times 3} & 1 \end{bmatrix}$$

A few basic homogeneous transformations arise as particular cases of (3.2.3)-(3.2.4). A rotation about the x axis by an angle θ can be represented by a 4x4 homogeneous matrix by letting $k = [1, 0, 0]^T$ and $d = 0$,

$$Rot(x, \theta) = \begin{bmatrix} 1 & 0 & 0 & 0 \\ 0 & \cos\theta & -\sin\theta & 0 \\ 0 & \sin\theta & \cos\theta & 0 \\ 0 & 0 & 0 & 1 \end{bmatrix}$$

Similarly, a rotation about the y axis by an angle ψ is

$$Rot(y, \psi) = \begin{bmatrix} \cos\psi & 0 & \sin\psi & 0 \\ 0 & 1 & 0 & 0 \\ -\sin\psi & 0 & \cos\psi & 0 \\ 0 & 0 & 0 & 1 \end{bmatrix}$$

and a rotation about the z axis by an angle φ is

$$Rot(z, \varphi) = \begin{bmatrix} \cos\varphi & -\sin\varphi & 0 & 0 \\ \sin\varphi & \cos\varphi & 0 & 0 \\ 0 & 0 & 1 & 0 \\ 0 & 0 & 0 & 1 \end{bmatrix}$$

A translation along x, y, and z by a, b, and c, respectively can then be obtained as a combination of three basic translations as follows

$$Trans(a, b, c) = \begin{bmatrix} 1 & 0 & 0 & a \\ 0 & 1 & 0 & b \\ 0 & 0 & 1 & c \\ 0 & 0 & 0 & 1 \end{bmatrix}$$

A sequence of two displacements T_1 and T_2 is naturally written as a product $T_1 T_2$. There are two different ways to attach physical meaning to this product, depending on whether we consider T_1 to *pre-multiply* T_2 or T_2 to *post-multiply* T_1. Viewing from right to left each transformation T_1 and T_2

represents a displacement with respect to the world reference frame. On the other hand, viewing from left to right, T_1 still represents a displacement with respect to the world, whereas T_2 represents a displacement with respect to the world frame after being transformed by T_1.

More generally, a sequence of actions $T_1 T_2 ... T_n$ provides simultaneously two types of information: looking from right to left transformations with respect to the world frame; or looking from left to right transformations with respect to transformed local coordinate frames.

III. THE DENAVIT-HARTENBERG MODEL AND ITS MODIFICATION

A. THE DENAVIT-HARTENBERG MODELING CONVENTION
A manipulator is a mechanical device that is capable of moving in different ways relative to its base (Figure 3.3.1).

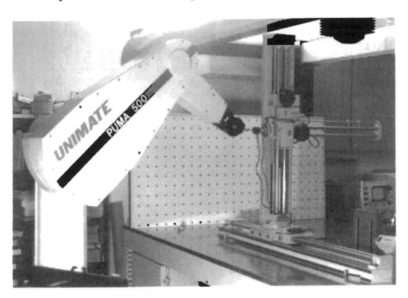

Figure 3.3.1 A manipulator

A manipulator usually consists of a group of rigid bodies or links connected together by joints. In most industrial manipulators, each link is connected to two other members. Therefore, each link has two axes (Figure 3.3.2). The Denavit-Hartenberg (D-H) link modeling characterizes the kinematic structure of the link in terms of physical link parameters. The first parameter a, that captures the link length information, is taken to be the *length of the common normal* of the two link axes. The second parameter α, known as the *twist angle*, is the angle between the two link axes (Figure 3.3.2).

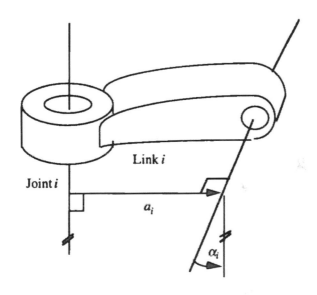

(a) A manipulator link

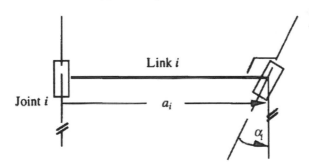

(b) Its symbolic representation

Figure 3.3.2 A manipulator link and its symbolic representation

The joint that connects two links makes one of the axes of one link coincide with one of the axes of the other link. It is assumed that each joint is either ideally revolute or ideally prismatic (Figure 3.3.3).

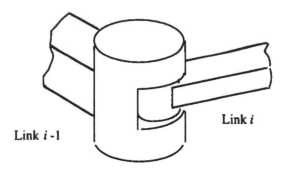

(a) A revolute joint

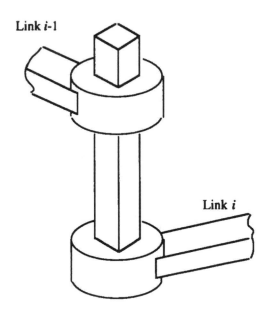

(b) A prismatic joint

Figure 3.3.3 Revolute and prismatic joints

As two links are connected at each joint axis, this axis has two perpendiculars to it - one for each link. The relative position of two such connected links is given by their *distance d* along the common axis and their *rotation angle* θ, measured in a plane normal to the common axis (Figure 3.3.4).

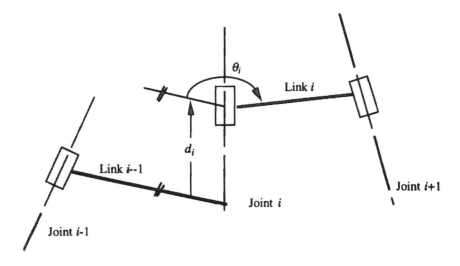

Figure 3.3.4 Neighboring links geometry

In summary, according to the D-H convention, there are two link parameters (a, α) which describe the link shape and two link parameters (d, θ) that describe the relative positions of two neighboring links. These latter parameters are essentially associated with joint i between links i and i-1. For a revolute joint, d is a fixed parameter and θ is a variable, and for a prismatic joint, d is a variable whereas θ is fixed. The variable parameter is referred to as *joint variable*.

The D-H modeling convention starts by assigning a Cartesian coordinate frame $\{x_i, y_i, z_i\}$, $i = -1, 0, \cdots, n$ for each link. The world, base and tool frames are denoted as the -1th, 0th and nth link frames, respectively. n is the number of degrees of freedom of the manipulator that, for a serial manipulator, equals the number of joints. For a revolute joint, the origin of frame i-1 is taken to be on joint i. On the other hand, for a prismatic joint, the origin can be assigned arbitrarily. It will be seen that in order to reduce the number of parameters that represent a prismatic joint, the origin can be placed on joint i+1. The positive direction of axis i defines the unit vector z_{i-1}.

The transformation from the world frame to the tool frame in the D-H convention is:

$$T_n = A_0 A_1 A_2 \ldots A_i \ldots A_{n-1} A_n \qquad (3.3.1)$$

where A_i is the transformation of link coordinate frame i with respect to frame *i*-1. Refer to Figure 3.3.5, in which joints are shown as revolute.

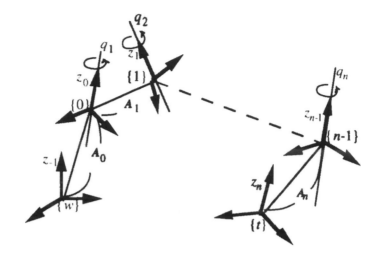

Figure 3.3.5 Coordinate assignment

The coordinate frame assignment for each of the internal links, $i = 0, ...,$ n-1 starts at $i = 0$ and proceeds serially from one link to its neighbor. These are established using three rules:

1. The z_{i-1} axis lies along the axis of motion of the ith joint.
2. The x_i axis is normal to the z_i axis and points in the direction of the common normal between axes i-1 and i.
3. The y_i axis completes a right-hand coordinate system.

The link transformation A_i is represented in terms of the four parameters $\{\theta_i\ d_i\ a_i\ \alpha_i\}$, thereafter referred to as the D-H parameters. The construction of A_i for internal links $i = 1, 2, ..., n$-1 is shown next. If the ith joint is revolute, the following transformations are required to bring link frame i-1 to link frame i (Figure 3.3.6),

1. Rotating frame i-1 about the z_{i-1} axis by an angle θ_i, so that the x axis of the moving frame is parallel to the x_i axis.
2. Translating the frame along the z_{i-1} axis a distance d_i, for the origin of the moving frame to reach the intersection point of the ith joint axis with the common normal.
3. Translating along x_i a distance a_i, so that the origin of the moving frame

coincides with that of frame i.

4. Rotating by an angle α_i about the rotated x_{i-1}. The moving frame is now coincident with frame i.

In summary,

$$A_i = Rot\ (z,\ \theta_i)Trans(0,\ 0,\ d_i)Trans(a_i,\ 0,\ 0)Rot\ (x,\ \alpha_i).$$

(3.3.2)

Or

$$A_i = \begin{bmatrix} \cos\theta_i & -\sin\theta_i\cos\alpha_i & \sin\theta_i\sin\alpha_i & a_i\cos\theta_i \\ \sin\theta_i & \cos\theta_i\cos\alpha_i & 0 & a_i\sin\theta_i \\ 0 & 0 & -\cos\theta_i\sin\alpha & d_i \\ 0 & 0 & 0 & 1 \end{bmatrix}$$

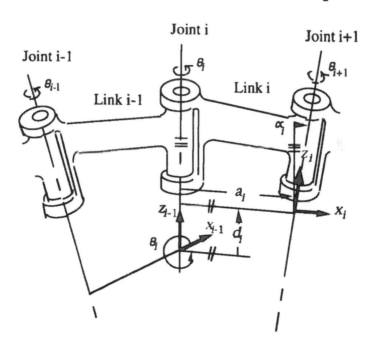

Figure 3.3.6 The D-H modeling convention for a revolute joint

In kinematic calibration, the geometric parameters to be identified in each link are d_i a_i α_i and the joint offset θ_i^{off}, where θ_i^{off} is the fixed portion of the joint variable θ_i. Similarly, if the ith joint is prismatic (Figure 3.3.7),

by definition $a_i = 0$. Therefore

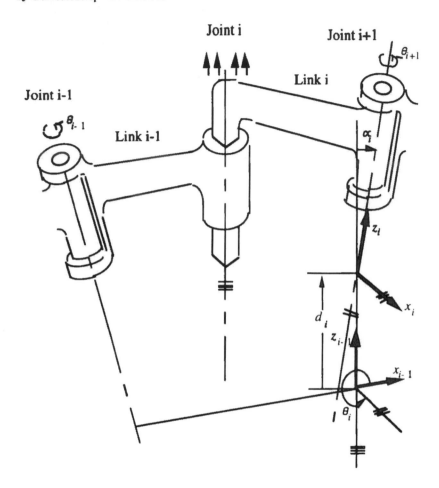

Figure 3.3.7 The D-H modeling convention for a prismatic joint

$$A_i = Rot\ (z,\ \theta_i)Trans(0,\ 0,\ d_i)Rot\ (x,\ \alpha_i).$$

(3.3.3)

In this case, the joint variable is d_i. The required parameters for calibration are α_i and θ_i.

In the case of consecutive parallel joint axes, the D-H convention stipulates that the common normal that satisfies $d_i = 0$ is selected.

The world and the tool coordinate systems are usually defined by the user. The base coordinate system, on the other hand, is often defined by the robot manufacturer. The original D-H convention restricted the assignment of the world and tool coordinate systems. For calibration this is unrealistic because this will cause these frames to vary depending on variations in the robot geometry. Allowing the world coordinate system $\{W\}$ to be defined and assigned arbitrarily and independently of the manipulator location, the fixed world-to-base transformation A_0, denoted also as the $BASE$ transformation, will depend in general on 6 independent parameters, reflecting the full 6 degrees of freedom in assigning the world coordinate frame.

Robot calibration often necessitates the definition of a so-called "flange frame" located on the mounting surface of the robot end-effector. $FLANGE$, the transformation from the $(n\text{-}1)$th link system to the flange system, contains the nth joint variable while $TOOL$, the transformation from the flange to the tool systems, is fixed. Thus

$$FLANGE\ TOOL\ \equiv A_n$$

In order to arbitrarily assign the world and tool frames, either the $BASE$ transformation or the $TOOL$ transformation shall have 6 independent link parameters. If 6 parameters are employed in the base transformation, one possible choice is

$$A_0 = Rot(z,\ \theta_0)Trans(0,\ 0,\ d_0)Rot(x,\ \alpha_0)Rot(y,\ \beta_0)Trans(a_0,\ b_0,\ 0) \quad (3.3.4)$$

Likewise

$$A_n = Rot(z,\ \theta_n)Trans(0,\ 0,\ d_n)Rot(x,\ \alpha_n)Rot(y,\ \beta_n)Trans(a_n,\ b_n,\ 0) \quad (3.3.5)$$

Equations (3.3.4)-(3.3.5) may not be suitable for all geometric setups. Particular selections of A_0 or A_n require in general some user ingenuity.

The D-H modeling convention that is now routinely taught in every basic robotics course has been by far the most widely used robot modeling convention. In particular, many robot manufacturers employ the D-H convention in implementing the robot forward and inverse kinematic routines that are part of the robot control software.

B. COMPLETENESS, PROPORTIONALITY AND SHORTCOMINGS OF THE D-H MODEL FOR ROBOT CALIBRATION

Robot calibration requires the identification of the manipulator link parameters in whatever kinematic modeling convention chosen by the user. At the beginning of the chapter the properties of completeness and proportionality were stated as "desired" in a kinematic model used for calibration.

To assess whether a given model is complete, there is a need to understand how many independent kinematic parameters are required in any complete model. The construction of any robot kinematic model involves the assignment of coordinate frames on the robot joint axes and the study of their relationship in an arbitrary robot position. In the case of a revolute joint, one places the origin of the frame in a specific location on the joint axis and the z-axis is defined along the joint axis. The origin location is irrelevant in the case of a prismatic joint. Therefore, there are four constraints associated with placing a coordinate frame on a revolute joint - two parameters have to specify the joint axis orientation and two parameters specify the origin place on the joint axis. There are only two constraints associated in placing a coordinate frame on a prismatic joint - the parameters that specify the joint axis orientation. In addition to allow arbitrary selection of the world and tool coordinate frame, one needs a total of 6 independent parameters, spread among the $BASE$ and $TOOL$ transformations.

The total number of independent kinematic parameters in a complete model of an n degrees-of-freedom manipulator is therefore

$$N = 4R + 2P + 6 \qquad (3.3.6)$$

where R and P are the numbers of revolute and prismatic joints, respectively.

Coming back to the case of the D-H model, the kinematic parameters that need to be identified are the D-H link parameters $\{\theta_i \; d_i \; a_i \; \alpha_i\}$ associated with the link transformations A_i of links preceded by revolute joints, the D-H parameters $\{\theta_j \; \alpha_j\}$ associated with the transformation A_j of links preceded by prismatic joints. In addition, A_0 is assumed to have 6 parameters. One can also choose A_n to have 6 independent parameters, in which case A_0 will have at most 4 parameters.

The joint offset d_j^{off}, associated with a prismatic joint j, is not to be identified. In the least squares sense, the effect of such an error parameter on the robot pose error is absorbed into other translational parameters, primarily those associated with A_0 (or A_n, depending upon the transformation that possesses 6 parameters).

Parameter identification methods will be discussed in detail later in the book. At this point it will do to recognize that there are in general two classes of identification techniques - those in which the parameters are identified directly, and those in which each additive error parameters with respect to the known nominal ones are identified using linearized accuracy models. The property of proportionality is important only for the latter class of techniques, as linearized accuracy models are valid only if the error parameters are indeed small.

The D-H model is not proportional for links that have two parallel revolute joint axes. The following example illustrates this phenomenon:

Example: Particular joint axis misalignment.

As a particular case of misaligned almost-parallel consecutive joint axes refer to Figure 3.3.8. In (b), a "worst case" scenario is assumed, in which the two almost-parallel joint axes lie on the same plane π. Since the two axes intersect, their common normal length a_i equals zero, compared to a certain finite value in the ideal case. Likewise, there is a jump discontinuity in d_i (from 0, by the D-H convention for parallel axes, to a very large number).

It is important to keep in mind that any kinematic parameter identification is to be done in a least squares sense. That is, the identified set of parameters account, in the least squares sense, for the robot accuracy errors. The number of identified parameters sometimes depends on the application. A robot user may perform full-scale calibration of all $4R + 2P + 6$ parameters at designated scheduled maintenance times, and may perform more frequently a partial calibration of the joint offsets only or the base transformation only, etc. This is perfectly feasible. Some identification methods may become ill-conditioned whenever redundant parameters are included (such as a prismatic joint offset), others will still produce some least-squares optimal solution. In practice, identification of redundant models has smaller domain of validity within the robot workspace.

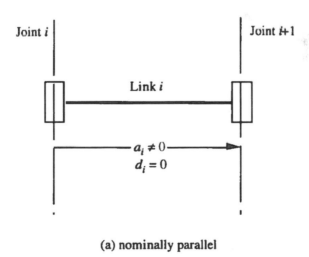

(a) nominally parallel

Figure 3.3.8 Two consecutive revolute joint axes

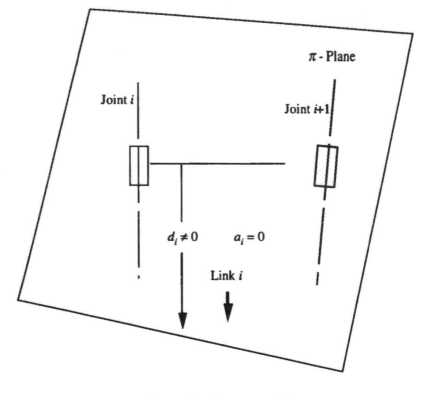

(b) actually almost-parallel

Figure 3.3.8 Two consecutive revolute joint axes (continued)

C. MODIFICATIONS TO THE D-H MODEL

A popular idea of making the D-H model proportional is known as Hayati's *Modified D-H model*. The following "IF statement", according to Hayati, is to be added to the D-H convention:

If the ith and $i+1$th joint axes are nominally parallel, the following transformation replaces the original D-H transformation,

$$A_i = Rot\ (z,\ \theta_i)Trans(a_i,\ 0,\ 0)Rot\ (x,\ \alpha_i)Rot\ (y,\ \beta_i).$$

for a revolute joint, or

$$A_i = Trans(0,\ 0,\ d_i)\ Rot\ (x,\ \alpha_i)Rot\ (y,\ \beta_i).$$

for a prismatic joint.

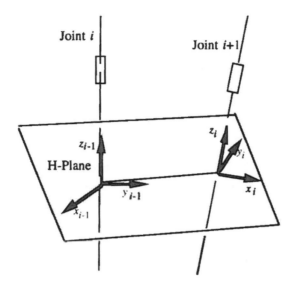

Figure 3.3.9 Hayati's modification to the D-H model for a revolute joint
(H-plane is perpendicular to joint i)

In other words, the coordinate frame $\{x_i, y_i, z_i\}$ is assigned as follows:
The origin of frame $\{x_i, y_i, z_i\}$ is placed at the intersection point of a plane
(termed the H-plane) that is perpendicular to joint axis i and passes through
the origin of frame $\{x_{i-1}, y_{i-1}, z_{i-1}\}$. The axis z_i coincides with joint axis
$i+1$.

The construction of a link transformation, between nominally parallel
revolute joint axes, takes the following steps:

1. The rotation $Rot\ (z,\ \theta_i)$ about z_{i-1} is to align x_{i-1} with the line connecting
   the two respective origins.
2. The translation by a_i along the rotated x_{i-1} aligns the two origins.
3. The final rotations $Rot\ (x,\ \alpha_i)Rot\ (y,\ \beta_i)$ align z_{i-1} to z_i.

Note that the x_i axis is the x_{i-1} axis after the above three-step
transformation. The rotation $Rot\ (z,\ \theta_i)$ and the translation $Trans(a_i,\ 0, 0)$ are
not needed for prismatic joints. With this convention, all link parameters are
continuous between the parallel and nearly-parallel cases. Hayati's convention
is to be used only for consecutive parallel joint axes.   If applied to
consecutively perpendicular joint axes, there would be a large discontinuity in
either α_i or β_i, between the perfectly-perpendicular and nearly perpendicular
cases.

IV.   THE CPC AND MCPC MODELS

A. A SINGULARITY-FREE LINE REPRESENTATION

The following line representation was introduced by Roberts (1988). A line B in 3-D space may be expressed in terms of four parameters by means of the following method. Let the orientation of the line be specified by two direction cosines in a suitable reference coordinate frame $\{x, y, z\}$. Let the position of the line be specified by a point P, the intersection of the line with a plane B that is perpendicular to the line and passes through the origin of $\{x, y, z\}$ (Figure 3.4.1). A 2-D Cartesian coordinate frame is defined on the B-plane, the origin of which is taken to be coincident with that of the reference frame. The remaining two parameters are taken to be the coordinate values, in that local frame, of Point P. Line B can then be represented by a 4-tuple $\{b_x, b_y, l_x, l_y\}$, where b_x and b_y are the x and y components of the direction unit vector b; l_x and l_y are the coordinates of Point P in the 2-D coordinate frame defined on the B-plane (Figure 3.4.1). Note that b_z, the z component of the direction unit vector is

$$b_z = (1 - b_x^2 - b_y^2)^{1/2} \qquad (3.4.1)$$

By definition (through adopting the plus sign in (3.4.1)) the direction unit vector b is forced to lie in the upper half-space defined by the x and y axes of the reference frame.

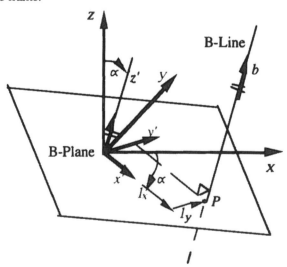

Figure 3.4.1 A line representation
(B-plane is perpendicular to the B-line)

A convention for choosing the coordinate axes of the local 2-D frame remains to be defined. The projection of the x axis, of the reference frame, onto the B-plane is taken to be the x axis of the local frame. Let z be a unit vector along the z axis of the reference frame. One way of implementing this projection is to rotate the reference frame by an angle of α about an axis k, where

$$\alpha \equiv \arccos(z \cdot b) = \arccos(b_z) \qquad (3.4.2)$$

and

$$k = \frac{z \times b}{\|z \times b\|} = \begin{bmatrix} \dfrac{-b_y}{\sqrt{b_x^2 + b_y^2}} \\[2mm] \dfrac{b_x}{\sqrt{b_x^2 + b_y^2}} \\[2mm] 0 \end{bmatrix} \qquad (3.4.3)$$

$\|\cdot\|$ denotes the Euclidean norm. The rotation matrix R is

$$R = Rot(k, \alpha) \qquad (3.4.4)$$

where k is a unit vector along the common normal and α is the familiar twist angle in the D-H convention. R is undefined from (3.4.3) when $b = z$. R however can be defined without explicit reference to $z \times b$. Substitution of (3.4.2) and (3.4.3) into the general rotation matrix formula of (3.2.3) yields

$$R = \begin{bmatrix} 1 - \dfrac{b_x^2}{1 + b_z} & \dfrac{-b_x b_y}{1 + b_z} & b_x \\[3mm] \dfrac{-b_x b_y}{1 + b_z} & 1 - \dfrac{b_y^2}{1 + b_z} & b_y \\[3mm] -b_x & -b_y & b_z \end{bmatrix} \qquad (3.4.5)$$

In the case of $b = z$, that is $[b_x, b_y, b_z]^T = [0, 0, 1]^T$, R becomes the identity matrix. Since b_z is by definition nonnegative, R is well defined for any direction of the line in 3-D space. Thus the unit basis vectors, x', y', z', for the new coordinate frame are given by the first, second and third columns of R, respectively.

Representing the above reference frame projection onto the B-plane through the component of the b vector is critical. It is very easy to go wrong at this step. For instance, representing R by $R = Rot(x, \alpha)Rot(y, \beta)$ where the first rotation by α brings the y axis to lie on the B-plane and the next rotation by β brings the x axis to the B-plane, would not yield a parametrically continuous model. More about this issue will be discussed later.

B. THE CPC MODEL

The CPC model synthesizes many attractive features of other kinematic models for robot calibration. The CPC model separates the joint motion from the link geometry. Each link transformation B_i has the following basic structure:

$$B_i = Q_i V_i, \qquad i = 0, 1, \cdots, n \qquad (3.4.6)$$

where Q_i and V_i are referred to as link motion matrix and shape matrix, respectively. It will be seen later in this section that the motion matrices only depend on the joint variables and the shape matrices are related only by the fixed link parameters. Furthermore, the link shape matrices, in the CPC model, are expressed in terms of the relative direction vector of each joint axis together with relative distance parameters. Roberts' line representation is adopted to ensure parametric continuity of the CPC model. Completeness of the model is achieved by adding two parameters to Roberts' line parameters. Post-multiplying the line representation's rotation matrix by $Rot(z, \beta)$ allows an arbitrary orientation of the new coordinate frame, and an additional translation by l_z along the z axis allows arbitrary placement of the frame origin.

Let the 4x4 homogeneous transformation T_n relating the position and orientation of an end-effector to the world coordinates be

$$T_n = B_0 B_1 B_2 \cdots\cdots B_{n-1} B_n \qquad (3.4.7)$$

In the CPC model, the position and orientation of the world and tool frames can be assigned arbitrarily, with the only restriction that the z axis of the world frame must not lie opposite to that of the base frame, and the z axis of the tool frame must not be opposite to that of the $(n-1)$th link frame. All other link frames are established based on the following convention:

1. The z_i axis must be on the $i+1$th joint axis for a revolute joint and parallel to it for a prismatic joint.
2. The coordinate frame $\{x_i, y_i, z_i\}$ forms an orthonormal right-hand system.

Link coordinate frame assignment follows Roberts' line representation convention. If however due to a physical constraint, the z axis of the world frame must point in an opposite or almost-opposite direction to the z axis of the base frame, an intermediate frame can be introduced to avoid model singularity. Similarly in the case when the z axes of the tool and the $(n-1)$th link frames are in opposite directions. A possible choice of such an intermediate frame is to have its z axis be perpendicular or almost perpendicular to the respective neighboring z axes.

In the case of a revolute joint, the link transformation matrix B_i is a function of five link parameters $\{b_{i,x}, b_{i,y}, b_{i,z}, l_{i,x}, l_{i,y}\}$ and one joint variable denoted as θ_i, while in the case of a prismatic joint, the joint variable is denoted as d_i, and the link parameter set is reduced to $\{b_{i,x}, b_{i,y}, b_{i,z}\}$. Let $l_i = [l_{i,x}, l_{i,y}, 0]^T$ and $b_i = [b_{i,x}, b_{i,y}, b_{i,z}]^T$. The assignment rules for link parameters and joint variables are as follows. If the ith joint is revolute, then (refer to Figure 3.4.2):

1. b_i is the direction unit vector of the $(i+1)$th joint axis represented in the $(i-1)$th link frame.
2. l_i represents the coordinates of the origin of the ith link frame in the rotated $(i-1)$th frame.
3. The zero position of joint variable θ_i (for $i = 1, \ldots, n$) corresponds to the zero reading of the ith joint position transducer.

Remark: The index of a joint axis is always greater by one than that of a link frame. That is why the $i+1$th joint axis is represented in the $(i-1)$th link frame by b_i.

If the ith joint is prismatic, $l_{i,x}$ and $l_{i,y}$ are set to zero (refer to Figure 3.4.3). The zero position of the joint variable d_i corresponds to the zero reading of the ith joint position encoder.

Define a 4x4 rotation matrix R_i whose upper-left 3x3 submatrix is as in (3.4.5). R_i is a function of the direction cosines $b_{i,x}, b_{i,y}$ and $b_{i,z}$ of the $i+1$th joint axis in the $(i-1)$th link frame. Specifically, for $i = 0, 1, \ldots, n$,

$$R_i = \begin{cases} Rot(k_i, \alpha_i) & \text{if } z_i \text{ is not parallel to } z_{i-1}, \\ I_{4\times4} & \text{if } z_i \text{ is parallel to } z_{i-1}, \end{cases}$$

with $k_i = e_3 \times b_i / \|e_3 \times b_i\|$ and $\alpha_i = \arccos(e_3 \cdot b_i)$, where $e_3 = [0, 0, 1]^T$. By (3.4.5),

$$R_i = \begin{bmatrix} 1 - \dfrac{b_{i,x}^2}{1 + b_{i,z}} & \dfrac{-b_{i,x} b_{i,y}}{1 + b_{i,z}} & b_{i,x} & 0 \\[2.5ex] \dfrac{-b_{i,x} b_{i,y}}{1 + b_{i,z}} & 1 - \dfrac{b_{i,y}^2}{1 + b_{i,z}} & b_{i,y} & 0 \\[2.5ex] -b_{i,x} & -b_{i,y} & b_{i,z} & 0 \\[1.5ex] 0 & 0 & 0 & 1 \end{bmatrix} \qquad (3.4.8)$$

where $b_{i,z} \geq 0$.

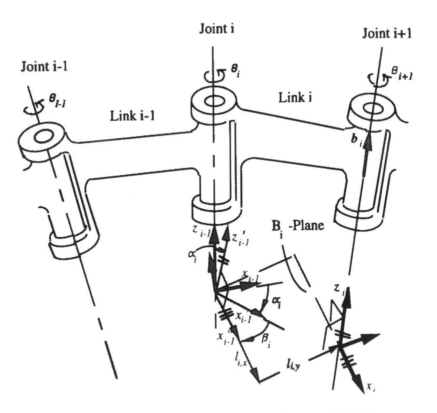

Figure 3.4.2   CPC modeling convention for a revolute joint

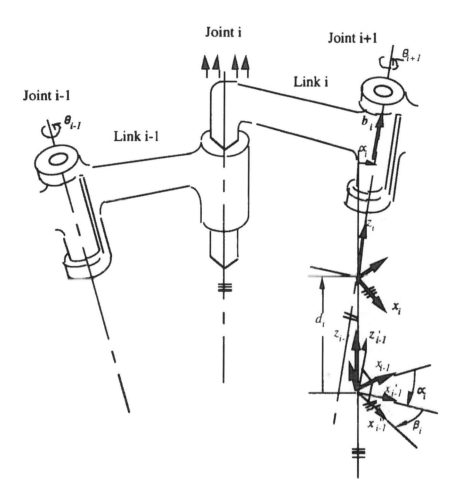

Figure 3.4.3 CPC modeling convention for a prismatic joint

Remark: During the iterations of a kinematic parameter identification process, $b_{i,z}$ is allowed to take negative values to describe the actual geometry of the manipulator as long as the condition $b_{i,z} > -1$ is met.

Assuming that the robot is at its zero position, the link transformation B_i is defined referring to a plane B_i which is perpendicular to the $(i+1)$th joint axis and passes through the origin of the $(i-1)$th link frame. The link transformation construction procedure is as follows:

1. The rotation R_i, as defined in (3.4.8), is performed. The resulting intermediate frame is denoted as $\{x_{i-1}', y_{i-1}', z_{i-1}'\}$. z_{i-1}' becomes parallel to z_i and the other two coordinate axes lie on the B_i plane.
2. The translation $Trans(l_{i,x}, l_{i,y}, 0)$ is performed. The resulting frame becomes coincident with the ith link frame.

The link transformation B_i is given by (3.4.6) where Q_i, the motion matrix, for $i = 1, 2, \cdots, n$, is

$$Q_i = \begin{cases} Rot\ (z, \theta_i) & \text{for a revolute joint} \\ \\ Trans(0, 0, d_i) & \text{for a prismatic joint} \end{cases} \qquad (3.4.9)$$

and V_i, the shape matrix, is

$$V_i = R_i Rot\ (z, \beta_i)Trans(l_{i,x}, l_{i,y}, l_{i,z}) \qquad i = 0, 1, \cdots, n \qquad (3.4.10)$$

By definition, $\theta_0 \equiv 0$ and $d_0 \equiv 0$. It is clear now that each shape matrix V_i, $i = 0, 1, \cdots, n$, is specified by all the fixed link parameters in the ith link transformation B_i, and is independent of the ith joint variable.

The CPC modeling convention allows arbitrary assignment of link coordinate frames. $Rot(z, \beta_i)$ is introduced to allow an arbitrary x-axis orientation of the ith link frame, and $Trans(0, 0, l_{i,z})$ is introduced to allow an arbitrary positioning of the ith link frame. The parameters β_i and $l_{i,z}$ are redundant for $i = 1, 2, ..., n-1$. Further, for prismatic joints, $l_{i,x}$ and $l_{i,y}$ are also redundant parameters. In a parameter identification process, each of these redundant parameters can be set to either zero or other constant value.

The assignment of link frames and link parameters for a "simple" robot (i.e, a robot in which two consecutive links are either perpendicular or parallel to one another) can be greatly simplified. For the sake of convenience, let $\beta_i = 0$, which is always so for internal links.   Two additional and highly convenient assignment rules are as follows:

1. If z_i is perpendicular to x_{i-1}, then let x_i have the same direction as x_{i-1}.
2. If z_i is parallel to x_{i-1}, then let y_i have the same direction as y_{i-1}.

For examples, refer to Sub-section D, of this section.

C. THE MCPC MODEL

The reasons for creating alternatives to the CPC model will become apparent later when kinematic error models are discussed. Meanwhile we will simply state the model construction convention. The modified CPC model uses four parameters to represent the internal link transformation of a robot. Like the CPC model, it too separates joint variables from link parameters. The difference between the models is that the MCPC model employs two angular parameters to model the rotation part of each link transformation. The procedure for the assignment of the MCPC internal link parameters is as follows (Figure 3.4.4):

1. A plane B_i is constructed to be perpendicular to the z_i axis and to pass through the origin of frame $\{i\text{-}1\}$.
2. Frame $\{i\text{-}1\}$ is rotated by an angle α_i about the x_{i-1} axis until the y axis of this frame lies on the B_i-plane. This y axis is now parallel to the y_i axis.
3. The moving frame is further rotated by an angle β_i about the resulting y axis so that the x axis of the moving frame lies on the B_i-plane. This x axis becomes parallel to the x_i axis. Consequently, the resulting z axis is also parallel to the z_i axis.
4. The moving frame is translated along the x_i and y_i axes by $l_{i,x}$ and $l_{i,y}$, respectively. The moving frame is finally coincident with $\{i\}$.

In summary, the ith link transformation in the MCPC convention is

$$B_i = Q_i V_i. \qquad i = 0, 1, \cdots, n \qquad (3.4.11)$$

where Q_i, the motion matrix, is given in (3.4.9). V_i, the shape matrix, for $i = 0, 1, \cdots, n$, is given by

$$V_i = Rot(x, \ \alpha_i)Rot(y, \ \beta_i)Rot(z, \ \gamma_i)Trans(l_{i,x}, \ l_{i,y}, \ l_{i,z}) \qquad (3.4.12)$$

By definition, $\theta_0 \equiv 0$ and $d_0 \equiv 0$. For the same reasons as in the CPC model, the additional redundant transformations represented in terms of parameters γ_i and $l_{i,z}$ are added for $i = 0, 1, ..., n\text{-}1$. Furthermore, for prismatic joint, $l_{i,x}$ and $l_{i,y}$ are also redundant.

Let $\gamma_i = 0$, which is always so for internal links. The same as in the CPC model, two additional convenient assignment rules for simple robots are as follows:

1. If z_i is perpendicular to x_{i-1}, then let x_i have the same direction as x_{i-1}.
2. If z_i is parallel to x_{i-1}, then let y_i have the same direction as y_{i-1}.

There are other similar MCPC models, depending on the parameters used to represent the projection of the reference frame x and y axes onto the B-plane. For instance, one may rotate first about the y axis and then about the x axis. It is also possible to rotate about z to bring the x or y axis to lie on the B-plane, then to rotate about the axis that has already been laid on this plane.

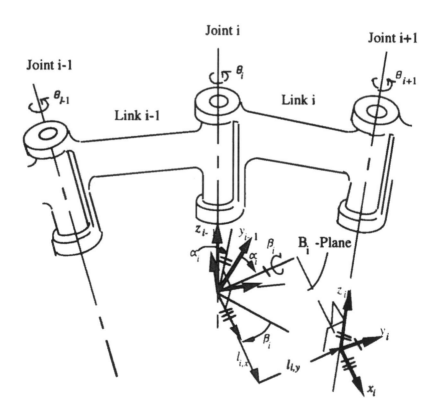

Figure 3.4.4 MCPC modeling convention for a revolute joint

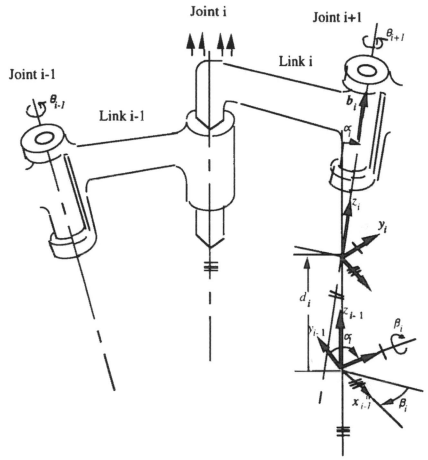

Figure 3.4.5 MCPC modeling convention for a prismatic joint

D. EXAMPLES

The following examples illustrate the CPC and MCPC modeling conventions.

Example 3.4.1 (The CPC Model):   Consider two internal and consecutive revolute joints (Figure 3.4.6). Let the z_i axis be nominally parallel to the z_{i-1} axis ($b_i = [0, 0, 1]^T$). Since z_i is perpendicular to x_{i-1}, the axis x_i has the same direction as x_{i-1} (refer to the simple rules given at the end of Section IV.B). Thus $l_i = [0, l_{i,y}^0, 0]^T$.

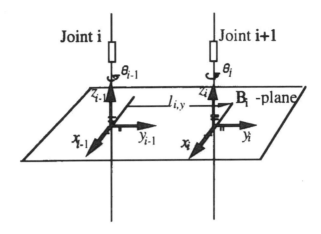

Figure 3.4.6   An example of two consecutive CPC link transformations

Example 3.4.2 (The MCPC Model): Consider the same geometry as in Example 3.4.1 (Figure 3.4.6). For the same reason, frame i has the same orientation as that of frame i-1 (refer to the simple rules given in the end of Section IV.C). Thus $\alpha_i = \beta_i = 0$ and $l_i = [0, l_{i,y}{}^0, 0]^T$.

Example 3.4.3 (The CPC Model): Let us continue the analysis of Example 3.4.1. Suppose now that joint i+1 becomes slightly misaligned such that $b_i = [0, e_y, (1 - e_y{}^2)^{1/2}]^T$ (Figure 3.4.7). The origin O_i, based on the CPC convention, is by $-e_y$. Therefore, by a linear approximation,

$$R_i = \begin{bmatrix} 1 & 0 & 0 \\ 0 & 1 & e_y \\ 0 & -e_y & 1 \end{bmatrix}$$

The new l_i is $[0, l_{i,y}{}^0 + e_y, 0]^T$. As can be seen, a small variation of the joint axis orientation results in small change of the CPC link parameters. Recall that in this case, since the $(i$-1)th and the ith joint axes become intersecting, the D-H parameter a_i (common normal length) jumps from the nominal value of $l_{i,y}{}^0$ to zero, and the D-H offset distance parameter d_i jumps from zero to the very large value of $l_{i,y}{}^0/e_y$.

The CPC model continues to be robust when two nominally-perpendicular consecutive joint axes become slightly nonperpendicular. However, the MCPC model may exhibit singularity behavior when two consecutive joint axes are perpendicular one to another.   This issue will be discussed in Section VII.

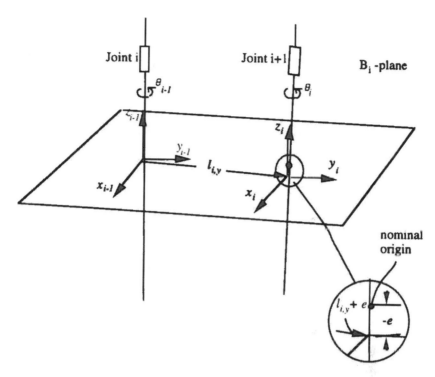

Figure 3.4.7   CPC modeling convention handles joint axes misalignment

Example 3.4.4 (The CPC Model): Modeling of joint offsets.

Whenever the reading of a joint variable is set to zero, whereas the actual (physical) joint angle is not at the zero position, the deviation from the zero joint position is called a "joint offset". In this example let us assume that the only geometric error of the ith link is its joint offset. In the D-H model, it is modeled by $d\theta_i$ (or dd_i). In the CPC model, the nominal value of the joint variables are always used to describe robot motion.   In this example, joint offsets can be accounted for through d_i.   Figure 3.4.8 illustrates such an example in which the nominal geometry of the link is the same as that given in Example 3.4.1 (in the figure the superscript "0" denotes a nominal entity). The joint offset here is modeled by $dl_{i,x}$ and $dl_{i,y}$.

If geometric errors involve both joint offsets and axis misalignment, it is important to stress that $d\theta_i$ (or dd_i) of the D-H model no longer represents the ith joint offset as the joint offset also depends on other parameter errors. Similarly for the CPC model.   However since joint offsets are redundant parameters, the identified independent CPC error parameters are sufficient for accounting, in the least squares sense, for pre-calibration pose errors. In addition, this set of independent parameters is sufficient for least-squares updating of nominal joint commands to minimize the post-calibration pose errors.

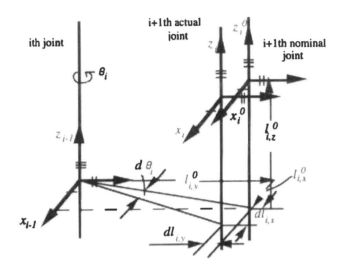

Figure 3.4.8 Modeling of a joint offset

Remark: In the case where the world and base z-axes are parallel and opposite, the *BASE* transformation is broken into a product of two transformations. Using the same line of reasoning, one may fix one of these transformations at its nominal value, and use six independent parameters from the other for kinematic identification and compensation.

V.   RELATIONSHIP BETWEEN THE CPC MODEL AND OTHER KINEMATIC MODELS

A. EXTRACTION OF CPC LINK PARAMETERS FROM LINK HOMOGENEOUS TRANSFORMATIONS

It is assumed for completeness of the discussion that the redundant CPC parameters are also included in the internal link transformations.

The objective of this section is to express the CPC model parameters in terms of the elements of the link transformation matrices B_i. Such formulas may be used to identify the *BASE* parameters for a calibrated robot whenever only the world frame changes its location. These formulas are also part of the proof of the parametric continuity of the CPC model (this will be presented in the next section). In the coming discussion, the index "i" is dropped for better clarity.

Denote by R^b and p^b the 3x3 rotational submatrix and the position vector of B, respectively. Then by (3.4.10)

$$R^b = Rot(z, \; \theta)RRot(z, \; \beta)$$
$$p^b = Rot(z, \; \theta)RRot(z, \; \beta)l + de_3$$

where by definition $\theta \equiv 0$ for a prismatic joint, $d \equiv 0$ for a revolute joint, and $e_3 \equiv [0, 0, 1]^T$. The joint rotation θ is not a CPC link parameter, thus $Rot(z, \; \theta)$ is known by taking a given joint command θ. Therefore

$$b \; = a^v \tag{3.5.1}$$

where a^v is the third column of $Rot(z, \; -\theta)R^b$. Given b, the parameters β and l can be solved for uniquely from

$$Rot(z, \; \beta) = R^T Rot(z, \; - \theta)R^b \tag{3.5.2}$$

and

$$l = Rot(z, -\beta)R^T Rot(z, \; -\beta)(p^b - de_3). \tag{3.5.3}$$

Hence, for a given value of the joint variable in B, the CPC parameters are continuous functions of the elements of B. This fact will be used to prove the parametric continuity of the CPC model.

B. MAPPING FROM THE D-H MODEL TO THE CPC MODEL

Transformations between the CPC model and the D-H model are useful whenever a robot controller already has its forward and inverse kinematics implemented in terms of the D-H model.

For a given set of D-H parameters, the matrices A_i, $i = 0,1, ..., n$, are computed using (3.3.2)-(3.3.3). The matrix values of B_i, the CPC link transformations, are then set to

$$B_i = A_i \qquad i = 0, 1, \cdots, n \tag{3.5.4}$$

The CPC link parameters for each link can then be found using (3.5.1)-(3.5.3).

C. MAPPING FROM THE CPC MODEL TO THE D-H MODEL

Given the CPC parameters, the D-H parameters must satisfy the following equality:

$$B_0 B_1 B_2 \ldots B_n = A_0 A_1 A_2 \ldots A_n .$$

Because A_0 is an arbitrary link frame transformation, let

$$A_0 = B_0 .$$

Assuming that all transformations A_i, $i = 0, \cdots, n-1$, have been found, then A_n, which is also an arbitrary link frame transformation, can be obtained through the following equation

$$A_n = \{A_1 A_2 \ldots A_{n-1}\}^{-1} B_1 B_2 \ldots B_n$$

The main problem is then to find A_1, A_2, ..., A_{n-1}, for a given transformation A_0. The solution method consists of three steps.

Step 1: The direction unit vector b_i of the joint axis and the origin p_i of the ith CPC link frame are transformed from the $(i-1)$th CPC link frame to the 0th link (base) frame. This can be done by the transformation T_{i-1} of (3.4.6). By doing so, ^0p_i and ^0b_i, the pose of the $(i+1)$th joint axis in the base frame, for $i = 1, 2, ..., n$, is obtained.

Remark: It is now straightforward to obtain the Zero-Reference model parameters from ^0p_i and ^0b_i for $i = 1, 2, ..., n$ (see Sub-section VIII).

Denote by ^0n_i the direction vector of the common normal between the ith and $i+1$th joint axes. Denote by s_i and w_{i+1} the intersection points of the common normal with the ith and $i+1$th joint axes, respectively. These too are represented in the base frame.

Step 2: ^0n_i, s_i and w_{i+1} are determined in terms of $\{^0p_{i-1}, {}^0b_{i-1}\}$ and $\{^0p_i, {}^0b_i\}$, assuming that $w_1 = 0$ (Figure 3.5.1). There are two cases.

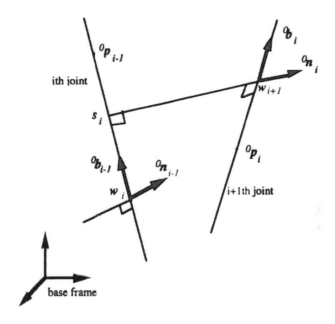

Figure 3.5.1 Determination of the intersection points of the common normal
with the two respective axes

Case 1: The ith and the $i+1$th axes are non-parallel.
By the definitions of ${}^0b_{i-1}$, ${}^0b_i$ and ${}^0n_i$,

$$^0n_i = {}^0b_{i-1} \times {}^0b_i \, / \|{}^0b_{i-1} \times {}^0b_i \|$$

The solution of s_i and w_{i+1} is obtained using standard vector analysis
formulas,

$$s_i = p_{i-1} + \frac{(p_i - p_{i-1}) \cdot ((b_i \cdot b_{i-1})b_i - b_{i-1})}{(b_i \cdot b_{i-1})^2 - 1} b_{i-1}$$

$$w_{i+1} = p_i + \frac{(p_i - p_{i-1}) \cdot (b_i - (b_i \cdot b_{i-1})b_{i-1})}{(b_i \cdot b_{i-1})^2 - 1} b_i$$

where the left superscript "0" is omitted for better clarity.

Case 2:   The ith and the $i+1$th axes are parallel.
    Let

$$s_i = {}^0p_i.$$

w_{i+1} is then the intersection point of the z_i axis with a plane that is perpendicular to z_{i-1} and passes through s_i, and it is given by

$$w_{i+1} = -{}^0b_i \cdot ({}^0p_i - s_i){}^0b_i + {}^0p_i$$

Also,

$${}^0n_i = (w_{i+1} - s_i)/ \|w_{i+1} - s_i\|$$

Step 3: The D-H parameters are determined in terms of ${}^0n_i$, s_i and w_{i+1}. The solution is

$$d_i = {}^0b_i \cdot (s_i - w_i)$$

If ${}^0n_{i-1} \times {}^0n_i \neq 0$, then

$$\theta_i = \text{sign}(k_1)\arccos({}^0n_{i-1} \times {}^0n_i)$$

where $k_1 = ({}^0n_{i-1} \times {}^0n_i) \cdot {}^0b_{i-1}$. Otherwise

$$\theta_i = 0.$$

If ${}^0b_{i-1} \times {}^0b_i \neq 0$, then

$$\alpha_i = \text{sign}(k_2)\arccos({}^0b_{i-1} \times {}^0b_i)$$

where $k_2 = ({}^0b_{i-1} \times {}^0b_i) \cdot {}^0n_{i-1}$. Otherwise

$$\alpha_i = 0.$$

Finally

$$a_i = {}^0n_{i-1} \cdot (w_{i+1} - s_i).$$

Remark: The mappings between the D-H and the MCPC models can be done in exactly the same way as outlined in the previous two sections.

VI.   PARAMETRIC CONTINUITY: GENERAL TREATMENT

The position and orientation of the $i+1$th robot joint axis in the $(i-1)$th link frame can be modeled by a point p_i on the $i+1$th axis and the direction unit vector b_i of the axis. Let A_i be the link homogeneous transformation between the $(i-1)$th and the ith link frames, in an arbitrary kinematic modeling convention. Let a_i be a vector that contains all 12 nontrivial elements of A_i, and let ρ_i be a vector of all the ith link parameters.

The vectors ρ_i and a_i are related in general through the functional relationship

$$\rho_i = \phi_i(a_i) \qquad i = 0, 1, ..., n$$

where ρ_i is an $m_i \times 1$ vector, and m_i is the number of link parameters. Likewise , a_i and the pose $\{p_i, b_i\}$ are related through

$$a_i = \psi_i(p_i, b_i) \qquad i = 0, 1, ..., n$$

where p_i and b_i are 3x1 vectors. Then

$$\rho_i = \phi_i (\psi_i(p_i, b_i)) = \eta_i (p_i, b_i) \qquad i = 0, 1, ..., n$$

Definition 3.6.1:   A manipulator kinematic model is said to be *parametrically continuous* if for every link, η_i is a continuous function of the respective joint axis poses $\{p_i, b_i\}$.

The vector function η_i is usually very complicated. On the other hand, each one of the functions, ϕ_i and ψ_i, is in general much simpler. Another version of the definition of parametric continuity will be based on the study of ϕ_i and ψ_i.

Definition 3.6.2:   A manipulator kinematic model is at its *model singularity* if either one of the following occurs:
1.   At least one of the elements of ρ_i cannot be written as a continuous function of a_i. That is, there exists an axis pose for which ϕ_i is a discontinuous function. This singularity is termed a *model singularity of the first type.*

2. At least one of the elements of a_i cannot be written as a continuous function of the pose $\{p_i, b_i\}$. That is, there exists an axis pose for which ψ_i is a discontinuous function. This singularity is termed a *model singularity of the second type*.

Remark: For a particular axis, p_i can be arbitrary. As long as there exists one p_i along this axis such that a_i is continuous in terms of p_i and b_i, we say that a_i is not singular at this pose.

Example 3.6.1: The D-H model.

It is not difficult to show that in the D-H modeling convention, ϕ_i is continuous for any manipulator. It can be observed from (3.3.2) that ψ_i becomes discontinuous whenever the direction unit vector b_i changes from $[0, 0, 1]^T$ to some neighborhood of $[0, 0, 1]^T$. The D-H model therefore has a model singularity of the second type.

Example 3.6.2: A z-y-x Euler angles model.

This example is constructed just for illustrating the existence of the first type of model singularities. The transformation A_i is formed as three translations $l_{i,x}$, $l_{i,y}$ and $l_{i,z}$ along the x, y and z axes of the (i-1)th link frame, followed by three rotations α_i, β_i and γ_i about the successively transformed z, y and x axes. The link transformation is

$$
A_i = \begin{bmatrix}
c\alpha_i c\beta_i & c\alpha_i c\beta_i s\gamma_i - s\alpha_i c\gamma_i & c\alpha_i s\beta_i c\gamma_i + s\alpha_i s\gamma_i & l_{i,x} \\
s\alpha_i c\beta_i & s\alpha_i c\beta_i s\gamma_i + c\alpha_i c\gamma_i & s\alpha_i s\beta_i c\gamma_i - c\alpha_i s\gamma_i & l_{i,y} \\
-s\beta_i & c\beta_i s\gamma_i & c\beta_i c\gamma_i & l_{i,z} \\
0 & 0 & 0 & 1
\end{bmatrix}
$$

Since the ith link frame can be assigned arbitrarily with respect to the (i-1)th link frame, there is no model singularity of the second type. The form of ϕ_i can be obtained by representing ρ_i in terms of the elements of A_i represented by the z-y-x Euler angles. As can be seen, if the x_i axis is non-parallel to the z_{i-1} axis, then $\alpha_i = \text{Atan2}(n^a_{i,y}, n^a_{i,x})$. Whenever the x_i axis is parallel to the z_{i-1} axis, α_i becomes undetermined from the equation. In this case, two angles, β_i and γ_i, are sufficient for the required transformation. The angle α_i may then be nominally set to zero. If, in particular, the x axis of the ith frame becomes slightly misaligned with the z axis of the (i-1)th frame, the α_i parameter will jump. This amounts to a model singularity of the first type.

Definition 3.6.3: A kinematic model is said to be *locally parametrically continuous* if any model singularity is outside of a closed subset of joint axis poses containing a nominal set of joint axis poses.

Definitions 3.6.1 and 3.6.3 differ in two respects: 1) A composite function of two discontinuous functions may theoretically be continuous; 2) Parametric continuity (in the sense of Definition 1) requires that there exist no model singularities in the entire space of all possible joint axis poses.

Theorem 3.6.1: The CPC model is locally parametrically continuous.
Proof: Let the pose of the $i+1$th joint axis be given by $\{p_i, b_i\}$. The elements of the rotation part of B_i are continuous functions of b_i since, by the CPC convention, $b_{i,z} \geq 0$. The 4th column of B_i can be set to p_i and is thus a continuous function of p_i. Therefore there exists no model singularity of the second type in any closed subset of joint axis poses containing nominal poses.

By (3.5.2)-(3.5.4), the CPC link parameters are continuous functions of the elements of B_i. Therefore there exists no model singularity of the first type. By Definition 3.6.3 the CPC model is thus a locally parametrically continuous model. □

We believe that Definition 3 will do for all practical purposes, and from there on the word "locally" will be omitted. The parametric continuity of the CPC model applies not only to the independent parameters but also to the redundant parameters.

Parametric continuity of a kinematic model is only a necessary condition for the application of error-model based kinematic identification. Conditions for the existence and nonsingularity of an identification Jacobian include differentiability of each B_i with respect to each ρ_i element and irreducibility of the error model (i.e., all error parameters have to be independent).

For kinematic identification methods that do not require the formation of an error model, the whole issue of parametric continuity is simply irrelevant.

VII. SINGULARITIES OF THE MCPC MODEL

As mentioned, the MCPC model is not free of singularities. First, the rotation part of its *TOOL* transformation is formed by three fixed angular parameters. As was analyzed in the last section, this kind of transformation has always one singular point. In addition to the wrist singularity, the internal link transformation has a singular point too. This is illustrated in the following example.

Let us assume that the $i+1$th joint axis is perpendicular to the ith joint axis. Let us also assume that x_{i-1} is nominally perpendicular to the $i+1$th joint axis. According to the MCPC modeling convention, the coordinate system $\{i\}$ is shown in Figure 3.7.1. The set of nominal link parameters $\{\alpha_i, \beta_i, l_{i,x}, l_{i,y}\}$ is $\{-90^0, 0, 0, 0\}$. If the $i+1$th joint axis is slightly misaligned, as shown in Figure 3.7.2, $\alpha_i, \beta_i, l_{i,x}$ and $l_{i,y}$ undergo small variations. Thus, there is no singularity in this case.

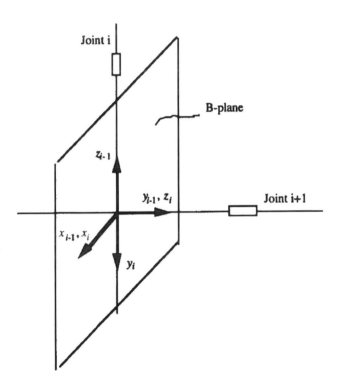

Figure 3.7.1 The MCPC coordinate frame assignment
when x_{i-1} is nominally perpendicular to z_i

Let us take the case where y_{i-1} is nominally perpendicular to the $(i+1)$th joint axis. The coordinate system $\{i\}$ in the MCPC convention is shown in Figure 3.7.3. If the $i+1$th joint axis is slightly misaligned, as shown in Figure 3.7.4, α_i may have a large jump, resulting in a model singularity.

Nominally, the representation of z_i in $\{i-1\}$ is $[-1, 0, 0]^T$. The set of nominal link parameters $\{\alpha_i, \beta_i, l_{i,x}$ and $l_{i,y}\}$ is $\{0, -90^0, 0, 0\}$. However, let us now assume that the actual z_i axis has a slight misalignment, but that it still passes through the origin of $\{i-1\}$. Its representation is $[(1 - e_1{}^2 - e_2{}^2)^{1/2},$ $e_1, e_2]^T$. It can be shown that the representation of y_i in this case becomes $[0,$ $e_1(e_1{}^2+e_2{}^2)^{1/2}, e_2(e_1{}^2+e_2{}^2)^{1/2}]^T$. To illustrate how large the jump in α_i is, let us consider the case of $e_1 = 0$ and $e_2 = 0.001$. Then $y_i = [0, 0, -1]^T$, and $\{\alpha_i,$ $\beta_i, l_i, l_{i,y}\} = \{-90^0, -90^0+\delta, 0, 0\}$, where δ is a small angle.

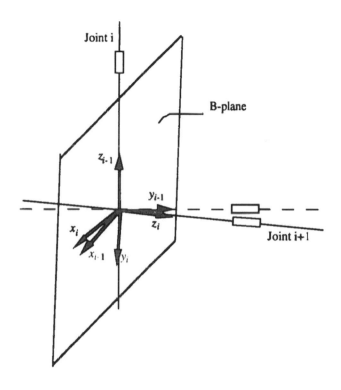

Figure 3.7.2 The MCPC coordinate frame assignment
when the $i+1$th joint is slightly misaligned from that given in Figure 3.7.1

Note that for a SCARA arm, the singularity illustrated in Figure 3.7.4 can never happen since there are no joint axes that are perpendicular to one another. For many other popular industrial manipulators, such as the PUMA 560 robot, this type of singularity can be bypassed by avoiding the case in which the x axis of the world coordinate frame is parallel to the z axis of the next joint axis. This z axis is also perpendicular to the first joint axis.

This type of singularity cannot be avoided whenever a manipulator features three mutually perpendicular consecutive axes, such as in the Cartesian IBM 7565 manipulator. This is a limitation of the MCPC model. Although modifications, such as reversing the order of rotations on x and y axes, can be adopted to get by the problem, the advantage of using the MCPC model in such a case is no longer significant over the modified D-H model. As a final comment, the type of singularity shown in this section is quite generic and in general can never be avoided whenever the rotation part of a link coordinate transformation is defined by any two angular parameters with respect to any axis.

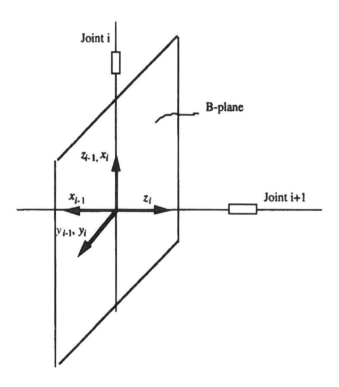

Figure 3.7.3 The MCPC coordinate frame assignment
when x_{i-1} is nominally parallel to z_i

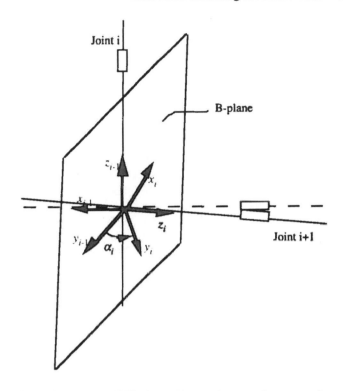

Figure 3.7.4 The MCPC coordinate frame assignment when
the $(i+1)$th joint axis is slightly misaligned from that given in Figure 3.7.3

VIII.   DISCUSSION AND REFERENCES

The general spatial displacement of a rigid body is the celebrated Chasles theorem (Chasles (1830)). For a thorough discussion of Chasles theorem and Rodrigues formula, the reader is referred to Murray, Li and Sastry (1994). Our own reference for deriving the homogeneous transformation from Rodrigues equation has been   Ravani (1985). Direct derivations of basic homogeneous transformations, the general rotation transformation and formulas for computing the axis of rotation and angle of rotation from numerical values of the rotation matrix can be found in Paul (1981).

The D-H convention for mechanisms was first introduced by Denavit and Hartenberg (1955), and became a "robotics standard" since the appearance of Paul's book. It should be noted though that many related conventions go by the name of Denavit-Hartenberg. Such conventions may differ in some of the details.   For instance, in the textbook by Craig (1989), the convention developed was that frame $\{x_i, y_i, z_i\}$ is attached to link i and has its origin lying on joint axis i. This results in a different link transformation.

$$A_i = Rot\ (x,\ \theta_{i-1})Trans(a_{i-1},\ 0,\ 0)Rot\ (z,\ \alpha_i)Trans(0,\ 0,\ d_i).$$

The standard textbook treatment, through the D-H modeling convention of prismatic joints, has been to ignore the difficulty of the redundant third parameter, namely the joint position offset.

The discovery by Mooring (1983) and Hayati (1983) that the D-H model is not proportional for parallel consecutive revolute joint axes has been one of the early sparks that started the intensive Robot Calibration research. Each of these researchers ended up working with different alternative models; refer to Mooring and Tang (1984), and Hayati and Mirmirani (1985). In fact, the number of kinematic models proposed, in the early days of Robot Calibration, almost equaled the number of researchers in this field.

The formula for the minimum number of independent parameters in a complete model is due to Everett, Driels and Mooring (1987). The book by Mooring, Roth and Driels (1991) explains in detail the origins and physical meaning of this formula. The same book also discusses in detail shortcomings of the D-H model such as the non-arbitrariness of the base frame selection and the manipulator zero position, which we decided not to repeat.

Mooring and Tang (1984) proposed the so-called Zero-Reference model that is outlined below (also refer to Suh and Radcliffe (1978), and Mooring, Roth and Driels (1991)). The model invokes the matrix version of Rodrigues' formula (equations (3.2.3)-(3.2.4)). Considering the right-to-left sequence of nominal pre-multiplying link transformations D_i

$$T = D_1 \ldots D_n T_0$$

Each transformation D_i is obtained from Rodrigues' formula by representing the ith joint axis of rotation and a point along each such axis with respect to the world frame. In quite a few practical cases, it is correct to assume that the robot nominal axes at its zero position are parallel to either the y or z axes of the world axes. To represent the world to tool transformation at the manipulator zero position, the actual link transformation D_i^a involves in general four independent physical error parameters, which assure that rotations are always performed with respect to actual axes of rotation.

One of the modified D-H models, coined the S-model by Stone (1987), was introduced primarily to assure that link coordinate frame will never fall outside of the manipulator structure. This is done by adding two parameters to the D-H model – an angle of rotation to allow arbitrary placement of the x axis and additional translation along the z axis, to allow more flexibility in placing the origin of the next frame. Since the S-model does not address the issues of completeness and proportionality of the D-H model, we chose not to discuss it within the chapter's text.

The CPC model is an evolution of several ideas that were "floating around" in the robot calibration literature. It is instructive to view the similarities to and distinctions from other models. As in Hayati's modification to the D-H model, a plane perpendicular to one of the joint axis is created. Hayati made the plane perpendicular to joint axis i-1, and the CPC model uses such a plane with respect to axis i. Like the Zero Reference model, a misaligned joint axis is modeled directly in terms of the components of its directional cosines, however in the Zero Reference model each joint axis is referred to the world coordinate system, whereas in the CPC model, each joint axis is referred to a local link coordinate frame.   The idea to separate the link transformation into a product of the "motion" and "shape" transformations was originally presented by Sheth and Uicker (1972) and adopted for robot calibration by Broderick and Cipra (1988). This partition allows the user to exclude the joint variable from the set of parameters that need to be identified.

What made the CPC model possible was a singularity-free line representation introduced by Roberts (1988) in the Computer Vision literature. Roberts surveyed commonly used line representations and concluded that all have either singularities or redundant parameters. He then presented a new line representation consisting of four parameters, which is the theoretical minimum. By way of specifying the orientation of a coordinate frame on the joint axis line, the only singularity in the line representation becomes removable.

The derivation of the CPC and MCPC models and their relationship to other models are based on Zhuang (1989), Zhuang, Roth and Hamano (1992b) and Zhuang, Wang and Roth (1993). Parametric continuity was first analyzed in Zhuang (1989) and expanded in Zhuang, Roth and Hamano (1992b). Singularities of the MCPC model were discussed in Zhuang and Roth (1996).

For those manipulators for which the MCPC model does not exhibit singularities, we found it to be more user-friendly than the CPC model because of its employing as link parameters of angular parameters, rather than direction vectors of joints. It will be seen in the next chapter that the MCPC error model is compact and convenient to construct. However, singularities in the MCPC model may prevent it from being applied to certain type of robots.

Calibration modeling and identification are not totally separate issues. Referring again to error model based kinematic identification, it is important to realize that for certain modeling conventions, including the CPC model, some of the error parameters may not have a direct physical meaning. A good example to that has been illustrated in Figure 3.4.8. More discussions on kinematic modeling and parameter identification were also given by Schroer (Bernhardt and Albright (1993)).

Chapter 4

POSE MEASUREMENT WITH CAMERAS

I.   INTRODUCTION

The measurement phase of robot calibration involves the collection of data needed for the kinematic parameter identification phase. In this book we focus only on model based kinematic identification. Such models, derived from the robot kinematic models, consist of sets of algebraic equations that relate the measured robot joint variables to sets of measured external variables. These variables could be the world coordinates of one or more points marked on the robot end-effector, or parameters related to certain constraint surfaces.

Let T_n and $T_n^{(0)}$ represent, respectively, the actual and nominal homogeneous transformation matrices of a robot end-effector coordinate frame with respect to its world coordinate frame. Given a set of measured joint positions in a particular robot configuration, the nominal transformation matrix $T_n^{(0)}$ can be computed using the nominal forward kinematic model. A 6x1 vector, which describes the position of the end-effector frame (through the position vector of $T_n^{(0)}$) and orientation (through three independent angles that characterize the rotation submatrix of $T_n^{(0)}$) in the world frame, is the *nominal robot end-effector's pose*. Likewise, we can define the *actual robot end-effector pose* (or in short the *robot pose*, as will be referred to from now on).

The transformation T_n provides the source for the algebraic equations used in the identification phase, in which the unknowns are the components of the kinematic parameter vector. The measured quantities are the joint variables and certain elements of the matrix T_n. For instance, if a single 3D end-effector point is measured in the world coordinate frame, the position vector of T_n, referred to as the end-effector "position", provides three algebraic equations that relate the endpoint position measurement to elements of the robot kinematic parameter vector. If multiple non-colinear end-effector points are measured at each robot measurement configuration (that is at each robot pose), one may compute the entire matrix T_n. From this extracted value of the matrix T_n, six independent algebraic equations can be obtained. These equations relate the pose measurement to all components of the robot

kinematic parameter vector.   Techniques for direct identification of robot
kinematic parameters are to be discussed in Chapter 6.

One may alternatively relate the pose error vector, defined in terms of the
difference between the measured and computed poses, to the kinematic
parameter error vector using a kinematic error model. The detailed discussion
of this issue is in Chapter 5. In both identification approaches, it is therefore
necessary to measure the robot end-effector pose or position in multiple robot
configurations.

Whenever a camera or a pair of cameras is used for robot endpoint
sensing, one can easily obtain a complete pose of the robot, because viewing
multiple target points is not much more difficult than viewing a single target
point.

Robot calibration by a camera system is potentially fast, automated, non-
invasive, user-friendly. Cameras can also provide full pose measuring
capability.   There are two typical setups for vision-based robot pose
measurement. The first is to fix cameras in the robot environment so that,
while the robot changes its configuration, the cameras can "view" a
calibration fixture mounted on the robot end-effector (Figure 4.1.1).   The
second setup is to mount a camera, or a pair of cameras, on the end-effector of
the robot manipulator (Figure 4.1.2). This setup will be referred to as a hand-
mounted camera setup.

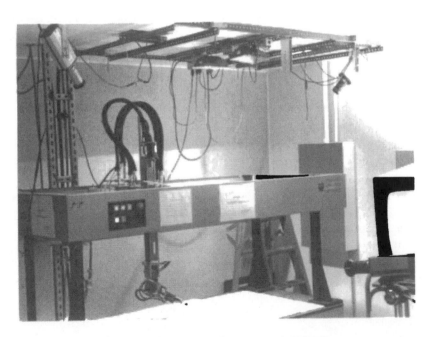

Figure 4.1.1 A stationary-camera setup

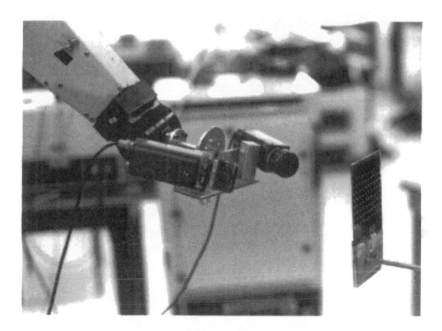

Figure 4.1.2   A hand-mounted camera setup

If the cameras in the system are calibrated prior to robot calibration, the locations of the hand-mounted calibration fixture in world coordinates for various robot measurement configurations can be computed by the vision system. The stationary-camera setup has two distinct advantages. First, it is non-invasive. The cameras are often placed outside the robot workspace, and need not be removed after robot calibration. Second, there is usually no need to identify the transformation relating the camera frame to the end-effector frame. This transformation in such a case is actually rather easy to compute in such a case. The major problem existing in all stationary camera setups is the accuracy of measurements. In order to have a large field-of-view for the stationary cameras, one has to sacrifice measurement accuracy. By using higher resolution cameras, the cost of the system, in particular that of its image processing part, may increase dramatically.

The moving camera approach can resolve the conflict between high accuracy and large field-of-view of the cameras. The cameras need only perform local measurements, whereas the global information on the robot end-effector pose is provided by a stationary calibration fixture. As the cameras are mounted on the robot hand, this method is necessarily invasive, which may prevent it from being used in certain applications. Another difficulty, which arises in this class of measurements, is that only the transformation from the world coordinate system to the camera coordinate system is computed. In many applications, the user may be mainly interested in the transformation from the world coordinate system to the tool coordinate system. Thus, a remaining task is to identify the transformation from the camera system to the tool system, which is in general a nontrivial task.

Methods for robot calibration using hand-mounted cameras can be classified into "two-stage" and "single-stage" methods. In a two-stage approach, the cameras are calibrated in advance. The calibrated cameras are then used to perform robot pose measurements. In a single-stage approach, the parameters of the manipulator and those of the cameras are jointly and simultaneously estimated. Depending upon the number of cameras mounted on the robot hand, these methods can be further divided into stereo-camera and monocular-camera methods. In the stereo-camera case, two cameras that have the same nominal optical characteristics are mounted on the robot hand. In the monocular case, only a single camera is utilized.

This chapter starts with a description of different system setups. It then discusses specific camera-based methods for pose collection.

II.   SYSTEM CONFIGURATIONS

A. STATIONARY CAMERA CONFIGURATIONS
This type of system setup requires the use of multiple cameras placed at fixed locations in the robot work cell (Figure 4.1.1). The major factors that determine the placement of the cameras are their field-of-view and the measurement accuracy.

The cameras have to be placed in locations that maintain the necessary field-of-view overlap. The actual camera positions often need to be selected empirically, because in practice it is not easy to avoid situations where the object is sometimes partially hidden by the manipulator links, cables or tools. Optimal placement of cameras requires a significant amount of trial and error.

There is a conflict between the accuracy of measurements and the common field-of-view of the cameras. On one hand, by reducing the tilt angles of the cameras, the accuracy of camera measurement in the z direction worsens compared to that in the other directions. In order to reduce such errors, one may use three cameras. The third camera is placed in an appropriate location to maintain a common field of view; it points in the direction perpendicular to those of the other two cameras. This strategy may not be feasible in many cases because when the third camera is perpendicular to the other cameras, it may not have a common field of view with the others, except when all view a single target point.

In the stereoscopic determination of an object location, there are various error sources, such as lens distortion, defocusing, electrical noise and spatial quantization. To reduce the effects of the lens distortion, one may choose to limit the useful interior image area to within 90% of the total available image area. Defocusing effects are minimized by carefully adjusting and by picking the brightest pixel located within the circular spot region. The effect of electrical noise can be reduced by using a better synchronization scheme. Thus the major contribution to the error in determining object location using cameras is due to spatial quantization inherent to any digital image processing systems. For CCD cameras the physical dimensions of a single sensing element (pixel) determine the resolution of the image. The following simplified analysis illustrates the order of magnitude of position measurement error due to spatial quantization.

Figure 4.2.1 depicts the case of a 3-D point P projected onto two 2-D images. Let the point projection onto the left image plane be at (x_l, y_l) and onto the right image plane be at (x_r, y_r). Due to quantization error, the stereo system will determine (x_l, y_l) and (x_r, y_r) with some error, which in turn causes error in the estimated location of P. Figure 4.2.2 illustrates this effect for errors caused by image quantization. Because of the finite resolution the estimated location of P can be anywhere in the shaded region surrounding the true location. The quantization error region often has the shape of an ellipsoid having its principal axis along the optical axes of the cameras if the two optical axes are parallel. This means that depth accuracy is severely affected by quantization and it could be improved by appropriate positioning of the cameras.

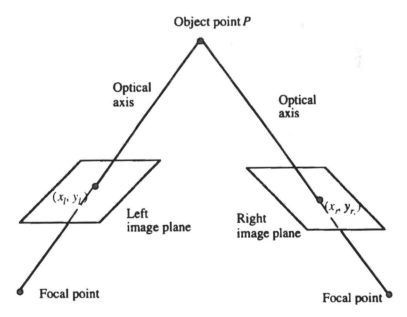

Figure 4.2.1 3D point projected onto 2D images in an ideal case (assuming that the number of pixels in the image plane is infinite)

The accuracy of depth determination improves with the increase of the separation between the cameras. This is also true in the case where the optical axes are not parallel. In practice, the best accuracy is obtained when the angle spanned by the rays at the object point is between 60° and 120°, with the cameras being approximately equidistant from the object point.

An approximated estimate of the errors in the object coordinates, for a highly simplified case (2D object, 1D images), is given by

$$\Delta x = (d/f)\Delta p$$

where Δx is the maximum 1D error in the object space due to the image quantization error, d is the distance from the object to the camera, f is the focal length, and Δp is the half of the 1D physical size of the image pixel.

For example, for the case where $d = 1000\ mm$, $f = 50.0\ mm$, and $\Delta p = 5mm/512/2 = 0.005mm$, the resulting error is $\Delta x = 0.1\ mm$.

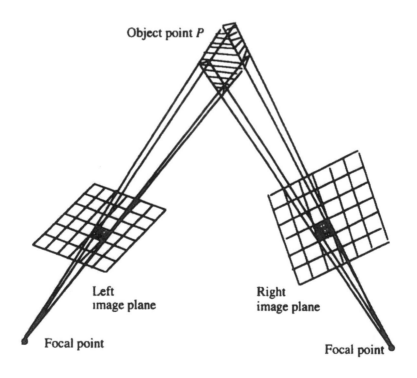

Figure 4.2.2 Error caused by the image quantization

B. HAND-MOUNTED CAMERA CONFIGURATIONS

Rather than keeping the cameras stationary in the environment, one may mount the cameras on the robot hand. Figures 4.2.3 and 4.2.4 depict two possible configurations. In Figure 4.2.3, a pair of stereo cameras is mounted on the robot hand. Poses of an object in the field of view of both cameras can be measured uniquely after the two cameras are calibrated. Figure 4.2.4 features a monocular camera mounted on the robot hand. In this case, full poses of any objects in the field of view cannot be uniquely determined from the image coordinates of the points measured by the monocular camera. See Section III.C for a discussion of camera recalibration. This is the only known proper way to measure robot poses in such a setup.

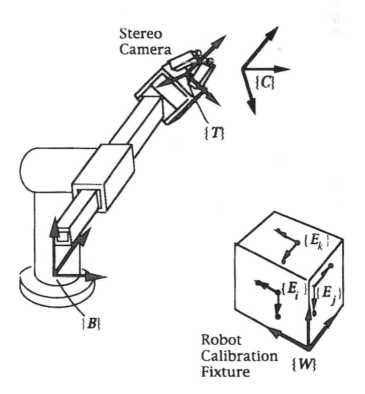

Figure 4.2.3   A stereo-camera measurement system
(definition of coordinate systems will be given in Section III)

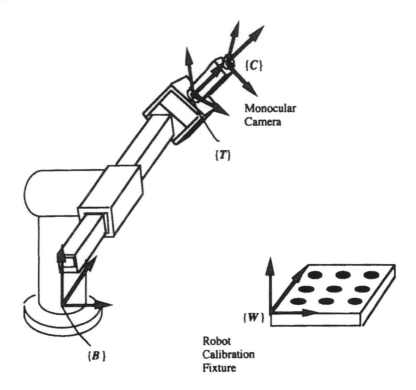

Figure 4.2.4 A monocular-camera measurement system

III. POSE MEASUREMENT WITH MOVING CAMERAS

There are two approaches in using moving cameras for robot calibration, which are categorized into one stage or two stage approaches. In a one-stage approach, the robot and its hand-mounted camera(s) are calibrated simultaneously. In a two-stage approach, on the other hand, the cameras (or camera) mounted on the robot hand are to be calibrated first. The calibrated cameras are then used to perform robot pose measurements. The issue of simultaneous calibration of the robot and its hand-mounted camera(s) is addressed in Chapter 7.   This section concentrates only on two stage approaches.

The difference between the stereo-camera case and the monocular camera case can be summarized as follows: In the stereo camera case, the cameras are calibrated only once. After camera calibration, the manipulator is moved to different measurement configurations.   At each configuration the hand-mounted stereo cameras compute the relative pose of the robot end-effector with respect to a precision calibration fixture, on which a world coordinate system is defined. On the other hand, in the monocular-camera case, the camera is recalibrated in each robot measurement pose.   At this particular robot measurement configuration, the calibrated camera model directly provides the pose of the camera with respect to the world coordinate system established using a precision calibration fixture.

A. COORDINATE SYSTEM ASSIGNMENT

Prior to describing the detailed measurement procedures, let us define the basic coordinate systems. These coordinate systems are given in terms of the stereo-camera setup (Figure 4.2.3). However, most definitions are applicable to the monocular-camera setup as well.

The coordinate frames are defined as follows:

1. $\{W\} = \{x_w, y_w, z_w\}$ — World coordinate system.   The world system is assigned in any convenient location.   It is often defined by the robot calibration fixture or calibration sensory system.   The coordinates of the precision points on the robot calibration fixture are assumed known in $\{W\}$.

2. $\{C\} = \{x_c, y_c, z_c\}$ — Camera coordinate system. Camera calibration is done separately from robot calibration. The camera coordinate system is the reference system used for camera calibration.   It thereafter moves together with the camera even though the calibration board may be long gone.   For instance, in Figure 4.2.3, $\{C\}$ may be coincident with $\{W\}$ in the stage of camera calibration.   After the cameras are calibrated, the manipulator is moved to various measurement configurations for pose collection.   At that time, $\{C\}$ and $\{W\}$ are no longer coincident.   This system should be distinguished from another coordinate system which lies on the focal point of the camera.

3. $\{B\} = \{x_b, y_b, z_b\}$ — Robot base coordinate system.   The nominal location of the base system is usually defined by the robot manufacturer. Robot users however often need to relocate the base system.

4. $\{T\} = \{x_t, y_t, z_t\}$ — Robot tool coordinate system.   The tool system is defined by the user in any convenient location.

The definition of $\{E_i\}$ is deferred to Section B.

The purpose of the robot calibration measurement is to determine, in each robot configuration, the robot tool position and orientation in $\{W\}$. Since the transformation from the camera to the tool coordinate systems is fixed, we may treat the camera coordinate system as an imaginary tool system.   More discussion about the identification of the camera-tool transformation can be found in Section IV.   In this section, we concentrate on the identification of the transformation from the world system to the camera system.

Recall that the 4x4 homogeneous transformation T_n relating the camera to the world coordinates is

$$T_n = A_0 A_1 \ldots\ldots A_{n-1} A_n$$

where A_0 is the 4x4 homogeneous transformation from the world coordinate system to the base coordinate system of the robot, A_i is the transformation from the $(i-1)$th to the ith link coordinate systems of the robot, and A_n is the transformation from the nth link coordinate system of the robot to the camera coordinate system.

A_i can be represented in terms of any proper kinematic modeling convention, such as the Denavit-Hartenberg convention, the Zero-Reference convention, the CPC convention, and more. T_n is a matrix function of the kinematic parameter vector ρ and the joint variable vector q. The main task of robot calibration is to estimate the robot kinematic parameter vector ρ given a set of robot pose measurements $T_n(\rho, q_i)$, where q_i denotes the ith robot measurement configuration.

Clearly, as far as the measurement phase is concerned, the leading problem is how to obtain robot poses in a large portion of the robot workspace. The next two subsections describe measurement procedures that obtain a set of robot poses $T_n(\rho, q_i)$ using stereo cameras or a monocular camera.

B. THE STEREO-CAMERA CASE

In a stereo-camera setup, the base line (i.e. distance) of the two cameras, which are rigidly mounted on the robot hand, has about the same length as the distance between the board and either of the cameras. In order for both cameras to share a large field-of-view, without compromising too much on accuracy, each camera had to be tilted approximately 30^0 with respect to the geometric center line of the two cameras.

To mathematically describe the calibration approach, the coordinate systems $\{E_i\}$ are defined next. $\{E_i\}$ is the ith coordinate frame on the precision robot calibration board (Figure 4.2.3). To measure the full pose of the robot end-effector, a group of at least three non-colinear precision points (as shown in Figure 4.2.3) needs to be viewed in each robot configuration. The positions of these points are measured and then used to construct a coordinate system $\{E_i\}$. The use of more calibration points can help in reducing the effect of measurement noise.

In the two stage approach, the stereo cameras are calibrated independently in the first stage, and the cameras are then used for robot pose measurement in the second stage.

Following the calibration of each camera, the 3D coordinates of any object point, viewed by the stereo cameras, are computed with respect to the camera coordinate system $\{C\}$ defined by the camera calibration board. Since $\{C\}$ is fixed with respect to the tool system $\{T\}$ of the robot, it moves with the robot hand from one calibration measurement configuration to another. Each pattern of the dots on the calibration board defines a coordinate system $\{E_i\}$, the pose of which is known in $\{W\}$. The pose of $\{E_i\}$ is also estimated with respect to $\{C\}$ in each robot calibration measurement. By that the pose of $\{C\}$ becomes known in $\{W\}$ at each robot measurement configuration. For a sufficient number of measurement configurations, the homogeneous transformation from $\{W\}$ to $\{C\}$ and thus the link parameters of the robot, can be identified.

The entities known with good precision are:

1. $^wT_{ei}$, the transformation from the world to the ith calibration frame, is known *a priori* since the world system is defined by the robot precision calibration board.
2. $^cT_{ei}$, the transformation from the camera to the ith calibration frames, is known through the measurements of the stereo cameras. Recall that after the cameras are carefully calibrated, these stereo cameras can provide full pose measurements of the ith coordinate system in terms of the camera system, by which $^cT_{ei}$ can be accurately computed.

The transformation from the world system to the camera system at this robot configuration can thus be accurately determined by

$$^wT_c(\rho, q_i) \equiv {}^wT_c = {}^wT_{ei}\,{}^{ei}T_c = {}^wT_{ei}\,{}^cT_{ei}^{-1} \quad i = 1, 2, ..., s.$$

Note that each robot pose ^wT_c is measured at a particular robot configuration, and is thus a function of the robot joint variable vector.

The measurement procedure for identifying ^wT_c, the transformation from the world to the camera frames, is summarized as follows:

The robot hand is first moved to a position where the ith group of points on the precision calibration board can be sensed by the cameras. Since the camera system is allowed to move, the distance between the cameras and the robot calibration board can be maintained to be within the focal distance. The coordinates of these points in both $\{W\}$ and $\{C\}$ are recorded, together with the robot joint variables. Note that the world coordinates of these points in each robot measurement configuration are used to construct $^wT_{ei}$, while the camera coordinates are used to compute $^cT_{ei}$. The recorded joint position vector, denoted by q_i, can be used to compute the nominal pose of the camera in this particular measurement configuration using the nominal forward kinematic model of the robot. The process is repeated until a sufficient number of robot-camera poses at various robot configurations is recorded.

The use of a cubic fixture for robot calibration appears to be a good practical choice. Having a number of precision calibration points on each face of the cubic fixture, as shown in Figure 4.2.3, allows the robot to be exercised in a sufficiently large portion of its dexterous workspace. In arbitrary application environments, precision calibration objects of any size and shape may be used, as long as the objects provide a large enough number of precision calibration dots, visible at sufficiently many different robot joint configurations.

The robot calibration precision fixture provides global information about the robot end-effector poses by enabling the robot to be exercised at many different configurations. Likewise the stereo cameras perform local measurements. The conflict between field-of-view and resolution is thus resolved. For instance, the field-of-view of the camera system can be as small as 50x50 mm², resulting in an overall accuracy of the measurement system of as high as 0.05 mm. This is at least 20 times better than the accuracy provided by stationary vision-based measurement systems that use common off-the-shelf cameras.

C. THE MONOCULAR-CAMERA CASE

In a monocular-camera setup, a single camera is rigidly mounted to the end-effector of the manipulator as shown in Figure 4.2.4.

The definitions of the various coordinate systems, given in Section 2, remain unchanged with the only exception that the location of the camera system may be different. This will be specifically stated in each of the following sections. The calibration task addressed in this section is still to identify the transformation from the world coordinate system to the camera coordinate system, or equivalently, to estimate the robot kinematic parameter vector ρ.

Lenz and Tsai proposed a hand-eye measurement technique for calibration of a Cartesian manipulator using a single moving camera. The key ideas were (1) to use a camera model that explicitly defines the camera pose, (2) to combine the robot and camera calibration fixtures, and (3) to recalibrate the camera at each robot measurement configuration, so that the calibrated camera pose can be directly treated as a robot end-effector pose.

In Lenz and Tsai's method, a single stationary precision planar calibration board is used to define the world coordinate system. The calibration board contains an array of precision dot points, whose world coordinates (i.e. the coordinates of the center of each dot) are assumed to be precisely known. At a given robot measurement pose, the image of the calibration board is taken and processed to extract the centroid of each precision dot. The world coordinates of the calibration points on the board, together with their image coordinates are thus known, which is in turn used to compute the intrinsic and extrinsic parameters of the camera model. The intrinsic parameters of the camera model include the focal length f, lens radial distortion coefficient α, etc. The extrinsic parameters of the camera model are the rotation matrix R and translation vector t that define the transformation from the camera to the world coordinate systems. Following this calibration step, the "pose" of the camera, which is just the position and orientation of the camera system in the world system, becomes known, as these entities are uniquely defined from the extrinsic parameters of the camera model.

As the robot moves from one configuration to another, the camera calibration process is repeated at each robot configuration, with the exception that the intrinsic parameters of the camera are treated as known quantities. This means that the poses of the camera, or equivalently the poses of the robot end-effector, are computed for each robot configuration. After the robot is moved to a sufficient number of measurement configurations, the robot parameter vector can be estimated using the corresponding poses.

An advantage of this method is that the resolution of the camera can be made very high, since its field-of-view can be made as small as 50x50 mm². Moreover, only one camera is needed for robot calibration, which greatly reduces the hardware and software complexity of the system and increases the computation speed. A limitation of this scheme however is that a full scale camera calibration process shall be performed whenever the robot changes its configuration. For calibration techniques suitable for this task, readers are referred to Chapter 2 of this book. Another limitation is that because the same precision calibration board is used for both camera and robot calibration, the set of robot measurement configurations may cover a relatively small portion of the robot joint space. The latter problem can be solved if more calibration boards are used and these calibration boards, whose relative positions are assumed known, are arranged in such a way that a large portion of the robot workspace can be covered. For instance, a cubic precision fixture can again be used. On each face of the cubic fixture, a set of calibration points is to be provided. The typical number of points has to be greater than 5x5 for a sufficient accuracy of the camera calibration.

IV. IDENTIFICATION OF THE RELATIONSHIP BETWEEN ROBOT END-EFFECTOR AND CAMERA

The methods, discussed thus far in this chapter, only obtain the geometric relationship between the world and the camera coordinate systems. Often in practice, the transformation from the tool to the world coordinate systems needs to be identified. Many researchers have addressed the issue by assuming that the robot geometry is accurately known, or at least relative motions of the tool coordinate system can be measured. This assumption is not valid in this case, as the transformations from any of the uncalibrated robot internal link coordinate systems to the tool frame are not known with sufficient precision. As the internal link coordinate systems, after robot calibration, become known with respect to the world system, the location of the tool system with respect to the world system remains unknown. Therefore many of the hand/eye calibration methods available in the literature may not be applicable here.

This problem is relatively easy to solve when stereo cameras are employed, as the stereo camera system enables the computation of the 3D coordinates of any unknown object point. Let a special fixture be mounted on the robot flange, on which there is a number of non-colinear calibration points. Suppose further that the stereo cameras can view simultaneously these calibration points. The 3D coordinates of these points can thus be computed in terms of the camera system, and a tool system can be established using the computed coordinates of these points. Consequently, the camera-tool transformation can be obtained.

It is difficult, on the other hand, to obtain the camera-tool transformation if only one camera is used. Of course one can use additional sensing devices to provide relative motion information of the robot tool. However, this invalidates the autonomy assumption that no measurements other than those provided by the camera/robot system are to be used in the calibration task. In this section, we provide a number of methods that can be used to identify the transformation between the robot tool frame and the camera frame.

A.   METHODS FOR STEREO CAMERAS

The homogeneous transformation ${}^cT_t$, from the camera frame to the tool frame, depends on the location of the tool frame. Without loss of generality, assume that the z axis of the tool frame coincides with the nth joint axis and that the x and y axes of the tool frame lie on the surface plane of the robot tool flange.

The nth joint axis can be identified directly by applying the "Circle Point Analysis" method. This is done as follows: Assume that the nth joint is revolute. All joints are to be fixed except for the last one. The nth joint is rotated continuously while the cameras snap images of a fixed point that resides on the precision calibration board. Moving cameras while taking pictures of a fixed point is equivalent to having fixed cameras take pictures of a moving point. Thus, after the nth joint completes a circle, the precision point has also moved a circle on the image plane; refer to Figure 4.4.1. Denote by ${}^cp_1$, ${}^cp_2$, ..., ${}^cp_k$ the coordinate values of the precision point in $\{C\}$ along the trajectory. Regression methods can be employed to identify, from $\{{}^cp_1, {}^cp_2, ..., {}^cp_{k.}\}$ the center of the rotation and the direction vector of the rotation axis, all expressed with respect to $\{C\}$. Similarly, the sliding trajectory of the axis is found, if the nth joint is prismatic.

The following procedure elaborates on the estimation of the axis of rotation and the center of rotation, given a measured circular trajectory.

Denote by $\{^{c}p_1,\ ^{c}p_2,\ ...,\ ^{c}p_k\}$ a set of trajectory points measured by the camera system. Ideally, these points are along a circle centered at $^{c}p_0$. The circle lies on Plane B, the common normal of which is ^{c}n.   The distance from the origin of the camera coordinate frame $\{C\}$ to Plane B is denoted as a. Refer to Figure 4.4.2. The problem is to find $^{c}p_0$ and ^{c}n given $\{^{c}p_1,\ ^{c}p_2,\ ...,$ $^{c}p_k\}$. To simplify the notation, whenever no ambiguity arises, the left-superscript "c" is dropped in the subsequent discussion for those vectors that are represented in $\{C\}$.

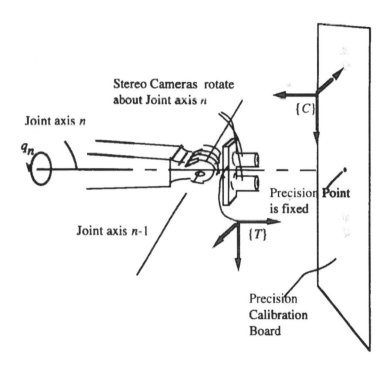

Figure 4.4.1 Measuring the position and orientation of $\{T\}$ in $\{C\}$

The *Circle Point Analysis Procedure* is described by the following 10 steps:

Step 1: Selection of three points, $p_1,\ p_{1+q},\ p_{1+2q},$ from $\{p_1,\ p_2,\ ...,\ p_k\}$, where q is the nearest integer to $k/3$.

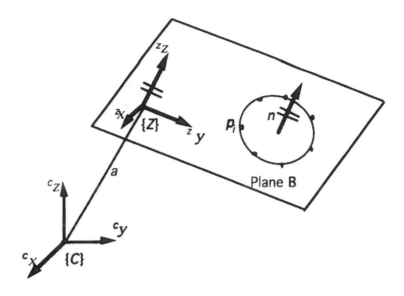

Figure 4.4.2 The rotation axis and rotation center in $\{C\}$ and $\{Z\}$

Ideally, the three points should form an equilateral triangle, in order to obtain a good initial estimate of n. If $\{p_1, p_2, ..., p_k\}$ do not complete a full circle, we need at least to make sure that p_1, p_{1+q} p_{1+2q} are not nearly colinear.

Step 2: Forming an initial guess n^0 for the normal of Plane B.

$$n = v_1 \times v_2 / \|v_1 \times v_2\| \tag{4.4.1}$$

where v_1 is the vector from p_{1+q} to p_1 and v_2 is the vector from p_{1+q} to p_{1+2q}

Step 3: Selection of the maximum magnitude element among the elements of $n^0 = [n_x{}^0, n_y{}^0, n_z{}^0]^T$.

This step is done to ensure that the equation of Plane B, given in Step 4 is valid.

Step 4: Forming the equation of Plane B:

$$n \cdot p_i + a = 0 \quad i = 1, 2, ..., k \tag{4.4.2}$$

Assuming for instance that $n_x{}^0 = \max\{n_x{}^0, n_y{}^0, n_z{}^0\}$, then Equation (4.4.2) can be rewritten as

$$y_i b + z_i c + e = x_i \quad i = 1, 2, ..., k \tag{4.4.3}$$

where $[x_i, y_i, z_i]^T \equiv p_i$, and $\{b, c, e\}$ are to be identified. Similar equations can be formulated whenever n_y^0 or n_z^0 are the maximum elements.

Step 5: Solving for $\{b, c, e\}$ from (4.4.3) using a least-squares algorithm. The least-squares solution is unique because $\{p_1, p_2, ..., p_k\}$ are not colinear.

Step 6: Finding n and a from $\{b, c, e\}$ as follows:

$$n \equiv [n_x, n_y, n_z]^T = [1/g, -b/g, -c/g]^T \tag{4.4.4}$$

and

$$a = e/g \tag{4.4.5}$$

where $g \equiv \{1+b^2+c^2\}^{1/2}$.

Step 7: Transformation, by a rotation R and a translation t, of the points $\{p_1, p_2, ..., p_k\}$ to a new coordinate frame $\{Z\}$ whose x-y plane is Plane B. R is the upper-left 3x3 sub-matrix given in the CPC model's homogeneous transformation matrix in which b_i is replaced by n, and

$$t = [0, 0, a]^T \tag{4.4.6}$$

The set $\{^Z p_1, ^Z p_2, ..., ^Z p_k\}$ denotes the representation of Points $\{p_1, p_2, ..., p_k\}$ in $\{Z\}$. These are obtained from the following equations,

$$^Z p_i = R^T p_i - t \quad i = 1, 2, ..., k. \tag{4.4.7}$$

The rotation R^T and the translation $-t$ bring $\{C\}$ to coincide with $\{Z\}$. Refer to Figure 4.4.2.

Step 8: Forming the circle equation in $\{Z\}$. The circle equation in the x-y plane of $\{Z\}$ can be written as

$$(^Z x_i - ^Z x_0)^2 + (^Z y_i - ^Z y_0)^2 = r^2 \quad i = 1, 2, ..., k \tag{4.4.8}$$

where $[^Z x_i, ^Z y_i, 0]^T \equiv ^Z p_i$, and $[^Z x_0, ^Z y_0, 0]^T \equiv ^Z p_0$. The latter is the center of the circle in $\{Z\}$. From (4.4.8),

$$2\,{}^{z}x_i\,{}^{z}x_0 + 2\,{}^{z}y_i\,{}^{z}y_0 - {}^{z}x_0{}^2 - {}^{z}y_0{}^2 + r^2 = {}^{z}x_i{}^2 + {}^{z}y_i{}^2 \quad i = 1, 2, ..., k$$

$$(4.4.9)$$

Let $h \equiv -{}^{z}x_0{}^2 - {}^{z}y_0{}^2 + r^2$, and $w_i \equiv {}^{z}x_i{}^2 + {}^{z}y_i{}^2$. Then (4.4.9) can be rewritten as,

$$2\,{}^{z}x_i\,{}^{z}x_0 + 2\,{}^{z}y_i\,{}^{z}y_0 + h = w_i \quad i = 1, 2, ..., k \qquad (4.4.10)$$

Step 9: Solving for $\{{}^{z}x_0,\ {}^{z}y_0,\ h\}$ from (4.4.10) using a least-squares technique. The least-squares solution is unique since $\{{}^{z}p_1,\ {}^{z}p_2,\ ...,\ {}^{z}p_k\}$ are not colinear.

Step 10: Transformation of ${}^{z}p_0$ back to $\{C\}$ using the following relationship,

$$p^0 = R({}^{z}p_0 + t) \qquad (4.4.11)$$

After the parameters of the last joint are estimated by the above procedure, the remaining task is to determine the origin and the coordinate axes of the tool frame in $\{C\}$. This can be done in several ways, none of which is absolutely better than the other. One method is to apply the circle point analysis method to the $(n\text{-}1)$th joint axis as well. This time all joint axes other than the $(n\text{-}1)$th axis are fixed. Assume that the $(n\text{-}1)$th axis is revolute. The scheme is similar to that given above, with the exception that more than one precision point should be used to increase the rotation range of the $(n\text{-}1)$th axis; refer to Figure 4.4.3. After the geometric features of the nth and $(n\text{-}1)$th axes are identified, the tool coordinate frame can be established. Because the actual distance between the center of the mounting surface and the $(n\text{-}1)$th joint axis is unknown, its nominal value may be used in the construction of ${}^{c}T_r$. The uncertainty in this distance will have influence on the robot calibration accuracy.

Another procedure, coined the Extended Circle Point Analysis procedure, for estimating the parameters of the $(n\text{-}1)$th joint axis is described next. In this procedure, more than one precision point on the precision calibration board are used. Assuming for simplicity that k different views are taken for each precision point, one has $\{{}^{c}p_{i,1},\ {}^{c}p_{i,2},\ ...,\ {}^{c}p_{i,k}\}$ for $i = 1, 2,..., r$, where r is the number of precision points used. Ideally, $\{{}^{c}p_{i,1},\ {}^{c}p_{i,2},\ ...,\ {}^{c}p_{i,k}\}$ lie on the ith circle (or contour) centered at a common center ${}^{c}p_0$. The circle lies on Plane B, the normal vector of which is ${}^{c}n$. The distance from the origin of the camera coordinate frame $\{C\}$ to Plane B is a. The problem is to find ${}^{c}p_0$ and ${}^{c}n$ given $\{\{{}^{c}p_{1,1},\ {}^{c}p_{1,2},\ ...,\ {}^{c}p_{1,k}\},\ ...,\ \{{}^{c}p_{r,1},\ {}^{c}p_{r,2},\ ...,\ {}^{c}p_{r,k}\}\}$.

The *Extended Circle Point Analysis Procedure* is given by the following 10 steps:

Step 1: Selection of three points, $^c p_{1,1}$, $^c p_{1,1+q}$, $^c p_{1,k}$, among $\{^c p_{1,1},$ $^c p_{1,2}, ..., ^c p_{1,k}\}$, where q is the nearest integer to $k/2$.

$\{^c p_{1,1}, ^c p_{1,2}, ..., ^c p_{1,k}\}$ can only form a small portion of a circle. Therefore the above is the best we can do to avoid colinearity of the three points.

Steps 2-7: The same as Steps 2-7 in the Circle Point Analysis procedure, except that in a few places the words "one trajectory" should be replaced by "a number of trajectories". This is done by changing a single index to double indices in (4.4.2), (4.4.3) and (4.4.7).

Step 8: Forming the equations of the circles in $\{Z\}$. The circle equations in the x-y plane of $\{Z\}$ can be written as

$$({}^z x_{j,i} - {}^z x_0)^2 + ({}^z y_{j,i} - {}^z y_0)^2 = r_j^2$$
$$j = 1, 2, ..., r, \, i = 1, 2, ..., k \quad (4.4.12)$$

where $[{}^z x_{j,i}, {}^z y_{,i}, 0]^T \equiv {}^z p_{j,i}$, and $[{}^z x_0, {}^z y_0, 0]^T \equiv {}^z p_0$. The latter is the center of the circle in $\{Z\}$ and r_j is the radius of the jth circle. From (4.4.12), the following relationships can be obtained,

$$2({}^z x_{j,i+1} - {}^z x_{j,i}){}^z x_0 + 2({}^z y_{j,i+1} - {}^z y_{j,i}){}^z y_0 = w_{j,i+1} - w_{j,i}$$
$$i = 1, 2, ..., k\text{-}1 \quad (4.4.13)$$

and

$$2({}^z x_{j,1} - {}^z x_{j,k}){}^z x_0 + 2({}^z y_{j,1} - {}^z y_{j,k}){}^z y_0 = w_{j,1} - w_{j,k} \quad (4.4.14)$$

for $j = 1, 2, ..., k$, where $w_{j,i} \equiv {}^z x_{j,i}^2 + {}^z y_{j,i}^2$.

These circles have a common center, therefore all the above equations can be used simultaneously to find the circle center.

Step 9: Solving for ${}^z x_0$ and ${}^z y_0$ from (4.4.13)-(4.4.14) using a least-squares technique. The least-squares solution is unique because $\{\{^c p_{1,1},$ $^c p_{1,2}, ..., ^c p_{1,k}\}, ..., \{^c p_{r,1}, ^c p_{r,2}, ..., ^c p_{r,k}\}\}$ are not colinear.

Step 10: The same as Step 10 in the Circle Point Analysis procedure.

Another method of determining the origin and the coordinate axes of the tool frame in $\{C\}$ involves extra measurements before the camera calibration fixture is removed. Since the field of view of the cameras is very small, the mounting surface can be within a range of 20 cm from the camera calibration fixture. In other words, the distances from some marked points on the mounting surface to a few (for instance, three) known points in $\{C\}$ can be measured by a gauge, refer to Figure 4.4.4. The coordinates of these marked points in {C} can then be calculated by elementary trigonometry. The direction and position of the x axis, defined by the user, is then determined. These geometric features are all represented in {C}. ^cT_t is then constructed. A disadvantage of this method is that it is labor intensive, because the precision measurements must be taken by the robot operator.

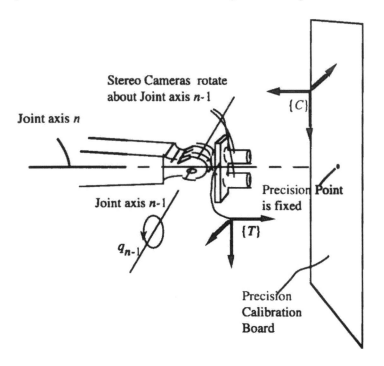

Figure 4.4.3 Measuring the position and orientation of
the $(n-1)$th link frame in {C}

The third method of finding the x axis of the tool frame requires a precisely-machined customized tool. The tool consists of two bars arranged in a T-shape; refer to Figure 4.4.5. One end of the round bar is to be inserted into the mounting surface of the robot and the other end of the bar is fixed in a right angle at the center of the rectangular bar. The distance a from the mounting surface of the tool flange to the center of the rectangular bar is known. There exist a few precision points on this bar that define the direction of the x axis of the tool frame. The geometry of the tool is determined in such a way that the cameras would be able to take pictures of the precision points on the tool. The direction vector of the x axis, defined by the user, is estimated from the measurements taken. The identified x axis is then shifted to the mounting surface by a distance a along the z axis of the tool frame.

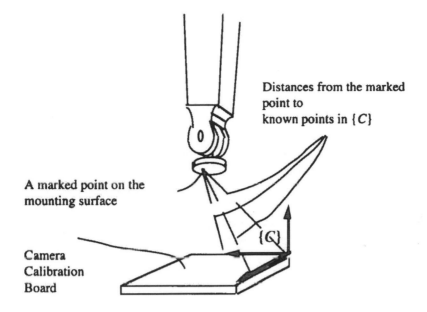

Figure 4.4.4 Measuring a marked point on the mounting surface in $\{C\}$

For all the above methods, there exists a common problem: That is, two actual axes may be neither orthogonal nor intersecting. For instance, in the first method, the identified nth and $(n\text{-}1)$th joint axes are likely to be non-intersecting and non-orthogonal. A solution approach to solving the problem is as follows: Let the z axis of the tool frame coincide with the identified nth joint axis. The common normal between the z axis and the $(n\text{-}1)$th joint axis (or the x axis of the tool frame, defined by the user in the second and the third methods) is computed. This line is then defined as the y axis of the tool frame.

Denote by $^{c}x_{t}$, $^{c}y_{t}$ and $^{c}z_{t}$ the direction vectors of the x, y and z axes of the tool frame, respectively; by $^{c}o_{t}$ the origin of the tool frame; by $^{c}z_{n\text{-}1}$ the direction vector of the z axis of the $(n\text{-}1)$th link frame; and by $^{c}p_{t}$ and $^{c}p_{n\text{-}1}$ two arbitrarily points on the nth and $(n\text{-}1)$th joint axes, respectively. All these vectors are represented in $\{C\}$. Then

$$^{c}y_{t} = {}^{c}z_{n\text{-}1} \times {}^{c}z_{t}$$

and

$$^{c}x_{t} = {}^{c}y_{t} \times {}^{c}z_{t}$$

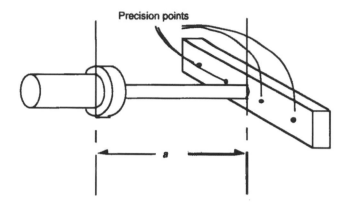

Figure 4.4.5 A customized tool designed for determining
the location of the tool frame in $\{C\}$

The origin of the tool frame, which is the intersection point of the common
normal with the z axis, is given by

$$
{}^c o_t = {}^c p_t + \frac{({}^c p_t - {}^c p_{n\text{-}1}) \cdot \left({}^c z_t - ({}^c z_t \cdot {}^c z_{n\text{-}1}) {}^c z_{n\text{-}1} \right)}{({}^c z_t \cdot {}^c z_{n\text{-}1})^2 - 1} \, {}^c z_t
$$

After ${}^c x_t$, ${}^c y_t$, ${}^c z_t$ and ${}^c o_t$ are determined, the transformation ${}^c T_t$ is readily
constructed.

B. A METHOD FOR MONOCULAR CAMERA WITH TWO VIEWS

The measurement schemes presented in the previous steps can be extended
to the case in which a single camera is used. Two views of a precision point
are required to compute the 3D position of the point in $\{C\}$.

The modification starts at the camera calibration stage. The monocular
camera, fixed on the tool mounting surface of the robot arm, is in a position
that is closer to one of the two stereo cameras discussed earlier (Position
One); say the one at the left-hand side (refer to Figure 4.4.6).

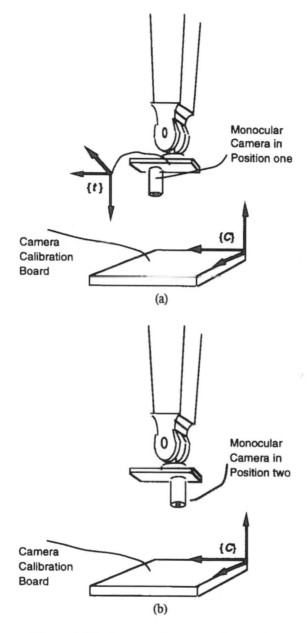

Figure 4.4.6 A monocular camera in two views

The nth joint position is, say, $q_{n,1}$.   After a camera calibration is performed at this location, the camera is moved to the position closer to the other one of the previously discussed stereo cameras, that is, the one at the right-hand side (Position Two). The displacement of the nth joint is $q_{n,1}$ + Δq_n. Another camera calibration is done for the second position of this camera. The identified camera models can now be used to conduct 3D measurement in $\{C\}$. The relative displacement of the nth joint Δq_n is recorded to place the monocular camera in two relatively fixed positions in the 3D measurement process.

A remaining task is to determine the transformation from the tool frame to the camera frame ^cT_t. The tool frame can be defined in terms of Position one of the camera. ^cT_t is the transformation from the tool frame to the camera frame when the camera is at Position One. Displacements of the nth joint at Position Two, therefore, are not used for the computation of the robot kinematic model. Keeping this in mind, all of the measurement and identification methods devised for the case of stereo cameras in part A of this section are applicable here except for the third method of identifying ^cT_t.

The single camera method has two obvious restrictions related to the last joint of the robot. First, its effective travel range is reduced by Δq_n. In general Δq_n equals 180 degrees. Consequently, the last joint may not be well calibrated. Second, the repeatability of the last joint influences directly the measurement accuracy.

V. SUMMARY AND REFERENCES

Pose measurement by using hand-mounted cameras is an effective method because of the following features:

1. With hand-mounted cameras, poses of the robot in a large portion of the robot workspace can be measured. Furthermore, cameras provide highly accurate pose measurements as their field of view can be made sufficiently small.
2. Standard off-the-shelf cameras can be used for robot calibration. Thus the cost of the measurement system is much lower, compared to laser tracking systems, theodolites, and high resolution motorized zoom-cameras.
3. It is possible to complete a pose computation while the robot changes its measurement configuration from one to another. Furthermore, the actions of the robot and the actions of cameras can be synchronized through proper communication protocols. Thus, autonomous robot calibration is possible.
4. In contrast to a single-beam laser tracking system and a theodolite, cameras provide full pose measurements, which allows the identification of all kinematic parameters of the robot.

The two-stage calibration methods also have the following distinct advantages:

1. Since the problem is decoupled, the computation involved in each stage is relatively simple. Also the concept is straightforward and therefore easy to understand.
2. Linear solution is possible in some cases. For instance, Lenz and Tsai calibrated a Cartesian robot using a linear least squares parameter estimation procedure. This further reduces the computational complexity of the parameter estimation algorithm and facilitates a near-real-time implementation of the calibration algorithm.

As to the one-stage calibration approaches discussed in this chapter, in addition to the features stated above, one has the following particular characteristics:

1. The camera and robot parameters are identified simultaneously. Thus the propagation errors which exist in multi-stage approaches are eliminated.
2. The approach allows the user to use a single object point for calibration reference, provided that the object point can be moved to three non-colinear locations.   One can simulate this by using a cube with one calibration point on each face. This arrangement provides a very high resolution to the cameras, and improves significantly their measurement accuracy.

Comparing with the stereo-camera based approaches, the monocular-camera methods has a larger field-of-view. Moreover, the processing speed using a monocular camera is faster. On the other hand, the stereo-camera setup provides absolute depth measurements.

The second issue is a comparison study between the stereo- and monocular-camera approaches in the one-stage case. It is clear that the monocular approach can be fast and can cover a large portion of robot workspace. It is yet to be demonstrated that the the stereo-camera approach has any significant advantage over the monocular approach, other than that the stereo cameras allow the computation of the camera-tool transformation.

An early experimental study of robot calibration using stereo cameras rigidly mounted to the robot hand is by Puskorius and Feldkamp (1987). A pin-hole model was first identified for each camera using a calibration block which had a plane of holes and which could be moved along a direction perpendicular to the plane of the holes. The cameras were therefore considered to be part of the robot last link. The calibration of the robot/camera system then followed using a single spherical target. Experimentation based on a similar setup was also conducted by Preising and Hsia (1991) and by Zhuang, Roth and Wang (1994).

A method for autonomous calibration of robots based on a hand-mounted stereo camera system was proposed by Bennett and Hollerbach (1991). The problem of calibrating a hand-mounted camera and a robot simultaneously has been formulated; its Identification Jacobian has been derived; and experimental studies have been conducted by Zhuang, Wang and Roth (1995).

Another vision-based pose measurement technique was reported by Driels and Pathre (1991). The vision-based theodolite is an automatic partial-pose measurement system. It uses rotation stages and measurements from the vision-system to determine the line of sight to a spherical illuminated target. After calibration, the system can continuously track the target with sufficient speed, reliability and accuracy. Since this approach requires a dedicated system hardware and only a partial pose can be measured, we did not elaborate on it further.

The Circle Point Analysis method for directly identifying robot joint axis parameters, by moving one axis at a time, was first proposed by Barker (1983) and then developed fully by Stone (1987) and Sklar (1988).

Chapter 5

ERROR-MODEL-BASED
KINEMATIC IDENTIFICATION

I. INTRODUCTION

The goal of kinematic identification is to estimate a kinematic parameter vector ρ that accounts in the least-squares sense for positioning and orientation errors of the robot. A common approach to the problem has been to define such a set of variables in terms of additive changes $d\rho$ to the robot nominal link parameter vector ρ^0 in a given kinematic model. Linearization of the robot forward kinematic equations about ρ^0, at the particular ith joint configuration q_i, provides an Identification Jacobian matrix $J_i = J(q_i, \rho^0)$. The Jacobian relates y_i, the vector of end-effector pose errors at the ith configuration to $d\rho$, the vector of independent kinematic parameter errors,

$$y_i = J_i \, d\rho \qquad (5.1.1)$$

Let the vector of all measured pose errors be $y = [y_1^T, y_2^T, ..., y_s^T]^T$. We define an aggregated Jacobian matrix

$$J = \begin{bmatrix} J_1 \\ J_2 \\ \cdot \\ \cdot \\ J_s \end{bmatrix}$$

where s is the number of measurements. The overall measurement equation for least squares estimation of $d\rho$ is then

$$y = J \, d\rho \qquad (5.1.2)$$

An iterative least squares kinematic identification procedure can go as follows: A number of robot end-effector pose measurements, together with the joint variables at each robot measurement configuration, is collected. For each robot pose, the pose error vector y_i and the Jacobian matrix J_i are computed using the measured poses and joint readings as well as the known robot geometric model. After (5.1.2) is obtained, a measure $\| y - J \, d\rho \|$ is minimized by the selection of the parameter error vector $d\rho$, where $\| . \|$ denotes the Euclidean norm. The kinematic parameter vector ρ is then updated. This procedure is repeated until either $\| y - J \, d\rho \|$ or $\| d\rho \|$ is small enough; that is, these measures fall within some prescribed small threshold values.

It is therefore clear that an error model that possesses certain "good" properties is vital to the success of error model based calibration. The nominal parameters of the robot are in general known from the machine design specifications. The actual values of the robot kinematic parameters often fall in the neighborhood of their nominal values if the kinematic model is parametrically continuous and the geometric structural changes of the robot are small. In order to apply a least-squares algorithm for parameter estimation, one needs to derive an error model by linearizing the kinematic model. For the error model to be valid, the kinematic model is required to be parametrically continuous and differentiable. Moreover, it is desired that the linearized error model associated with the kinematic model be irreducible; that is, all the error parameters in the linearized error model be independent.

This chapter opens in Section II with mathematical preliminaries and discussion of some generic properties of error models. Finite difference approximation to the Identification Jacobian is then outlined in Section III, as a quick practical way of deriving error models. A generic kinematic error model is derived in Section IV, followed by study of observability issues of robot kinematic parameters. Analytical derivations of error models for the D-H, CPC and MCPC modeling conventions are detailed in Sections V-VII, respectively. The chapter concludes with discussion about the advantages and disadvantages of these error models.

II.   DIFFERENTIAL TRANSFORMATIONS

Certain properties of error models are independent of the choice of a particular kinematic modeling convention.

Let T_n denote the homogeneous transformation matrix that relates the tool frame of the manipulator to the world frame,

$$T_n = A_0 A_1 A_2 \ldots A_{n-1} A_n$$

Each link transformation A_i relates two coordinate frames located on two consecutive joint axes. Let T be an arbitrary homogeneous transformation matrix, written as

$$T = \begin{bmatrix} R & p \\ 0_{1\times3} & 1 \end{bmatrix}$$

where R is a 3x3 orthonormal matrix and p is a 3x1 vector.

The *additive differential transformation dT* of T is defined as

$$dT \equiv T - T^0 \qquad (5.2.1)$$

where T^0 is a matrix function of the nominal kinematic parameters and T is that of the actual kinematic parameters. The *right multiplicative differential transformation* of T, denoted as ΔT^u, is defined as

$$T^0 \Delta T^u \equiv T$$

Similarly, the *left multiplicative differential transformation* of T, denoted as ΔT^l, is defined as

$$\Delta T^l T^0 \equiv T$$

The superscripts "u" and "t" indicate that the respective entity is associated with either the U_i or the T_i transformation matrices defined later in Section IV.

The following trivial relationships show the connection between the additive and multiplicative differential transformations.

$$\Delta T^u = I + (T^0)^{-1} dT$$

and

$$\Delta T^l = I + dT (T^0)^{-1}$$

Useful versions of multiplicative differential transformations may be defined directly from dT:

$$\delta T^u \equiv (T^0)^{-1} dT$$

and

$$\delta T^l \equiv dT (T^0)^{-1}$$

Obviously,

$$\Delta T^u = I + \delta T^u$$

and

$$\Delta T^l = I + \delta T^u$$

Both differential matrices δT^u and δT^l have the mathematical structure shown below. For simplicity we let δT denote either one of these matrices. Then

$$\delta T = \begin{bmatrix} 0 & -dz & dy & dx \\ dz & 0 & -dx & dy \\ -dy & dx & 0 & dz \\ 0 & 0 & 0 & 0 \end{bmatrix} \qquad (5.2.2)$$

where $d^T \equiv [dx,\ dy,\ dz]^T$ represents a set of translational errors and $\delta^T \equiv [\delta x,\ \delta y,\ \delta z]^T$ are a set of rotational errors. These error vectors are defined with respect to either the base coordinates when the left multiplicative differential transformation convention is adopted, or with respect to the tool coordinates when the right multiplicative differential transformation convention is used.

The right and left multiplicative differential transformations are related through

$$\Delta T^l = T^0 \Delta T^u (T^0)^{-1}$$

Similarly,

$$\delta T^l = T^0 \delta T^u (T^0)^{-1}$$

The choice of whether to use right or left multiplicative differential transformation is not model-independent. To ensure that joint displacements are always performed about *actual* axes, care should be exercised in selecting a differential transformation formalism. In the following discussion, for the sake of convenience, assume that all joints are revolute, and A_i can be modeled in the following two alternative forms:

$$A_i^0 = \text{Rot}(z,\ \theta_i)V_i \qquad i = 0, 1, \cdots, n \qquad (5.2.3)$$

for some homogeneous "shape" transformation V_i, where $\theta_0 \equiv 0$; or

$$A_i^0 = V_i'\text{Rot}(z,\ \theta_{i+1}) \quad i = 0, 1, \cdots, n \qquad (5.2.4)$$

for some homogeneous "shape" transformation V_i', where $\theta_{n+1} \equiv 0$. In (5.2.3)-(5.2.4), V_i and V_i' are related only to the ith nominal link parameters, excluding the joint variable. Examples of modeling conventions of the type (5.2.3) are the original D-H model and the CPC model. An example of modeling conventions of the type (5.2.4) is Craig's version of the D-H model (Craig (1989)).

Assuming first that left multiplicative differential transformations are adopted to model kinematic errors, and that A_i^0 is modeled as in (5.2.3), then

$$A_i = \Delta A_i^t A_i^0 = \Delta A_i^t \; \text{Rot}(z, \; \theta_i) V_i$$

In this case, the joint rotation θ_i is about the differentially transformed ith "imaginary" axis; see Fig. 5.2.1. On the other hand, if A_i^0 is modeled as in (5.2.4), then

$$A_i = \Delta A_i^t A_i^0 = \Delta A_i^t V_i \; \text{Rot}(z, \; \theta_{i+1})$$

In this case, the joint displacement θ_{i+1} is about the $(i+1)$th actual axis; see Fig. 5.2.2. Similar arguments apply to the use of right multiplicative differential transformations. In summary, to ensure that the joint rotations are always performed about *actual* axes, the following rule must be observed: Whenever A_i is in the form of (5.2.4), the left multiplicative differential transformation formalism should be adopted to model kinematic errors. Likewise, whenever A_i is in the form of (5.2.3), the right multiplicative differential transformation formalism is to be chosen. The same guidelines are true for robots featuring prismatic joints.

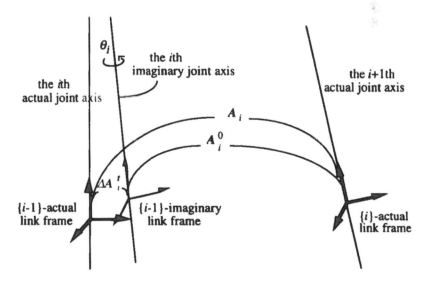

Figure 5.2.1 Modeling kinematic errors with right multiplicative differential transformation

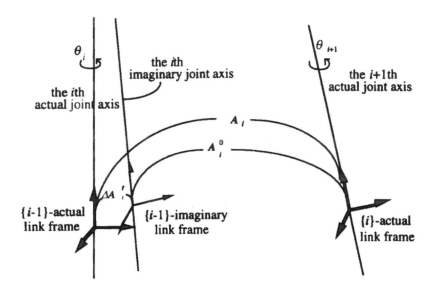

Figure 5.2.2 Modeling kinematic errors with
left multiplicative differential transformation

A highly useful mathematical tool that helps to significantly simplify the
notation is to represent a vector $v = [v_x, v_y, v_z]^T$ by a skew-symmetric matrix
Ω_v, as follows:

$$\Omega_v \equiv \Omega(v) = \begin{bmatrix} 0 & -v_z & v_y \\ v_z & 0 & -v_x \\ -v_y & v_x & 0 \end{bmatrix} \qquad (5.2.5)$$

For example, the differential transformation δT (for either δT^u or δT^t) has the
following structure,

$$\delta T = \begin{bmatrix} \Omega_\delta & d \\ 0_{1\times 3} & 1 \end{bmatrix} \qquad (5.2.6)$$

An operator, S, defined as $S(\delta T) = [\delta^T, d^T]^T$, extracts the six nontrivial
elements from any matrix of the form (5.2.6)

Another useful relationship that holds between a three dimensional vector

δ' and its skew-symmetric matrix representation $\Omega_{\delta'}$ is as follows: Let R be any 3x3 orthonormal matrix. Then the matrix equality

$$\Omega_{\delta'} = R^T \Omega_\delta R \tag{5.2.7}$$

implies the vector equality

$$\delta = R^T \delta. \tag{5.2.8}$$

III. FINITE DIFFERENCE APPROXIMATION TO KINEMATIC ERROR MODELS

Given the analytic expression of T, the forward kinematic transformation that relates the end-effector coordinate frame to the world coordinate frame, the kinematic error model can be found, either analytically through linearization of the kinematic model, or numerically by a finite difference approximation.

A finite difference approximation to a kinematic error model is straightforward to construct and for many practical applications can be used to replace a differential error model without loss of accuracy or numerical stability. This section provides the necessary tools for a robot calibration user to implement such an error model in any kinematic modeling convention.

Because this book employs primarily the D-H, CPC and MCPC models, it is assumed that the right multiplicative differential transformation of a homogeneous transformation T, $\delta T = T^{-1} dT$, is adopted for finite difference error modeling. It is straightforward to modify the result given in this section to the situation in which the left multiplicative differential transformation has to be used.

Recall that the objective of kinematic identification is to minimize the discrepancy between the computed robot pose T and the measured pose T^m by choosing the kinematic parameter vector. Ideally, $T(\rho, q) = T^m$, where $\rho = \rho^0$. The need for calibration arises whenever these two entities are not equal. An iterative procedure starts with the known value of ρ, and continues with the updated values of ρ. In each iteration, $d\rho$ is to be determined such that the position and orientation error components, the six nontrivial elements of the differential error transformation $T^{-1}(T(\rho + d\rho, q) - T^m)$ are minimized. In the first iteration, we take $\rho = \rho^0$, and we seek $d\rho^0$. In the second iteration, we let $\rho^1 = \rho^0 + d\rho^0$, and we seek $d\rho^1$, and so on.

Expanding $T(\rho + d\rho, q)$ in to a power series yields the following first order approximation:

$$T(\rho, q)^{-1}\left\{[T^m - T(\rho, q)] - \frac{\partial T}{\partial \rho}d\rho\right\} \cong 0 \qquad (5.3.1)$$

Invoking the operator S, defined in the previous section, the pose error vector y can be expressed as

$$y \equiv \begin{bmatrix} \delta \\ d \end{bmatrix} \equiv S\left\{T(\rho, q)^{-1}[T^m - T(\rho, q)]\right\} \qquad (5.3.2)$$

Note that $T^{-1}(\rho, q)(T^m - T(\rho, q))$ is in the form of (5.2.6).
From (5.3.1), and up to a first order approximation,

$$y = Jd\rho \equiv S\left\{T^{-1}\frac{\partial T}{\partial \rho}\right\}d\rho \qquad (5.3.3)$$

The finite difference approximation of the error model is obtained by computing the Identification Jacobian matrix J as

$$S\left\{T^{-1}\frac{\partial T}{\partial \rho}\right\} \cong \left[S\left\{T^{-1}\frac{T(\rho_1, \cdots, \rho_i + \Delta\rho_i, \cdots, \rho_m; q) - T(\rho, q)}{\Delta\rho_i}\right\}\right]_{i = 1, 2, \ldots, m} \qquad (5.3.4)$$

where $[M]_i$ denotes the ith column of a matrix M. Substituting (5.3.4) in to (5.3.3), yields the finite difference approximation to the kinematic error model.

For most practical applications, when using the metric unit systems, a choice of $\Delta\rho_i$ between 10^{-5} to 10^{-7} is sufficient. To effectively use the error model, the kinematic model adopted for modeling the manipulator should be complete and parametrically continuous.

Let us summarize the construction of the Identification Jacobian J as a step by step procedure. For that, let $j = 1, 2, \ldots, s$ denote the jth robot measurement configuration, and let $i = 1, 2, \ldots, m$ denote the ith robot kinematic parameter vector component.

Step 1:   Considering the first measurement configuration, set $j = 1$.
Step 2:   Using the joint variable vector q_j and the identified kinematic vector ρ, compute the world-to-tool homogeneous transformation $T(\rho, q_j)$. In the first iteration, we use the nominal kinematic vector.
Step 3:   Considering the first element of the kinematic parameter vector, set $i = 1$.

Step 4:   Selecting a sufficiently small increment $\Delta\rho_i$, we update ρ_i into $\rho_i + \Delta\rho_i$. Using the joint variable vector and the modified kinematic parameter vector, we compute the perturbed world-to-tool transformation $T(\rho + \Delta\rho_i, q_j)$, where $\Delta\rho_i = [0, ..., 0, \Delta\rho_i, 0, ..., 0]^T$.

Step 5:   Computation of the additive differential transformation:

$$dT_{i,j} = T(\rho + \Delta\rho_i, q_j) - T(\rho, q_j)$$

Step 6:   Computation of the right multiplicative differential transformation

$$\delta T_{i,j} = T^{-1}(\rho, q_j) dT$$

Step 7:   Determination of the pose error vector $y_{i,j}$ as follows:

$$y_{i,j} = S(\delta T_{i,j})$$

Step 8:   Computation of the ith column of the matrix J_j. This is done by dividing y_i by $\Delta\rho_i$.

Step 9:   Incrementing $i \rightarrow i + 1$. If $i > m$, the algorithm goes to Step 10. Otherwise it goes to Step 3.

Step 10:   If $j > 1$, we stack matrix J_j under J_{j-1}.

Step 11:   Incrementing $j \rightarrow j + 1$. If $j \leq s$, the algorithm returns to Step 2. Otherwise the procedure ends. The stacked matrices constitute the finite difference approximation of the Identification Jacobian

$$J = \begin{bmatrix} J_1 \\ J_2 \\ \vdots \\ J_s \end{bmatrix}$$

IV. GENERIC LINEARIZED KINEMATIC ERROR MODELS

This section focuses on the portion of the manipulator kinematic error model that is independent of the choice of a specific modeling convention. The derivation of kinematic error model becomes much simpler once this portion of the model is utilized.

A. LINEAR MAPPINGS RELATING END-EFFECTOR CARTESIAN ERRORS TO CARTESIAN ERRORS OF INDIVIDUAL LINKS

Let $y^u = [d^{uT}, \delta^u T]^T = [dx^u, dy^u, dz^u, \delta x^u, \delta y^u, \delta z^u]^T$ be the vector of Cartesian errors of the end-effector using the right multiplicative differential transformation and $x^u = [d^u{}_0{}^T, \delta^u{}_0{}^T, \cdot\cdot, d^u{}_n{}^T, \delta^u{}_n T]^T = [dx^u{}_0, dy^u{}_0, dz^u{}_0, \delta x^u{}_0, \delta y^u{}_0, \delta z^u{}_0, \ldots, dx^u{}_n, dy^u{}_n, dz^u{}_n, \delta x^u{}_n, \delta y^u{}_n, \delta z^u{}_n]^T$ be the vector of Cartesian errors of every link frame of the robot using the right multiplicative differential transformation. Let $U_i \equiv A_i A_{i+1} \ldots A_{n-1} A_n$ for $i = 0, 1, \ldots, n$ with $U_{n+1} \equiv I$.

Theorem 5.4.1: Assume that a given kinematic model T_n is parametrically continuous and differentiable. Then the linearized relationship between the Cartesian errors of individual links and those of the end-effector is given by

$$y^u = L^u x^u \tag{5.4.1}$$

where y^u is a 6x1 vector and x^u is a 6(n+1)x1 vector, and the linear mapping L^u is

$$L^u = \begin{bmatrix} R^u{}_1{}^T & -R^u{}_1{}^T \Omega^u{}_{p,1} & \cdots & R^u{}_n{}^T & -R^u{}_n{}^T \Omega^u{}_{p,n} & I_{3\times3} & 0_{3\times3} \\ 0_{3\times3} & R^u{}_1{}^T \cdots & & 0_{3\times3} & R^u{}_n{}^T & 0_{3\times3} & I_{3\times3} \end{bmatrix} \tag{5.4.2}$$

where $\Omega^u{}_{p,i}$, $i = 1, \cdot\cdot, n$, is a skew-symmetric matrix constructed from the elements of the fourth column vector $p^u{}_i$ of U_i. The matrix $R^u{}_i$ is the rotation matrix of U_i.

Proof: By straightforward differentiation (Paul (1981)),

$$\delta T^u{}_n = \sum_{i=0}^{n} U_{i+1}^{-1} \delta A^u{}_i U_{i+1} \tag{5.4.3}$$

Therefore

$$\Omega^u{}_\delta = \sum_{i=0}^{n} R^u{}_{i+1}{}^T \Omega^u{}_{\delta,i} R^u{}_{i+1}$$

where Ω^u_{δ} is the upper-left 3x3 submatrix of δT^u_n and $\Omega^u_{\delta,i}$ is the upper-left 3x3 submatrix of δA^u_i.

By (5.4.2) and (5.2.7)-(5.2.8)

$$\delta^u = \sum_{i=0}^{n} R^u_{i+1}{}^T \, \delta^u{}_i \tag{5.4.4}$$

The last three rows of L^u in (5.4.2) are thus obtained, noting that $R^u_{n+1} = I_{3\times3}$. Also from (5.4.3),

$$d^u = \sum_{i=0}^{n} R^u_{i+1}{}^T \Omega^u_{d,i} \, p^u_{i+1} + R^u_{i+1}{}^T d^u_i$$

$$= \sum_{i=0}^{n} -R^u_{i+1}{}^T \Omega^u_{p,i+1} \, \delta^u_i + R^u_{i+1}{}^T d^u_i \tag{5.4.5}$$

using the fact that $\Omega^u_{\delta,i} \, p^u_{i+1} = \delta^u_i \times p^u_{i+1}$. The first three rows of L^u in (5.4.2) are thus obtained noting that $R^u_{n+1} = I_{3\times3}$ and $\Omega^u_{p,n+1} = 0_{3\times3}$.   []

Similar relationships exist for left multiplicative differential transformation models. Let $y^l = [d^{lT}, \delta^{lT}]^T = [dx^l, dy^l, dz^l, \delta x^l, \delta y^l, \delta z^l]^T$ be the vector of Cartesian errors of the end-effector using the left multiplicative differential transformation and $x^l = [d^l_0{}^T, \delta_0{}^T, \ldots, d^l_n{}^T, \delta_n{}^T]^T = [dx^l_0, dy^l_0, dz^l_0, \delta x^l_0, \delta y^l_0, \delta z^l_0, \ldots, dx^l_n, dy^l_n, dz^l_n, \delta x^l_n, \delta y^l_n, \delta z^l_n]^T$ be the vector of cartesian errors of every link frame of the robot using the left multiplicative differential transformation. Let $T_i \equiv A_0 \ldots A_{i-1} A_i$ for $i = 0$, 1, ..., n with $T_{-1} \equiv I$.

Theorem 5.4.2: Assume that a given kinematic model T_n is parametrically continuous and differentiable. The linearized relationship between the Cartesian errors of individual links and those of the end-effector is given by

$$y^l = L^l x^l \tag{5.4.6}$$

where y^l is a 6x1 vector and x^l is a $6(n+1)$x1 vector, and the linear mapping L^l is

$$L^t = \begin{bmatrix} I_{3\times3} & 0_{3\times3} & R'_0 & \Omega^t_{p,0}R'_0 & \cdots & R'_{n-1} & \Omega^t_{p,n-1}R'_{n-1} \\ 0_{3\times3} & I_{3\times3} & 0_{3\times3} & R'_0 & \cdots & 0_{3\times3} & R'_{n-1} \end{bmatrix} \qquad (5.4.7)$$

The following cautionary remark is made whenever the matrices L^u or L^t are used as building blocks in the construction of the Identification Jacobian. Whenever the link transformation A_i is in the form

$$A_i = Q_i V_i$$

where Q_i is the motion matrix and V_i is the shape matrix, L^u should be adopted to model kinematic errors. Otherwise q_n may never appear in the Identification Jacobian; consequently the kinematic errors associated with the nth link become unidentifiable. Similarly, whenever A_i is in the form

$$A_i = V_i Q_i$$

L^t should be chosen; otherwise q_0 may never appear in the Identification Jacobian.

The structure of the matrices L^u and L^t is independent of the choice of a particular link modeling convention. From now on for simplicity the superscripts u and t will be dropped.

B. LINEAR MAPPING RELATING CARTESIAN ERRORS TO LINK PARAMETER ERRORS

It is assumed that the link kinematic model is differentiable. Then

$$x = K d\rho \qquad (5.4.8)$$

where $d\rho$ is an $m \times 1$ parameter error vector, and x is a $6(n+1) \times 1$ vector, representing the Cartesian errors of every robot link frame. By (5.4.1) or (5.4.6) as well as (5.4.8), the pose error vector y is related to the parameter deviations through

$$y = LK d\rho. \qquad (5.4.9)$$

The matrix structure of K depends on the particular choice of the kinematic modeling convention. For a given kinematic model, K can be derived through $\delta T^u = (T^0)^{-1} dT$ or $\delta T^t = dT(T^0)^{-1}$, depending on the choice of error model convention. More specifically,

$$K = \text{diag}(K_0, K_1, .., K_n)$$

where K_i relates m_i parameter errors to the 6 link Cartesian error in the ith link, and m_i is the number of link parameters in A_i.

Example: A version of the modified Denavit-Hartenberg modeling convention can be defined by post-multiplying A_i with Rot(y, β_i), where A_i is as in the Denavit-Hartenberg modeling convention.

Notice that the case of $i = n$ is excluded, as the modified Denavit-Hartenberg convention may not be used to model the nth link transformation.

$$
K^u_i =
\begin{bmatrix}
-s\beta_i c\alpha_i & c\beta_i & a_i s\beta_i s\alpha_i & 0 & 0 \\
s\alpha_i & 0 & a_i c\alpha_i & 0 & 0 \\
c\beta_i c\alpha_i & s\beta_i & -a_i c\beta_i s\alpha_i & 0 & 0 \\
0 & 0 & -s\beta_i c\alpha_i & c\beta_i & 0 \\
0 & 0 & s\alpha_i & 0 & 1 \\
0 & 0 & c\beta_i c\alpha_i & s\beta_i & 0
\end{bmatrix}
$$

The matrix LK is an Identification Jacobian of the manipulator. If a large enough number of pose measurements is taken, the identification of dp is possible by applying least squares methods.

C. ELIMINATION OF REDUNDANT PARAMETERS

A necessary condition for the Identification Jacobian LK to be full rank is that the elements of dp are independent. If dp contains dependent elements, it is always possible to find a linear mapping M, where M relates dp with z, an m'x1 vector, and m' is the number of independent parameters in the kinematic model,

$$dp = Mz \tag{5.4.10}$$

M has the following form,

$$M = \text{diag}(M_0, M_1, .., M_n)$$

where M_i is an $m_i \times m'$ matrix. Here m_i is the number of parameters and m_i' is the number of independent parameters in A_i. Consequently KM is block diagonal.

Finally, combining (5.4.9) with (5.4.10) yields

$$y = LKM z = Jz \qquad (5.4.11)$$

where $J = LKM$ transforms the independent parameters of the robot to its pose errors.

Example: Again the modified D-H model is used to illustrate the construction process. Whenever the $i+1$th joint axis is not nominally parallel to the ith axis, the angle β_i is redundant. Thus

$$M_i = \begin{bmatrix} 1 & 0 & 0 & 0 \\ 0 & 1 & 0 & 0 \\ 0 & 0 & 1 & 0 \\ 0 & 0 & 0 & 1 \\ 0 & 0 & 0 & 0 \end{bmatrix} \qquad i = 0, 1, \cdots, n\text{-}1$$

yielding a reduced-order parameter error vector $[dl_i, da_i, d\theta_i, d\alpha_i]^T$. Whenever the $(i+1)$th joint is parallel to the ith joint, the parameter d_i is redundant. Thus

$$M_i = \begin{bmatrix} 0 & 0 & 0 & 0 \\ 1 & 0 & 0 & 0 \\ 0 & 1 & 0 & 0 \\ 0 & 0 & 1 & 0 \\ 0 & 0 & 0 & 1 \end{bmatrix} \qquad i = 0, 1, \cdots, n\text{-}1$$

yielding a reduced-order parameter error vector $[da_i, d\theta_i, d\alpha_i, d\beta_i]^T$.

The irreducibility of a linearized error model is defined as follows:

Definition 5.4.1: Consider a parametrically continuous and differentiable kinematic model. Its linearized error model is *irreducible* if all link parameters are independent. Otherwise it is *reducible*.

The concept of irreducibility is often applied to the construction of a robust linearized error model. Model reduction is often done analytically. An alternative way of eliminating dependent parameters is through Singular Value Decomposition of the Identification Jacobian.

D. OBSERVABILITY OF KINEMATIC PARAMETERS

In this section, we define the observability of kinematic parameters first, followed by some case studies. We then introduce two important observability measures.

Let us go back to Equation (5.1.2). The Jacobian matrix in (5.1.2) is the augmented Jacobian matrix consisting of s Jacobian matrices given in (5.4.11).

Definition 5.4.2:   The kinematic error parameters are said to be *observable* if $J^T J$ is full rank.

The observability depends on both the kinematic modeling convention as well as the selection of measurement configurations.

Theorem 5.4.3:   Assume that the right multiplicative differential transformations are used to model kinematic errors. If the orientational errors of a manipulator end-effector are not measured, then all orientational parameters of A_n are unobservable. In this case, the number N of observable kinematic error parameters must satisfy the following inequality

$$N \leq 4n - 2p + 6 - o_n \qquad (5.4.12)$$

where o_n is the number of independent orientational parameters in A_n, n is the number of degrees of freedom, and p is the number of prismatic joints in the robot. If in addition the last joint of the manipulator is revolute and the origin of the tool frame lies on the last joint axis, then the number N of observable kinematic error parameters must satisfy the following inequality

$$N \leq 4n - 2p + 6 - o_n - ot_{n-1} \qquad (5.4.13)$$

where $ot_{n-1} = \min\{o_{n-1}, t_{n-1}\}$; o_{n-1} and t_{n-1} are the number of independent orientational and translational parameters in A_{n-1}, respectively.

Proof: If the orientational errors of the manipulator are not measured, then the last three rows in the linear mapping L^u (given in (5.4.2)) are deleted. It is clear that in this case, the columns of L^u corresponding to the orientational parameters of the nth link transformation are always zero. Thus the orientational parameters of A_n are unobservable. Combining that with the fact that $N \leq 4n - 2p + 6$, one obtains (5.4.12).

If in addition the last joint of the manipulator is revolute and the origin of the tool frame lies on the last joint axis, then $p^u_{n,x}$ and $p^u_{n,y}$ are zero and $p^u_{n,z} = p^a_{n,z}$. Thus

$$\Omega^u_{p,n} = \begin{bmatrix} 0 & -p^a_{n,z} & 0 \\ p^a_{n,z} & 0 & 0 \\ 0 & 0 & 0 \end{bmatrix}$$

$\Omega^u_{p,n}$ is independent of the joint variables. Therefore, the columns of $R^u_n{}^T \Omega^u_{p,n}$ depend linearly on the columns of $R^u_n{}^T$. This implies that either the orientational parameters, corresponding to the columns of $- R^u_n{}^T \Omega^u_{p,n}$, or the translational parameters, corresponding to the columns of $R^u_n{}^T$, are unobservable. Combining the result with (5.4.12) yields (5.4.13).   []

Remark: Theorem 5.4.3 explains why only 25 parameters are independent when the PUMA arm was calibrated using only positioning errors of the end-effector, as was done by Chen and Chao (1986). Other researchers have also observed the same phenomenon; for instance, refer to (Whitney, Lozinski and Rourke (1986) and to Everett and Hsu (1988)).

Theorem 5.4.4:   Assume that the left multiplicative differential transformations are used to model kinematic errors. If the orientational errors of a manipulator end-effector are not measured, then all orientational parameters of A_0 are unobservable. In this case, the number N of observable kinematic error parameters must satisfy the following inequality

$$N \leq 4n - 2p + 6 - o_0 \qquad (5.4.14)$$

where o_0 is the number of independent orientational parameters in A_0.

Proof: Similar to the proof of the first part of Theorem 5.4.3.

Theorem 5.4.5:   Assume that the left multiplicative differential transformations are used to model kinematic errors. If the 0th link frame is parallel to the world frame, then the translational parameters in A_0 are unobservable. In this case, the number N of observable kinematic error parameters must satisfy the following inequality

$$N \leq 4n - 2p + 6 - t_{01} \qquad (5.4.15)$$

where $t_{01} = \min\{t_0, t_1\}$, and t_i is the number of independent translational parameters in A_i. If in addition the orientational errors of the end-effector are not measured, then

$$N \leq 4n - 2p + 6 - o_0 - t_{01} \qquad (5.4.16)$$

where o_0 is the number of independent orientational parameters in A_0.

Proof: When the base frame is parallel to the world frame, R'_0 is an elementary matrix. Thus R'_0 is independent of the joint variables (i.e., it is not a function of q_1) and the columns related to R'_0 in L' depend linearly on the first three columns of L'. In this case, either the translational error parameters corresponding to R'_0 or to the first three columns of L' are unobservable. Inequality (5.4.16) is obtained after combining (5.4.14) with (5.4.15). []

Theorems 5.4.3-5.4.5 are true for any choice of kinematic modeling conventions. The next simple fact reveals the relationship between the observability of kinematic parameters and the irreducibility of linearized error models.

Theorem 5.4.6: Consider a parametrically continuous and differentiable kinematic model. The parameters in the linearized error model are unobservable if the linearized error model is reducible.

Proof: Denote the Identification Jacobian matrix associated with the reducible error model as LK. Since the model is reducible, there exists a linear mapping M, where M is an $m' \times m$ matrix relating the nonindependent parameter vector to the independent parameter vector. Let $J = LKM$. Notice that the column rank of LK is the same as that of J. Since $m' < m$, $\text{Dim}((LK)^T LK) > \text{Dim}(J^T J)$, where $\text{Dim}(\cdot)$ denotes the dimension of a matrix. However $\text{Rank}((LK)^T LK) = \text{Rank}(J^T J)$. Thus $(LK)^T LK$ is singular. By Definition 5.4.1 the proof is complete. []

Observability characterizes the possibility of identifying kinematic parameters by examining the singular values of the Jacobian matrix J. Observability measures are usually defined in terms of the singular values of J. By singular value decomposition, J can be written as

$$J = U\Sigma V^T$$

where $U^TU = I_s$, $V^TV = I_m$, and $\Sigma = \text{diag}(\sigma_1, \sigma_2, ..., \sigma_m)$. Here $q{\times}m$ is the size of J, and σ_i, $i = 1, 2,\cdots, m$, are the singular values of J^TJ. The matrix U consists of q orthonormalized eigenvectors associated with the eigenvalues of JJ^T, whereas the matrix V consists of m orthonormalized eigenvectors associated with the eigenvalues of J^TJ.

An observability measure O is defined as follows (Menq and Borm 1988):

$$O(J) = (\sigma_1 \sigma_2...\sigma_m)^{1/m} q^{1/2} \qquad (5.4.17)$$

If all values of σ_i are available, the measure $O(J)$ is conveniently computed by the following equation,

$$O(J) = q^{-1/2} exp\{m^{-1}\sum_{i=1}^{m} ln\sigma_i\}$$

provided that none of the singular values σ_i is zero. O can also be written as

$$O(J) = (\det(J^TJ))^{1/2m} q^{1/2}$$

It was pointed out by Menq and Borm (1988) that whenever O is zero, the error parameters are unobservable from the measurements performed at the selected configurations. As O increases, the contribution of geometric errors to the overall robot positioning errors dominates the effects of nongeometric and other unmodeled errors; consequently better estimation of the error parameters is expected. Experimental studies were performed also by Borm and Menq (1991) to demonstrate the importance of the observability measure to the estimation of kinematic parameters. Optimal measurement configurations for robot calibration have been investigated through simulation studies, using the above observability measure.

Despite its being a good practical index for investigating the observation strategy in robot calibration, the observability measure defined in (5.4.17) has a few drawbacks. First it is not scale invariant. This can be illustrated in the following way. Let J be multiplied by a scalar k. Then all singular values of J are multiplied by the same scaling factor, therefore

$$O(kJ) = (k\sigma_1 k\sigma_2...k\sigma_m)^{1/m} q^{1/2}$$

$$= k(\sigma_1\sigma_2...\sigma_m)^{1/m}q^{1/2}$$
$$= kO(J)$$

Another problem associated with Menq and Borm's observability measure is that two extremely dissimilar matrices may have similar values of the observability measure. For instance, let the singular values of J be $\{10^2, 10, 1, 10^{-1}, 10^{-2}\}$. According to (5.4.17), the observability measure has a value of $q^{-1/2}$, which is identical to that of the identity matrix. Obviously such a matrix behaves numerically very differently from the identity matrix.

If instead the condition number of J is used as an observability index, the problems mentioned above are eliminated. Driels and Pathre (1989) studied through simulations the observation strategy of robot calibration experiments by using the condition number of J, defined in terms of its maximum and minimum singular values,

$$\text{Cond}(J) = \sigma_{max}/\sigma_{min} \qquad (5.4.18)$$

Obviously
$$\text{Cond}(kJ) = \text{Cond}(J)$$

where k is a scalar. This means that the condition number is scale invariant. Also,

$$\text{Cond}(J) > 1$$

$\text{Cond}(J) = 1$ is the ideal situation for good observability of error parameters. If $\text{Cond}(J)$ is large, then the columns of J are nearly dependent, therefore a poor estimation of error parameters is inevitable. If $\text{Cond}(J) = \infty$, then the error parameters are not fully observable.

From a computational view point, the condition number is somewhat more expensive, compared to the observability measure defined in (5.4.18). This is because either matrix inversion or singular value decomposition requires more computation time than the determinant computation of the matrix. In (Forsythe, Malcolm and Moler (1977); Golub, Charles and Loan (1983)) and other texts on matrix computation, efficient algorithms for computing the condition number can be found.

The matrix J could become singular at any iteration step during the numerical process even if the kinematic model itself has no singularity. Some robust minimization techniques, such as the Levenberg-Marquardt algorithm, have to be applied to solve the problem, refer to ((Mooring and Padavala (1989) and to Bennett and Hollerbach (1988)).

An observation similar to that of Theorem 5.4.6 is given next:

Theorem 5.4.7:   Consider a parametrically continuous and differentiable kinematic model. If the linearized error model is reducible, then

$$\text{Cond}(J) = \infty$$

and

$$O(J) = 0.$$

Proof: If the error model is reducible, the columns of J are not linearly independent. In this case $\sigma_{min} = 0.$ []

If the error model is reducible, observability measures are always zero or infinity, respectively, no matter how many measurements are taken and how well the configurations are chosen. In practice, even if parameter errors are not independent, a robust algorithm can be applied to estimate the geometric errors, as long as good configurations are chosen and enough measurements are taken. This may suggest that the values of the observability measure or the condition number may be meaningless in such a case.   Whenever the observability measure or condition number are used for planning a robot calibration experiment, the irreducibility of the linearized error model has to be verified first.

The technique of singular value decomposition can be employed to eliminate dependent parameters in an error model.   Those singular values that are close to zero correspond to dependent parameters.   By nullifying these singular values, the dimension of the parameter vector is reduced. For a subset selection using the technique of singular value decomposition, readers are referred to (Golub, Charles and Loan (1983).   However, care should be taken in doing so because the dependence of parameters is not the only source for the Jacobian matrix J to become singular.

Observability measures are defined based on the identification Jacobian, therefore these do not take into account the nongeometric nature of some error sources, such as gravity loading and backlash.   Least-squares techniques can be used to effectively reduce the influence of nongeometric errors.   The observability indices fail to characterize the efficiency of such a process.   A common sense approach in designing a robot calibration experiment may often produce better results than designs resulting from the application of these mathematical indices.   As an example, errors in the first three joints of the PUMA arm play a dominant role in determining the positioning accuracy of the manipulator.   The PUMA can be in either the right-arm configuration or the left-arm configuration. If kinematic parameters are identified by using measurements taken only from the right-arm configuration, the positioning errors may not be reduced using the resulting model if the robot had been programmed in the left-arm configuration. This is because gravity effects on the arm are amplified in such a case. On the other hand, if kinematic parameters are identified with measurements taken evenly from both types of arm configurations, the errors due to the weight of the arm can be averaged out, resulting in much accurate compensation results. Both observability measures fail to indicate the great difference between the above two schemes.

V. THE D-H ERROR MODEL

The derivation of the linearized D-H error models follows two steps. First, a linear mapping relating Cartesian errors to D-H parameter errors of individual links is derived. The results are then plugged into the generic error model described in the previous section to obtain the D-H error model.

A. LINEAR MAPPING RELATING CARTESIAN ERRORS TO D-H PARAMETER ERRORS OF INDIVIDUAL LINKS

Let $[d_i^T, \delta_i^T]^T$ be a Cartesian error vector of the ith link, where $[d_i^T, \delta_i^T]$ $\in R^6$. Let $[d\theta_i, dl_i, da_i, d\alpha_i]^T$ be the corresponding D-H parameter error vector.

Let us derive the translation part first. Let R_i and p_i denote the upper-left 3x3 rotation matrix and 3x1 position vector of A_i in the D-H model, respectively. The translation error vector d_i can then be written as

$$d_i = -R_i^T dp_i$$

The position vector p_i in the D-H model is

$$p_i = [a_i cos\theta_i, a_i sin\theta_i, d_i]^T$$

By straightforward differentiation, it can be shown that

$$d_i \;\; = \;\; k_{i,1} d\theta_i \; + k_{i,2} dd_i \;\; + k_{i,3} da_i$$

where

$$k_{i,1} = [0, a_i cos\alpha_i, -a_i sin\alpha_i]^T \qquad (5.5.1)$$
$$k_{i,2} = [0, sin\alpha_i, cos\alpha_i]^T \qquad (5.5.2)$$
$$k_{i,3} = [1, 0, 0]^T. \qquad (5.5.3)$$

Let us now derive the orientation part. Note that $\delta R_i = R_i^T dR_i$ is a skew-symmetric matrix,

$$\delta R_i = Rot(x, \alpha_i)^T Rot(z, \theta_i)^T dRot(z, \theta_i) Rot(x, \alpha_i) + Rot(x, \alpha_i)^T dRot(x, \alpha_i)$$

Thus, the vector δ_i of three nontrivial elements in δR_i can be written as

$$\delta_i = Rot(x, \alpha_i)^T e_z d\theta_i + e_x d\alpha_i$$

where $e_z = [0, 0, 1]^T$, and $e_x = [1, 0, 0]^T$. However $k_{i,2} = Rot(x, \alpha_i)^T e_z$, and $k_{i,3} = e_z$. In summary, we have,

$$\delta_i = k_{i,2} d\theta_i + k_{i,3} d\alpha_i \qquad (5.5.4)$$
$$d_i = k_{i,1} d\theta_i + k_{i,2} dd_i + k_{i,3} da_i \qquad (5.5.5)$$

where the coefficients $k_{i,1}$, $k_{i,2}$ and $k_{i,3}$ are given in Equations (5.5.1)-(5.5.3).

B. LINEARIZED D-H ERROR MODEL

It is assumed that the structure of transformations A_i, $i = 0, 1, ..., n$, is based on the standard D-H modeling convention. Recall that $[d^T, \delta^T]^T$ is a Cartesian error vector representing the pose errors of the manipulator. Let dp be the corresponding D-H parameter error vector.

Denote

$$U_i = \begin{bmatrix} n_i^u & o_i^u & a_i^u & p_i^u \\ 0 & 0 & 0 & 1 \end{bmatrix}$$

where n^u_i, o^u_i, a^u_i and p^u_i are 3x1 vectors.

The D-H error model is given in the following theorem:

Theorem 5.5.1: The linearized D-H error model can be written as

$$\delta = R_\theta d\theta + R_\alpha d\alpha \qquad (5.5.6)$$
$$d = T_\theta d\theta + T_d dd + T_a da \qquad (5.5.7)$$

where

$$d\theta = [d\theta_0, d\theta_1, ..., d\theta_n]^T,$$

$$d\alpha = [\, d\alpha_0, \, d\alpha_1, \, \ldots, \, d\alpha_n \,]^T,$$

$$dd = [\, dd_0, \, dd_1, \, \ldots, \, dd_n]^T,$$

$$da = [\, da_0, \, da_1, \, \ldots, \, da_n]^T,$$

and R_θ, R_α, T_θ, T_d and T_a are all $3\times(n+1)$ matrices, the components of which are functions of link parameters and joint variables. Specifically, the ith column of these matrices are:

$$[R_\theta]_i = \begin{bmatrix} n^u_{i+1} \cdot k_{i,2} \\ o^u_{i+1} \cdot k_{i,2} \\ a^u_{i+1} \cdot k_{i,2} \end{bmatrix}$$

$$[R_\alpha]_i = \begin{bmatrix} n^u_{i+1} \cdot k_{i,3} \\ o^u_{i+1} \cdot k_{i,3} \\ a^u_{i+1} \cdot k_{i,3} \end{bmatrix}$$

$$[T_\theta]_i = \begin{bmatrix} (k_{i,2} \times p^u_{i+1} + k_{i,1}) \cdot n^u_{i+1} \\ (k_{i,2} \times p^u_{i+1} + k_{i,1}) \cdot o^u_{i+1} \\ (k_{i,2} \times p^u_{i+1} + k_{i,1}) \cdot a^u_{i+1} \end{bmatrix}$$

$$[T_\alpha]_i = \begin{bmatrix} (k_{i,3} \times p^u_{i+1}) \cdot n^u_{i+2} \\ (k_{i,3} \times p^u_{i+1}) \cdot o^u_{i+2} \\ (k_{i,3} \times p^u_{i+1}) \cdot a^u_{i+2} \end{bmatrix}$$

$$[T_d]_i = [R_\theta]_i$$
$$[T_a]_i = [R_\alpha]_i$$

where $k_{i,j}$ for $j = 1, 2, 3$ are given in (5.5.3)-(5.5.5).

Proof: Substituting (5.5.1) into (5.4.4) yields

$$\delta = \sum_{i=0}^{n} R^u_{i+1}{}^T \{k_{i,2}d\theta_i + k_{i,3}d\alpha_i\}$$

Substituting (5.5.1) and (5.5.2) into (5.4.5), yields

$$d = \sum_{i=0}^{n} -R^u{}_{i+1}{}^T \Omega^u{}_{p,i+1} \{k_{i,2}d\theta_i + k_{i,3}d\alpha_i\}$$

$$+ R^u{}_{i+1}{}^T \{k_{i,1}d\theta_i + k_{i,2}dd_i + k_{i,3}da_i\}$$

$$= \sum_{i=0}^{n} R^u{}_{i+1}{}^T \{-\Omega^u{}_{p,i+1} k_{i,2} + k_{i,1}\}d\theta_i - R^u{}_{i+1}{}^T \Omega^u{}_{p,i+1}k_{i,3}d\alpha_i$$

$$+ R^u{}_{i+1}{}^T k_{i,2}dd_i + R^u{}_{i+1}{}^T k_{i,3}da_i$$

using the relationship $\Omega^u{}_{p,n+1} = 0$ and $\Omega^u{}_{p,i+1} \times k_{i,j} = p^u{}_{i+1} \times k_{i,j}$ for $j = 1$, 2. Equations (5.5.6) and (5.5.7) are obtained by rearrangement of the above equations.

Remark: The above D-H error model has $4R + 2P + 4$ error parameters, where R and P are the number of revolute and prismatic joints in the manipulator, respectively. This is because only 4 parameters are employed in the 0th link transformation.

VI.   THE CPC ERROR MODEL

The CPC error model is obtained following a similar path to that of the D-H error model derivation, except that an additional step is introduced to eliminate the redundant error parameters in the CPC error model.

A. LINEAR MAPPING RELATING CARTESIAN ERRORS TO INDEPENDENT CPC PARAMETER ERRORS OF INDIVIDUAL LINKS

The derivation of the linearized error mapping of individual links follows two steps.   First, a linear mapping K_i, relating the Cartesian errors to the CPC link parameter errors of each individual link, is obtained.   The redundant parameter errors are then eliminated by another linear mapping M_i.

1. Linear Mapping Relating Cartesian Errors to CPC Parameter Errors of Individual Links

Let $[d_i{}^T, \delta_i{}^T]^T$ be a Cartesian error vector of the ith link and $[dl_i, db_i, d\beta_i]^T$ be the corresponding CPC parameter error vector, where $[d_i{}^T, \delta_i{}^T] \in R^6$ and $[dl_i, db_i, d\beta_i]^T \in R^7$.   Prior to a linear mapping from the CPC parameter error space to the Cartesian error space, several preliminary derivations are introduced first.

Let R_i denote the upper-left 3x3 submatrix of R_i in the CPC model. Then $R_i{}^T dR_i$, is a skew-symmetric matrix. In a straightforward derivation, the vector of three nontrivial elements of $R_i{}^T dR_i$, denoted by δ_i, can be written as

$$\delta_i = k^r_{i,x} db_{i,x} + k^r_{i,y} db_{i,y} + k^r_{i,z} db_{i,z} \quad i = 0, 1, \cdots, n \quad (5.6.1)$$

where the coefficients $k^r_{i,x}$, $k^r_{i,y}$ and $k^r_{i,z}$ are 3x1 vectors given by

$$k^r_{i,x} = [b_{i,x} b_{i,y} w_i, \ 1-b_{i,y}{}^2 w_i, \ (1+b_{i,z}+b_{i,x}{}^2) b_{i,y} w_i{}^2]^T \quad (5.6.2)$$

$$k^r_{i,y} = [-1+b_{i,y}{}^2 w_i, \ -b_{i,x} b_{i,y} w_i, \ -(1+b_{i,z}-b_{i,y}{}^2) b_{i,x} w_i{}^2]^T \quad (5.6.3)$$

$$k^r_{i,z} = [b_{i,y}, \ -b_{i,x}, \ b_{i,x} b_{i,y} b_{i,z} w_i{}^2]^T. \quad (5.6.4)$$

Here $w_i \equiv 1/(1+b_{i,z})$, for $i = 0,1, \ldots, n$. Equations (5.6.1)-(5.6.4) are kept for later use.

Let us now derive the linear mapping K_i. A right multiplicative differential transformation is to be used to model the error transformation. From $B_i = Q_i V_i$,

$$\delta B_i = B_i{}^{-1} dB_i = V_i{}^{-1} dV_i.$$

Then

$$
\begin{aligned}
\delta B_i &= Trans(-l_{i,x}, -l_{i,y}, -l_{i,z}) Rot(z_i, -\beta_i) R_i{}^{-1} d\{ R_i Rot(z_i, \beta_i) Trans(l_{i,x}, l_{i,y}, l_{i,z}) \} \\
&= Trans(-l_{i,x}, -l_{i,y}, -l_{i,z}) Rot(z_i, -\beta_i) R_i{}^{-1} dR_i Rot(z_i, \beta_i) Trans(l_{i,x}, l_{i,y}, l_{i,z}) \\
&\quad + Trans(-l_{i,x}, -l_{i,y}, -l_{i,z}) Rot(z_i, -\beta_i) dRot(z_i, \beta_i) Trans(l_{i,x}, l_{i,y}, l_{i,z}) \\
&\quad + Trans(-l_{i,x}, -l_{i,y}, -l_{i,z}) dTrans(l_{i,x}, l_{i,y}, l_{i,z}) \quad (5.6.5)
\end{aligned}
$$

It is now convenient to break δB_i into two parts: the upper-left 3x3 submatrix that contains the elements of the orientational error vector δ_i, and the last column that is the positioning error vector d_i. Then

$$\delta_i = \delta_{i,1} + \delta_{i,2} + \delta_{i,3} \quad (5.6.6)$$

and

$$d_i = d_{i,1} + d_{i,2} + d_{i,3} \quad (5.6.7)$$

where $\delta_{i,j}$ and $d_{i,j}$ are the orientational and positioning error vector corresponding to the jth additive term of the right-hand side of (5.6.5). By using the relationships (5.2.8) and (5.2.9) and noting that the effects of the forward and backward translations on the orientational errors cancel out, we have

$$\delta_{i,1} = Rot(z_i, -\beta_i)_{3x3} \, \delta_i \tag{5.6.8}$$

$$\delta_{i,2} = k_{i,\beta} d\beta_i \tag{5.6.9}$$

$$\delta_{i,3} = 0 \tag{5.6.10}$$

where δ_i is given in (5.6.1), and

$$k_{i,\beta} = [0, 0, 1]^T. \tag{5.6.11}$$

By straightforward manipulation of (5.6.5) and the use of (5.6.8)-(5.6.10), we obtain

$$d_{i,1} = \delta_{i,1} \times l_i \tag{5.6.12}$$

$$d_{i,2} = \delta_{i,2} \times l_i \tag{5.6.13}$$

$$d_{i,3} = d_i. \tag{5.6.14}$$

Plugging (5.6.8)-(5.6.14) into (5.6.6) and (5.6.7) yields

$$\delta_i = k_{i,x} db_{i,x} + k_{i,y} db_{i,y} + k_{i,z} db_{i,z} + k_{i,\beta} d\beta_i \tag{5.6.15}$$

$$d_i = d_i + k_{i,x} \times l_i db_{i,x} + k_{i,y} \times l_i db_{i,y} + k_{i,z} \times l_i db_{i,z} + k_{i,\beta} \times l_i d\beta_i \tag{5.6.16}$$

for $i = 0,1, \cdots, n$, where

$$k_{i,j} = Rot(z_i, -\beta_i)_{3x3} k'_{i,j} \quad j \in \{x,y,z\}. \tag{5.6.17}$$

For robot internal links, β_i is often set to zero, thus $k_{i,j} = k'_{i,j}$. Equations (5.6.15) and (5.6.16) express the Cartesian errors in terms of the CPC parameter errors for an individual link. The coefficients in these two equations define the linear mapping K_i.

2. Linear Mapping Relating Cartesian Errors to Independent CPC Parameter Errors of Individual Links

The feasibility of constructing an irreducible error model in the CPC modeling convention is analyzed first. The linearized error model is obviously reducible if it is obtained by plugging K directly into (5.4.9) because the parameter errors are not independent. However, a linear mapping M can be constructed to transform the linearized error model to an irreducible one. Among the orientational parameters $\{b_{i,x}, b_{i,y}, b_{i,z}\}$, one is redundant and can be recovered from the other two.   Among the translational parameters $\{l_{i,x}, l_{i,y}, l_{i,z}\}$, $l_{i,z}$ is introduced to allow an arbitrary assignment of the ith link frame along the $(i+1)$th joint axis. It is therefore redundant for $i = 0, 1, .$ . ., n-1, except for $l_{n,z}$. Likewise β_i, $i = 0, 1, \ldots, n$-1, is redundant, except for β_n. Both $l_{n,z}$ and β_n are introduced to provide 6 degrees of freedom to the tool frame. Thus four independent parameters remain for each revolute joint and two independent parameters for a prismatic joint.   The above analysis clearly shows that the linearized error model can be made irreducible.

In the next few paragraphs, the subscript "i" is dropped for convenience.

It seems straightforward to eliminate an error parameter from the set $\{db_{i,x}, db_{i,y}, db_{i,z}\}$ in the error model. For instance, if db_x and db_y are chosen as the independent parameter errors, db_z can be expressed as follows

$$db_z = -\frac{b_x db_x + b_y db_y}{b_z} \qquad (5.6.18)$$

After substituting the above into (5.6.15) and (5.6.16), δ_i is only related to db_x and db_y. However, (5.6.18) has a singular point $b_z = 0$. The model is no longer differentiable at this point.   Therefore the linearized error model is no longer valid at such a pose. Physically this happens when two consecutive axes are perpendicular to each other. Similar problems will appear if other parameter errors are chosen as independent ones in the set of $\{db_x, db_y, db_z\}$. The following two rules for selecting two independent parameters out of the set $\{db_x, db_y, db_z\}$ can be used to keep the CPC model differentiable:

1.   The error parameter that corresponds to $\max\{|b_x|, |b_y|, |b_z|\}$ is eliminated.
2.   The direction vector b must be kept a unit vector at every step of a kinematic parameter estimation process.

The transformation M, relating the dependent parameter errors with the independent CPC parameter errors, can now be easily written down. Due to the simplicity of the mapping, one can directly find $K_i M_i$, the mapping relating the Cartesian error vector $[d^T_i, \delta^T_i]^T$ to the independent parameter error vector.

Let j and k indicate that $b_{i,j}$ and $b_{i,k}$ in b_i are chosen as independent parameters, and let h indicate that $b_{i,h}$ in b_i is to be eliminated. $j, k, h \in \{x, y, z\}$ and $j \neq k \neq h$. δ_i is only related to the orientational parameters,

$$\delta_i = k\,m_{i,j,h}db_{i,j} + k\,m_{i,k,h}db_{i,k} \qquad i = 0, 1, \ldots, n\text{-}1 \quad (5.6.21)$$

and

$$\delta_n = k\,m_{n,j,h}db_{n,j} + k\,m_{n,k,h}db_{n,k} + k_{n,\beta}d\beta_n \qquad (5.6.22)$$

where $k\,m_{i,j,h}$ and other similar coefficients are 3x1 vectors. d_i is in general a function of both orientational and translational parameters,

$$d_i = dl_i + k\,m_{i,j,h}xl_i db_{i,j} + k\,m_{i,k,h}xl_i db_{i,k} \qquad i = 0, 1, \ldots, n\text{-}1$$
$$(5.6.23)$$

and

$$d_n = dl_n + k\,m_{n,j,h}xl_n db_{n,j} + k\,m_{n,k,h}xl_n db_{n,k} + k_{n,\beta}xl_n d\beta_n$$
$$(5.6.24)$$

where $dl_i = [dl_{i,x}, dl_{i,y}, 0]^T$ for $i = 0, 1, \ldots, n\text{-}1$ and $dl_n = [dl_{n,x}, dl_{n,y}, dl_{n,z}]^T$. Equations (5.6.21)-(5.6.24) express the Cartesian errors in terms of the independent CPC parameter errors for individual links. The coefficients in this set of equations define the linear mapping $K_i M_i$.

The coefficients of (5.6.21)-(5.6.24) can be obtained by straightforward derivations. Assume for example that $db_{i,x}$ and $db_{i,y}$ are chosen as the independent parameter errors. Then

$$\delta_i = k\,m_{i,x,z}db_{i,x} + k\,m_{i,y,z}db_{i,y}$$
$$d_i = dl_i + k\,m_{i,x,z}xl_i db_{i,x} + k\,m_{i,y,z}xl_i db_{i,y}.$$

Substituting (5.6.18) into (5.6.15) and (5.6.16) and comparing the result with (5.6.21) and (5.6.24), one obtains,

$$k\,m_{i,x,z} = Rot(z_i,\ \beta_i)_{3\times3} \times [-b_{i,x}b_{i,y}w_i/b_{i,z},\ 1 + b_{i,y}^2 w_i/b_{i,z},\ b_{i,y}w_i]^T,$$

$$k\,m_{i,y,z} = Rot(z_i,\ \beta_i)_{3\times3} \times [-1 - b_{i,y}^2 w_i/b_{i,z},\ b_{i,x}b_{i,y}w_i/b_{i,z},\ -b_{i,x}w_i]^T,$$

for $i = 0, 1, \ldots, n$, where again $w_i \equiv 1/(1+b_{i,z})$. Similarly if $db_{i,x}$ and $db_{i,z}$ are chosen as the independent parameter errors. Then

$$\delta_i = k\,m_{i,x,y}db_{i,x} + k\,m_{i,z,y}db_{i,z}$$
$$d_i = dl_i + k\,m_{i,x,y}xl_i db_{i,x} + k\,m_{i,z,y}xl_i db_{i,z}.$$

where

$$k\,m_{i,x,y} = Rot(z_i,\ \beta_i)_{3\times3} \times [b_{i,x}/b_{i,y},\ 1,\ (1 - b_{i,z})/b_{i,y}]^T,$$

$$k\,m_{i,z,y} = Rot(z_i,\ \beta_i)_{3\times3} \times [b_{i,z}/b_{i,y} + b_{i,y}w_i,\ -b_{i,x}w_i,\ b_{i,x}b_{i,z}w_i/b_{i,y}]^T,$$

for $i = 0, 1, \ldots, n$. Finally if $db_{i,y}$ and $db_{i,z}$ are chosen as the independent parameter errors. Then

$$\delta_i = k\,m_{i,y,x}db_{i,y} + k\,m_{i,z,x}db_{i,z}$$
$$d_i = d_i + k\,m_{i,y,x}xl_i db_{i,y} + k\,m_{i,z,x}xl_i db_{i,z}.$$

where

$$k\,m_{i,y,x} = Rot(z_i,\ \beta_i)_{3\times3} \times [-1,\ b_{i,y}/b_{i,x},\ -(1 - b_{i,z})/b_{i,x}]^T,$$

$$k\,m_{i,z,x} = Rot(z_i,\ \beta_i)_{3\times3} \times [b_{i,y}w_i,\ -b_{i,z}/b_{i,x} - b_{i,x}w_i,\ -b_{i,y}b_{i,z}w_i/b_{i,x}]^T,$$

for $i = 0, 1, \ldots, n$. Note that for $i = n$, the term related to $d\beta_n$ has to be added to both δ_n and d_n, as given in (5.6.9).

B. THE LINEARIZED CPC ERROR MODEL

The CPC error model can be obtained by utilizing the results obtained in Sections VI.A and IV.

Theorem 5.6.1: The linearized CPC error model can be written as

$$\delta = Rb_j db_j + Rb_k db_k + k_{n,\beta} d\beta_n \qquad (5.6.25)$$
$$d = Tb_j db_j + Tb_k db_k + Tl_x d_x + Tl_y dl_y + [Tl_z]_n dl_{n,z} + k_{n,\beta} xl_n d\beta_n$$

$$(5.6.26)$$

where

$$db_j = [\,db_{0,j},\ db_{1,j},\ \ldots,\ db_{n,j}\,]^T,$$

$$db_k = [\,db_{0,k},\ db_{1,k},\ \ldots,\ db_{n,k}\,]^T,$$

$$d_x = [\,dl_{0,x},\ dl_{1,x},\ \ldots,\ dl_{n,x}\,]^T,$$

$$d_y = [\,dl_{0,y},\ dl_{1,y},\ \ldots,\ dl_{n,y}\,]^T,$$

and Rb_j, Rb_k, Tb_j, Tb_k, Tl_x, and Tl_y are all $3\times(n+1)$ matrices whose components are functions of link parameters and joint variables. Specifically, the ith column of these matrices are:

$$[Rb_j]_i = \begin{bmatrix} n^u_{i+1} \cdot k\, m_{i,j,h} \\ o^u_{i+1} \cdot k\, m_{i,j,h} \\ a^u_{i+1} \cdot k\, m_{i,j,h} \end{bmatrix}$$

$$[Rb_k]_i = \begin{bmatrix} n^u_{i+1} \cdot k\, m_{i,k,h} \\ o^u_{i+1} \cdot k\, m_{i,k,h} \\ a^u_{i+1} \cdot k\, m_{i,k,h} \end{bmatrix}$$

$$[Tb_j]_i = \begin{bmatrix} ((p^u_{i+1} + l_i) \times n^u_{i+1}) \cdot k\, m_{i,j,h} \\ ((p^u_{i+1} + l_i) \times o^u_{i+1}) \cdot k\, m_{i,j,h} \\ ((p^u_{i+1} + l_i) \times a^u_{i+1}) \cdot k\, m_{i,j,h} \end{bmatrix}$$

$$[Tb_k]_i = \begin{bmatrix} ((p^u_{i+1} + l_i) \times n^u_{i+1}) \cdot k\, m_{i,k,h} \\ ((p^u_{i+1} + l_i) \times o^u_{i+1}) \cdot k\, m_{i,k,h} \\ ((p^u_{i+1} + l_i) \times a^u_{i+1}) \cdot k\, m_{i,k,h} \end{bmatrix}$$

$$[Tl_x]_i = [\, n^u_{i+1,x},\ o^u_{i+1,x},\ d^u_{i+1,x}\,]^T$$

$$[Tl_y]_i = [\, n^u_{i+1,y},\ o^u_{i+1,y},\ d^u_{i+1,y}\,]^T$$

and

$$[Tl_z]_n = [\, n^u_{n,z},\ o^u_{n,z},\ d^u_{n,z}\,]^T.$$

The coefficients $k\, m_{i,\,*,\,*}$, where each subscript $*$ represents either x, y or z, are all given at the end of Section 5.5.B.

Proof: Substituting (5.6.21) and (5.6.22) into (5.4.4) results in

$$\delta = \sum_{i=0}^{n} R^u_{i+1}{}^T \{k\, m_{i,j,h} db_{i,j} + k\, m_{i,k,h} db_{i,k}\} + k_{n,\beta} d\beta_n$$

since $R^u_{n+1} = I$. Substituting (5.6.23) and (5.6.24) into (5.4.5), yields

$$d = \sum_{i=0}^{n} -R^u_{i+1}{}^T \Omega^u_{p,i+1} \{k m_{i,j,h} db_{i,j} + k m_{i,k,h} db_{i,k}\}$$

$$+ \sum_{i=0}^{n} R^u_{i+1}{}^T \{k m_{i,j,h} \times l_i db_{i,j} + k m_{i,k,h} \times l_i db_{i,k} + dl_i\} + k_{n,\beta} \times l_i d\beta_n$$

$$= \sum_{i=0}^{n} R^{u}_{i+1}{}^{T} \{ km_{i,j,h} \times (p^{u}_{i+1} + l_{i}) db_{i,j} + km_{i,k,h} \times (p^{u}_{i+1} + l_{i}) db_{i,k} + dl_{i} \}$$

$$+ k_{n,\beta} \times l_{i} d\beta_{n}$$

using the relationship $\Omega^{u}_{p,n+1} = 0$, $\Omega^{u}_{p,i+1} \times k\, m_{i,j,h} = p^{u}_{i+1} \times k\, m_{i,j,h}$, and $\Omega^{u}_{p,i+1} \times k\, m_{i,k,h} = p^{u}_{i+1} \times k\, m_{i,k,h}$. Equations (5.6.25) and (5.6.26) are obtained from rearrangement of the above equations.   []

The CPC error model possesses $4R + 2P + 6$ error parameters. It is an irreducible error model.

C. LINEARIZED *BASE* AND *TOOL* ERROR MODELS

In this section, the error models for the *BASE* and *TOOL* transformations is presented in the CPC modeling convention. Similar treatment can be applied for the MCPC model.

1. Linearized *BASE* Error Model

Several methods exist for determining the nominal *BASE* transformation (denoted also as $^{w}T_{b}$ or B_{0}). Obviously, if a tool frame can be established on an end-effector such that three points from each of the three axes of the tool frame, respectively, can be measured in terms of the world frame, then the determination of the nominal *BASE* becomes trivial.

Alternatively, the nominal *BASE* can be found by using the 3-measurements scheme given next. Taking three non-colinear points in the world frame, one is able to establish a tool frame having three orthonormal vectors along its coordinate axes; say $^{w}n_{t}$, $^{w}o_{t}$, and $^{w}a_{t}$. Furthermore, one of the points, the intersection of the three vectors, can be chosen as the origin of the tool frame represented in the world frame; say $^{w}p_{t}$. Thus the homogeneous transformation matrix $^{w}T_{t}$ has the rotation submatrix $^{w}R_{t} = [^{w}n_{t}, \, ^{w}o_{t}, \, ^{w}a_{t}]$ and the position vector $^{w}p_{t}$. Similarly, the forward kinematics of the robot model provides the coordinates of the three points in the base frame provided that their joint readings are recorded. Similar procedures of constructing $^{w}T_{t}$ are then applied to find $^{b}T_{t}$. From the solutions $^{w}T_{t}$ and $^{b}T_{t}$, we have

$$^{w}T_{b} = {}^{w}T_{t} \, {}^{b}T_{t}^{-1}.$$

As mentioned earlier, the nominal *BASE* obtained above suffers a drawback. That is, the transformation on the whole is inaccurate due to measurement uncertainties. Thus updating $BASE$ is a necessary step if the nominal $BASE$ does not satisfy the required positioning accuracy of the robot.

The 0th link parameters can be initialized from the nominal $BASE$. By using an error model that maps the 0th link parameter error vector to the 0th Cartesian error vector, these parameters can be iteratively updated.

Six independent error parameters should be used if $BASE$ is to be identified. Recall that the same number of independent error parameters are used to model the nth link errors in the CPC error model. The formulas for modeling errors in $BASE$ can be obtained simply by changing the subscript "n" to "0" in (5.6.22) and (5.6.24). The above error model needs to be modified if the orientational errors are not measured.

The transformation from the tool frame to the world frame can be written as

$$T_n = B_0 U_1$$

where U_1 is the transformation relating the tool frame to the base frame. Assuming that the pose errors of the end-effector are only caused by the deviations of the 0th link parameters, then

$$d T_n = d B_0 U_1 = B_0 \, \delta B_0 U_1.$$

Thus

$$\delta B_0 = B_0^{-1} d T_n U_1^{-1}. \tag{5.6.27}$$

$d T_n$ is found by measuring the poses of the manipulator end-effector. Thus the right-hand side of (5.6.5) becomes known. The above equation serves as a measurement equation in the case in which only the CPC $BASE$ parameters are to be found.

If positioning error measurements are the only ones available, the above measurement equation cannot be used. This is because the fourth columns of the right-hand side of (5.6.27) contains unknown orientation information. The following method is designed to overcome this difficulty. From (5.6.27),

$$\delta B_0 U_1 = B_0^{-1} d T_n. \tag{5.6.28}$$

The fourth column of the right-hand side of the above equation becomes known. Denote that column by $d_0{}'$. Then from (5.6.28),

$$d_0{}' = \delta_0 \times p^u{}_1 + d_0 \tag{5.6.29}$$

where $p^u{}_1$ is the nontrivial part of the fourth column of U_1. Substituting d_0 and δ_0, given in (5.6.22) and (5.6.24), into (5.6.29) yields

$$d_0' = d_0 + k\, m_{0,j,h} \times (p^u{}_1 + l_0) db_{0,j} + k\, m_{0,k,h} \times (p^u{}_1 + l_0) db_{0,k}$$
$$+ k_{0,\beta} \times l_0 d\beta_0$$

where $d_0 = [dl_{0,x}, \; dl_{0,y}, \; dl_{0,z}]^T$. The above is the linearized $BASE$ error model when the orientational errors are not available.

2. Linearized $TOOL$ Error Model

Tool dimensions can often be measured very accurately off-line before it is attached to the robot. If mounting errors of the end-effector are negligible, the $TOOL$ transformation can be treated as error-free. In that case, the robot need not be calibrated after a tool is changed. $TOOL_{old}$, the old transformation relating the tool frame to the flange frame, is replaced by $TOOL_{new}$, the new transformation relating the tool frame to the flange frame. Both $TOOL_{old}$ and $TOOL_{new}$ are determined from the given geometry of the end-effectors.

However, if the mounting errors are significant, $TOOL$ cannot be considered error-free. Thus, calibrating $TOOL$ is necessary in order to guarantee the accuracy of the robot. $TOOL$ may also need to be calibrated later while the entire robot is calibrated if the orientational errors are not identified. Assume that the nominal $TOOL$ is determined off-line in the CPC modeling convention, which is equivalent to introducing an extra link transformation in the CPC model. $TOOL$ can be updated through updating the parameters in its error model.

The linearized $TOOL$ error model is exactly in the form of (5.6.22) and (5.6.24) with the exception that the parameters in the formulas are from $TOOL$ instead of $BASE$. One merit of the CPC model is that both error relationships in $BASE$ and $TOOL$ can be modeled in the same manner as those in the internal links.

Up to six independent parameters in $BASE$ are observable, if proper measurements are taken, even if only positioning errors of the end-effector are provided. However, three parameters in the $TOOL$ are not observable, when orientational errors of the end-effector are not given. This is because the rank of the Jacobian matrix, defined by the coefficients in (5.6.29), is three as $k\, m_{n,j,h}$, $k\, m_{n,k,h}$ and $k_{0,\beta}$ are not functions of joint variables. In this case, increasing the number of measurements will not resolve the rank-deficiency problem.

VII. THE MCPC ERROR MODEL

The derivation of the MCPC error model is obtained in the same manner as that of the D-H error model. Therefore in this section we only provide the final results.

A. LINEAR MAPPING RELATING CARTESIAN ERRORS TO MCPC PARAMETER ERRORS OF INDIVIDUAL LINKS

Recall that $[d_i^T, \delta_i^T]^T$ is a Cartesian error vector of the ith link. Let $l_i = [l_{i,x}, l_{i,y}, 0]^T$ and $d_i = [dl_{i,x}, dl_{i,y}, 0]^T$ for $i = 0, 1, \ldots, n-1$. Let $l_n = [l_{n,x}, l_{n,y}, l_{n,z}]^T$ and $d_n = [dl_{n,x}, dl_{n,y}, dl_{n,z}]^T$. Also let $d\alpha_i$, $d\beta_i$, and $d\gamma_i$ be the MCPC orientation parameter errors. Then, following the derivation of the D-H and the CPC error models, one can show that

$$\delta_i = k_{i,\alpha} d\alpha_i + k_{i,\beta} d\beta_i \tag{5.7.1}$$

$$d_i = dl_i + dk_{i,\alpha} \times l_i d\alpha_i + dk_{i,\beta} \times l_i d\beta_i + dk_{i,\gamma} \times l_i d\gamma_i \tag{5.7.2}$$

for $i = 0, 1, \cdots, n-1$, and

$$\delta_n = k_{n,\alpha} d\alpha_n + k_{n,\beta} d\beta_n + dk_{n,\gamma} d\gamma_n \tag{5.7.3}$$

$$d_n = dl_n + dk_{n,\alpha} \times l_n d\alpha_n + dk_{n,\beta} \times l_n d\beta_n + dk_{n,\gamma} \times l_n d\gamma_n \tag{5.7.4}$$

where

$$k_{i,\alpha} = Rot(y, -\beta_i))_{3\times3} [1, 0, 0]^T$$

$$k_{i,\beta} = [0, 1, 0]^T$$

$$k_{n,\alpha} = Rot(z, -\gamma_n))_{3\times3} Rot(y, -\beta_g)_{3\times3} [1, 0, 0]^T$$

$$k_{n,\beta} = Rot(z, -\gamma_b))_{3\times3} [0, 1, 0]^T$$

$$k_{n,\gamma} = [0, 0, 1]^T.$$

B. THE LINEARIZED MCPC ERROR MODEL

Theorem 5.7.1: The linearized MCPC error model can be written as

$$\delta = R_\alpha d\alpha + R_\beta d\beta + k_{n,\gamma} d\gamma_n \tag{5.7.5}$$

$$d = T_\alpha d\alpha + T_\beta d\beta + Tl_x dl_x + Tl_y dl_y + [Tl_z]_n dl_{n,z} + k_{n,\gamma} l_n d\gamma_n$$

$$(5.7.6)$$

where

$$d\alpha = [\, d\alpha_0, \, d\alpha_1, \, \ldots, \, d\alpha_n \,]^T,$$

$$d\beta = [\, d\beta_0, \, d\beta_1, \, \ldots, \, d\beta_n \,]^T,$$

$$d_x = [\, dl_{0,x}, \, dl_{1,x}, \, \ldots, \, dl_{n,x} \,]^T,$$

$$d_y = [\, dl_{0,y}, \, dl_{1,y}, \, \ldots, \, dl_{n,y} \,]^T,$$

and R_α, R_β, T_α, T_β, Tl_x, and Tl_y, are all $3 \times (n+1)$ matrices whose components are functions of link parameters and joint variables. Specifically, the ith column of these matrices are:

$$[R_\alpha]_i = \begin{bmatrix} n^u_{i+1} \cdot k_{\alpha,i} \\ o^u_{i+1} \cdot k_{\alpha,i} \\ a^u_{i+1} \cdot k_{\alpha,i} \end{bmatrix}$$

$$[R_\beta]_i = \begin{bmatrix} n^u_{i+1} \cdot k_{\beta,i} \\ o^u_{i+1} \cdot k_{\beta,i} \\ a^u_{i+1} \cdot k_{\beta,i} \end{bmatrix}$$

$$[T_\alpha]_i = \begin{bmatrix} ((p^u_{i+1} + l_i) \times n^u_{i+1}) \cdot k_{\alpha,i} \\ ((p^u_{i+1} + l_i) \times o^u_{i+1}) \cdot k_{\alpha,i} \\ ((p^u_{i+1} + l_i) \times a^u_{i+1}) \cdot k_{\alpha,i} \end{bmatrix}$$

$$[T_\beta]_i = \begin{bmatrix} ((p^u_{i+1} + l_i) \times n^u_{i+1}) \cdot k_{\beta,i} \\ ((p^u_{i+1} + l_i) \times o^u_{i+1}) \cdot k_{\beta,i} \\ ((p^u_{i+1} + l_i) \times a^u_{i+1}) \cdot k_{\beta,i} \end{bmatrix}$$

$$[Tl_x]_i = [\, n^u_{i+1,x}, \, o^u_{i+1,x}, \, d^u_{i+1,x} \,]^T$$

$$[Tl_y]_i = [\, n^u_{i+1,y}, \, o^u_{i+1,y}, \, d^u_{i+1,y} \,]^T$$

and

$$[Tl_z]_n = [\ n^u_{n,z}, \ o^u_{n,z}, \ d^u_{n,z}]^T.$$

Proof: Similar to the proof of the D-H error model.

VIII   SUMMARY AND REFERENCES

Kinematic error models are used in an error-model based kinematic identification procedure.   To be able to effectively identify the error parameters, the kinematic model chosen for the construction of the error model should be complete and parametrically continuous.   Furthermore, the error model should be irreducible.

There are two main ways to derive a differential kinematic error model. Wu (1984) used the differential transformation method based on analytic differentiation of the kinematic model.   Bennett and Hollerbach (1991) made use of the vector cross product method proposed by Whitney (1972), namely, using the geometric interpretation of each column of the Jacobian.   Readers are referred to Wu (1984) for a complete derivation of the D-H error model.   In this book because the D-H error model is derived by utilizing the generic error model, the derivation process is shortened somewhat.   The derivation of the CPC error model and the MCPC error model can be found respectively in (Zhuang and Roth (1992)) and (Zhuang, Wang and Roth (1993)).   Paul's textbook (1981) is an excellent reference on the mathematical background of differential transformations.

Versions of the generic kinematic error models can be found in (Wu (1984)) and (Everett and Suryohadiprojo (1988)).   The version presented in this chapter can be found in (Zhuang (1989)) and (Zhuang, Roth and Hamano (1992a)).

The linearized D-H error model, without any modifications, may run into singularity problem.   So does the linearized MCPC error model.   A way to avoid singularity in the MCPC model is to introduce a fixed transformation so that the two consecutive joint axes are always parallel after the fixed transformation.   This can be done by using a proper modeling convention. Similarly, for the D-H error model, a fixed transformation can be introduced so that the two consecutive joint axes are always perpendicular.   The disadvantage of doing so is that more fixed kinematic parameters are needed to model the robot manipulator.

The CPC error model is singularity free.   It has a sufficient number of parameters to span the entire geometric error space of the manipulator.   Yet it is irreducible.   However, the CPC error model is more complex in terms of the number of lines of code in the software implementation, comparing with the D-H and the MCPC error models.

As mentioned, the finite difference kinematic error model computed numerically is easy to implement and possesses similar numerical properties comparing with the differential error model. On the other hand, a differential error model facilitates theoretical investigation of some important issues such as identifiability of kinematic parameters and singularity of the Identification Jacobian.

Chapter 6

KINEMATIC IDENTIFICATION:
LINEAR SOLUTION APPROACHES

I. INTRODUCTION

Kinematic identification is the process by which all kinematic model parameters of a robot manipulator are identified given a set of end-effector pose measurements and the corresponding joint position measurements.

Determining kinematic parameters directly from the kinematic model given the measurement data, although appealing both theoretically and practically, is difficult since pose components of the robot end-effector are in general nonlinear functions of the robot link parameters. In Chapter 5, linearization techniques were used to construct kinematic error models relating robot pose errors to its kinematic parameter errors, based on which kinematic parameter deviations are found by iteratively solving linear least squares problems. This technique usually allows effective implementation and requires a relatively small number of pose measurements. Extensive off-line simulations are often required to determine for each individual robot, the number of and the optimal choice of measurement configurations, to ensure a reasonably small condition number of the "Identification Jacobian". For the linearized error model to be valid the nominal model of the robot has to be sufficiently close to the actual model. Although certain nonlinear optimization algorithms may be applied to handle large parameter deviations, their convergence properties are difficult to be established.

An alternative class of robot kinematic identification techniques consists of extracting the kinematic parameters from identified orientation and position of the joint axes obtained by tracking one or more target points located on the robot while moving each robot joint one at a time through a significant portion of the joint travel. The technique is attractive for studying joint imperfections. It allows an estimation of joint features (such as center of rotation, plane of rotation etc.) from relatively noisy measurements of the target positions. A disadvantage of the method is that in order to increase the robustness of the identification, a relatively large number of target position measurements need to be taken to perform least squares curve fitting to a target trajectory resulting from joint motion of each identified joint axis.

This chapter presents a number of linear solutions for the identification of unknown robot kinematic parameters. These solution methods employ the Complete and Parametrically Continuous (CPC) model. One reason for choosing the CPC model is that there exists a sequence of pose measurements for which all CPC link parameters except for one appear linearly in the system of equations to be solved. Another reason is that the redundant parameters in the CPC model can be systematically eliminated, a crucial step in the solution method.

The chapter is organized in the following manner. The problem of direct kinematic parameter identification is defined and a solution strategy is outlined in Section II. A hybrid solution for all-revolute robots is presented in Section III. "Hybrid" means here that a recursive linear procedure is devised for obtaining orientation parameters and a batch linear procedure is used for determining translation parameters. The recursive portion of the procedure is modified in Section IV to find all parameters, including orientation and translation parameters recursively. In Section V, the hybrid procedure is further extended to deal with general serial manipulators. Simulation results for the hybrid procedure are given in Section VI. The chapter ends with discussion and references.

II.   PROBLEM FORMULATION AND A SOLUTION STRATEGY

The following notation will be used thereafter in this chapter:

q_i —— the ith joint variable, $q_i \in \{\theta_i, d_i\}$

q —— the joint variable vector, $q \equiv [q_1, q_2, ..., q_n]^T$

p —— the link (kinematic) parameter vector, $p \equiv [b_0{}^T, l_0{}^T, ..., b_n{}^T, l_n{}^T, \beta_n]^T$ for the CPC model

^iT_j —— homogeneous transformation relating link frame $\{i\}$ to link frame $\{j\}$

^iR_j —— the upper-left 3x3 sub-matrix (the Orientation part) of ^iT_j.

$Rot(z, \theta_i)$ —— the upper-left 3x3 sub-matrix of $R o t(z, \theta_i)$

R_i —— the upper-left 3x3 sub-matrix of R_i given in the CPC model.

^ip_j —— the first 3 elements of the 4th column (the translation vector) of ^iT_j

$^ir_{jx}, {}^ir_{jy}$ and $^ir_{jz}$ —— the first, second, and third columns of ^iR_j

r_{ix}, r_{iy}, r_{iz} —— the first, second, and third columns of R_i

The CPC parameters b_0, b_1, ..., b_n and β_n will be referred to as "CPC orientation parameters", whereas the CPC parameters l_0, l_1, ..., l_{n-1} and l_n will be called "translation parameters".

By using the above notation, the CPC forward kinematic model T_n can be broken into the rotation and position equations

$$^wR_n = {}^wR_0 \prod_{i=1}^{n} Rot(z, \theta_i)R_i \tag{6.2.1}$$

$$^wp_n = \left\{ \sum_{i=1}^{n} \left[{}^wr_{ix}\, l_{i,x} + {}^wr_{iy}\, l_{i,y} + {}^wr_{(i-1)z}\, d_i \right] + {}^wR_n\, l_n \right\} \tag{6.2.2}$$

where by definition, $d_i \equiv 0$ if the respective joint is revolute, and $\theta_i \equiv 0$ if the ith joint is prismatic. The superscript "w" stands for reference to the world frame.

Finally, to specify a particular value of q_i, an additional subscript is introduced. For example, $q_{i,1}$ and $q_{i,2}$ denote two different displacements of the ith joint.

This chapter addresses the following robot kinematic parameters identification problem:

Problem: Given that a sufficiently large number of poses $\{^wT_n\}$ of the robot end-effector have been measured, and that the joint positions $\{q\}$ at the respective measurement configurations have been recorded, can we find from (6.2.1)-(6.2.2) a linear solution of the robot kinematic parameter vector p ?

In particular: (i) Should there be a specific relationship among the pose measurements, or can these be random? (ii) If a linear solution is possible, is it unique?

A few observations can be made regarding equations (6.2.1)-(6.2.2):

1. The position equation (6.2.2) is a system of linear equations in the components of the translation parameter vectors l_i. If the orientation parameters are known, the coefficient vectors in (6.2.2) become known. l_i can thus be solved linearly from (6.2.2), if the coefficient matrix is nonsingular.

2. The rotation equation (6.2.1) is independent of the translation parameters and the prismatic joint variables. Therefore, prismatic joint motions do not affect (6.2.1).

The above observations suggest a multi-stage procedure to find linear solution for all kinematic parameters. In the simplest case of an all-revolute manipulator, the orientation parameters are solved first using the robot orientation measurements. The estimated orientation parameters along with both robot position and orientation measurements are then used to solve for the translation parameters.

The remaining problem now is how to determine CPC orientation parameters linearly from (6.2.1). This can be done by restricting the motion pattern of all revolute joints in the robot in a suitable way. More specifically, starting from an arbitrary robot configuration, the robot revolute joints should be moved one at a time sequentially from the first joint to the last joint. By this measurement strategy, a set of rotation matrix equations in the form

$$R_{Ai}R_{i-1} = R_{i-1}R_{Bi} \qquad (6.2.3)$$

can be derived from Equation (6.2.1), where R_{Ai} and R_{Bi} are known and R_{i-1} are to be determined. It will be shown (in Section III) that under certain conditions, R_{i-1} can be solved uniquely from (6.2.3). The following example illustrates admissible motion sequences for calibration measurement.

Example: Consider a R-P-R robot shown in Fig. 6.2.1. Joints 1 and 3 are revolute, and thus have to move sequentially. Joint 2 is prismatic and therefore can move together with any other joint. In the particular sequence shown in Fig. 6.2.1b, joint 1 moves first from its initial position $\theta_{1,1}$ to a new position $\theta_{1,2}$. Joints 2 and 3 then move together to new positions $d_{2,2}$ and $\theta_{3,2}$, respectively.

III. A HYBRID LINEAR SOLUTION METHOD
FOR ALL-REVOLUTE MANIPULATORS

It is assumed in this section that all joints of the manipulator are revolute. The solution procedure consists of two stages. In the first stage, the CPC orientation parameters are solved one joint at a time using the robot orientation measurements only. All CPC translation parameters are then determined simultaneously using the estimated orientation parameters and the robot position measurements.

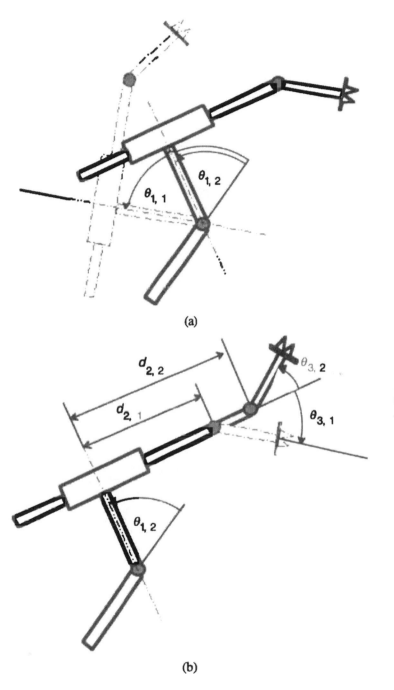

Figure 6.2.1 An example of possible robot motion sequences

A. SOLUTION FOR THE ORIENTATION PARAMETERS

As said in Section II, the measurement procedure starts at an initial robot configuration $q_0 \equiv (\theta_{1,1}, \theta_{2,1}, \cdots, \theta_{n-1,1}, \theta_{n,1})$. It then proceeds by *displacing* one joint at a time, as follows: $q_1 \equiv (\theta_{1,2}, \theta_{2,1}, \cdots, \theta_{n-1,1}, \theta_{n,1})$, $q_2 \equiv (\theta_{1,2}, \theta_{2,2}, \cdots, \theta_{n-1,1}, \theta_{n,1})$, $\cdots\cdots$, $q_{n-1} \equiv (\theta_{1,2}, \theta_{2,2}, \cdots, \theta_{n-1,2}, \theta_{n,1})$, and $q_n \equiv (\theta_{1,2}, q_{2,2}, \cdots, \theta_{n-1,2}, \theta_{n,2})$.

The rotation matrices R_0, R_1, $\cdots$, R_{n-1} and $^{n-1}R_n$ (in this order) are solved for recursively. Specifically, R_{i-1} is determined in Step i using R_0, R_1, $\cdots$, R_{i-2}, which are obtained in the preceding steps. Step i involves the measurement configurations q_{i-1} and q_i and the pose orientation measurements $^wR_n(q_{i-1})$ and $^wR_n(q_i)$. From (6.2.1),

$$^wR_n(q_{i-1}) = R_y R_{i-1} Rot(z, \theta_{i,1}) R_x \qquad (6.3.1)$$

$$^wR_n(q_i) = R_y R_{i-1} Rot(z, \theta_{i,2}) R_x \qquad (6.3.2)$$

where R_y is a computable rotation matrix using previously identified rotation matrices, and R_x is a fixed unknown rotation matrix. Eliminating R_x from (6.3.1)-(6.3.2) yields the equation

$$R_y^T \, ^wR_n(q_{i-1}) \, ^wR_n(q_i)^T R_y R_{i-1} = R_{i-1} Rot(z, \theta_{i,1} - \theta_{i,2})$$

Let $C_i \equiv R_y^T \, ^wR_n(q_{i-1}) \, ^wR_n(q_i)^T R_y$ and $\Delta\theta_i \equiv \theta_{i,1} - \theta_{i,2}$. Then

$$C_i R_{i-1} = R_{i-1} Rot(z, \Delta\theta_i) \qquad i = 1, 2, \dots n \qquad (6.3.3)$$

For all the practical purposes, it is assumed that $|\Delta\theta_i| < 2\pi$. Matrix C_i depends on the current and previously measured end-effector poses and on the previously identified rotation matrices R_0, $\cdots$, R_{i-2}. The matrix C_i is therefore fully known. Equation (6.3.3) is thus in the form of (6.2.3). The next fact reveals the condition for a unique solution from (6.3.3) for the unknown rotation matrix R_{i-1}.

Theorem 6.3.1: For $|\Delta\theta_i| < 2\pi$, a solution R_{i-1} of (6.3.3) always exists. Equivalently, the direction vector b_{i-1} of joint i represented in $\{i-1\}$ can be solved from

$$(C_i - I)b_{i-1} = 0 \qquad (6.3.4)$$

where I is a 3x3 identity matrix. For $b_{i-1,z} > 0$, the solution is unique. For $b_{i-1,z} = 0$, the sign of b_{i-1} cannot be determined.

Proof: Equation (6.3.4), obtained directly from the third columns of (6.3.3) is equivalent to (6.3.3) in the sense that b_{i-1} contains all the information about R_{i-1}. For (6.3.4) to have a nonzero solution, $Rank(C_i - I)$ must be less than three. By (6.3.3), $Rot(z, \Delta\theta_i)$ is obtained from C_i through a similarity transformation R_{i-1}. Thus $Rot(z, \Delta\theta_i)$ and C_i have the same eigenvalues $\{\lambda_1, \lambda_2, \lambda_3\}$ which are $\{cos\Delta\theta_i + jsin\Delta\theta_i, \quad cos\Delta\theta_i - jsin\Delta\theta_i, \Delta\theta_i\}$. Consequently, $C_i - I$ is similar to $diag(cos\Delta\theta_i + jsin\Delta\theta_i - 1, cos\Delta\theta_i - jsin\Delta\theta_i - 1, 0)$ and thus has a rank of 2 since $|\Delta\theta_i| < 2\pi$. The vector b_{i-1} is solved from (6.3.4) up to an undetermined scale factor, which can be determined using the fact that b_{i-1} is a unit vector and that $b_{i-1,z} \geq 0$. However whenever $b_{i-1,z} = 0$, the sign of b_{i-1} cannot be determined.   []

Note that whenever $b_{i-1,z} = 0$, the sign of b_{i-1} can be determined by additional steps. b_{i-1} is first substituted into (6.3.1) and (6.3.2) by assuming an arbitrary sign for b_{i-1}. If both sides of the two equations are consistent, the sign of b_{i-1} is retained, otherwise the sign is reversed.

Following the estimation of $R_0, R_1, \cdots, R_{n-1}$, any one of the previously obtained end-effector rotation equations will do to determine $^{n-1}R_n$,

$$^{n-1}R_n = {}^wR_{n-1}{}^T {}^wR_n$$

where $^{n-1}R_n = R_n Rot(z, \beta_n)$. The vector b_n, the third column of $^{n-1}R_n$, is thus obtained and so is R_n. The transformation $Rot(z, \beta_n)$ can now be computed numerically

$$Rot(z, \beta_n) = R_n{}^T {}^{n-1}R_n$$

from which the value of β_n is uniquely determined as $\beta_n = atan2(sin(\beta_n), cos(\beta_n))$ where $sin(\beta_n)$ and $cos(\beta_n)$ are the (2,1) and (1,1) elements of $Rot(z, \beta_n)$, respectively.

Remarks:

1. The rotation matrix R_y in (6.3.1) needs not be the same as R_y in (6.3.2), as long as these matrices are computable. Physically, this means that joints 1, 2, ..., i-1 can be placed at any arbitrary fixed positions while the ith revolute joint is moved to two different positions. All derivations remain valid by this modification. The measurement procedure stated at the beginning of this section can thus be generalized as follows: To compute the CPC orientation parameters of the robot revolute joints, the pose measurements of the robot are taken while *releasing* each revolute joint one at a time and sequentially from the first joint to the last joint.

2. The matrix C_i represents a relative rotation of the manipulator induced by a relative rotation of the ith axis motion. Equation (6.3.3) relates changes in robot orientation to changes $\Delta\theta_i$ in the joint coordinates. In this way, the effects of pose measurement static errors and joint position offsets can be greatly reduced. However, the last rotation matrix $^{n-1}R_n$ is computed using absolute pose orientation measurements. Therefore static errors may particularly affect this rotation matrix.

3. If measurements from three or more robot configurations are provided, then each pair of configurations contributes a measurement equation similar to (6.3.3). A least squares algorithm can then be applied to estimate b_{i-1}. Such an over-determined system of equations is necessary in practice to reduce the influence of measurement noise.

4. From each rotation equation, an estimate of $^{n-1}R_n$ can be obtained. One can then have s estimates of $^{n-1}R_n$ from s pose measurements. In the presence of measurement noise, these estimates are not necessarily identical. The quaternions approach presented in Chapter 8 can be used to fit a best rotation matrix $^{n-1}R_n$ from s estimates.

5. It is unnecessary to restrict joints to move one at a time as far as the solution of translation parameters is concerned. This is because least squares algorithms can be applied to solve for unknown parameters from (6.3.3) even if joints are moved randomly.

B. SOLUTION FOR THE TRANSLATION PARAMETERS

After b_0, b_1, $\cdots$, b_n and β_n become known, the CPC translation parameters can be found from (6.2.2). Unlike the recursive method of determining the CPC rotational parameters, all the translation parameters are computed simultaneously. It is therefore important to check the rank of the coefficient matrix defined by (6.2.2).

Multiplying both sides of (6.2.2) by R_0^T yields

$$R_0^T {}^w p_n = \sum_{i=0}^{n-1} \left[{}^0 r_{ix} l_{i,x} + {}^0 r_{iy} l_{i,y} \right] + {}^0 R_n l_n)$$

(6.3.5)

where the terms related to d_i have been set to zero as we are discussing all-revolute robots.   The left-hand side of (6.3.5) is known.   The remaining problem is to linearly solve for the translation parameters $l_{i,x}$ and $l_{i,y}$, $i = 0$, $1, \cdots, n-1$, as well as l_n.

Following the measurement procedure described in the beginning of Section III,   the minimum number of required measurement configurations is $n+1$ and if this exact minimum number is taken then measurement i indeed corresponds to the motion of joint i.   Let y_i be the left-hand side of (6.3.5) evaluated at the ith robot configuration, let $y \equiv [y_1^T, y_2^T, \ldots, y_{n+1}^T]^T$, and denote the $2n+3$ vector of unknown translation parameters by $x \equiv [l_{0,x}, l_{0,y},$ $\cdots, l_{n-1,x}, l_{n-1,y}, l_{n,x}, l_{n,y}, l_{n,z}]^T$. The $n+1$ measurement equations of the type (6.3.5) can be written in a matrix form

$$y = G x$$

(6.3.6)

where, referring to the notation explained in Section II, the coefficient matrix G is

$$
G = \begin{bmatrix}
I_{3\times 2} & {}^0 r_{1x}(q_0) & {}^0 r_{1y}(q_0) & \cdots\cdots & {}^0 r_{(n-1)x}(q_0) & {}^0 r_{(n-1)y}(q_0) & {}^0 R_n(q_0) \\
I_{3\times 2} & {}^0 r_{1x}(q_1) & {}^0 r_{1y}(q_1) & \cdots\cdots & {}^0 r_{(n-1)x}(q_1) & {}^0 r_{(n-1)y}(q_1) & {}^0 R_n(q_1) \\
 & & & \vdots & & & \\
I_{3\times 2} & {}^0 r_{1x}(q_{n-1}) & {}^0 r_{1y}(q_{n-1}) & \cdots\cdots & {}^0 r_{(n-1)x}(q_{n-1}) & {}^0 r_{(n-1)y}(q_{n-1}) & {}^0 R_n(q_{n-1}) \\
I_{3\times 2} & {}^0 r_{1x}(q_n) & {}^0 r_{1y}(q_n) & \cdots\cdots & {}^0 r_{(n-1)x}(q_n) & {}^0 r_{(n-1)y}(q_n) & {}^0 R_n(q_n)
\end{bmatrix}
$$

Theorem 6.3.2: For $|\Delta \theta_i| < 2\pi$, the robot translation parameter vector x can be uniquely determined from (6.3.6).

Proof: The $2n+3$ equations of the form (6.3.6) are consistent.   The theorem then follows immediately from the following lemma.

Lemma 6.3.3: $rank(G) = 2n+3$, where $2n+3$ is the number of columns of G.

Proof: The particular measurement sequence described earlier, in which each joint is moved one at a time, greatly simplifies the structure of matrix G. Specifically,

$$^0r_{ix}(q_j) = {^0r_{ix}(q_i)} \text{ if } j > i$$

$$^0r_{iy}(q_j) = {^0r_{iy}(q_i)} \text{ if } j > i$$

That is, the rotation matrix ^0R_i is independent of all measurement configurations that come after measurement i. This allows us by elementary matrix transformations to bring G into a block upper triangular matrix G_3. Without loss of generality, assume that $q_0 = 0$. One can now verify by direct multiplication that

$$G_3 = G_1 G_2$$

where

$$G_1 = diag(I, [Rot(z, \theta_{1,2}) - I]_{3 \times 2}, [^0R_1 \cdot \{Rot(z, \theta_{2,2}) - I\}]_{3 \times 2},$$
$$..., [^0R_{n-1} \cdot \{Rot(z, \theta_{n,2}) - I\}]_{3 \times 2})$$

and

$$G_2 = \begin{bmatrix} [I]_{3 \times 2} & [R_1]_{3 \times 2} & [R_1 R_2]_{3 \times 2} & \cdots & [R_1 \cdots R_{n-1}]_{3 \times 2} & [R_1 \cdots R_n Rot(z, b_n)]_{3 \times 3} \\ & [R_1]_{2 \times 2} & [R_1 R_2]_{2 \times 2} & \cdots & [R_1 \cdots R_{n-1}]_{2 \times 2} & [R_1 \cdots R_n Rot(z, b_n)]_{2 \times 3} \\ & & [R_2]_{2 \times 2} & \cdots & [R_2 \cdots R_{n-1}]_{2 \times 2} & [R_2 \cdots R_n Rot(z, b_n)]_{2 \times 3} \\ & & & \ddots & & \\ & 0 & & & [R_{n-1}]_{2 \times 2} & [R_{n-1} R_n Rot(z, b_n)]_{2 \times 3} \\ & & & & & [R_n Rot(z, b_n)]_{2 \times 3} \end{bmatrix}$$

where $[R]_{i \times j}$ denotes an upper-left $i \times j$ submatrix of R. Note that G_1 is an $M \times N$ matrix and G_2 an $N \times N$ matrix, where $M = 3n + 3$, and $N = 2n + 3$.

We next observe the following trivial fact: If $\theta_{i,1} \neq \theta_{i,2} + 2k\pi$, for any integer k, then

$$Rank([^0R_{i-1} \cdot \{Rot(z, \theta_{i,1}) - Rot(z, \theta_{i,2})\}]_{3 \times 2}) = 2$$

for $i = 1, 2, \cdots, n$.

By Sylvester's Inequality,

$$Rank(G_1) + Rank(G_2) - N \leq Rank(G) \leq Min(Rank(G_1), Rank(G_2)) \quad (6.3.7)$$

However, since G_1 is block-diagonal and by (6.3.7), we have that $Rank(G_1) =$ $2n+3$. Performing a sequence of elementary transformations on G_2 results in a block diagonal matrix. Then $Rank(G_2) = 2n+3$. This completes the proof.

$$\square$$

Remarks:
1. Unlike the identification of orientation parameters, absolute position measurements are used for the identification of the robot translation parameter vector. Static errors of the measuring system will affect the estimated robot translation parameters.
2. Since the identified orientation parameters are utilized, errors in the orientation parameters will propagate to the translation parameters. As all translation parameters are computed simultaneously, there will be no further error propagation.
3. Again it is straightforward to apply a least squares technique to solve (6.3.6) whenever more than $n+1$ measurements are provided. This remark applies also to a general robot.

IV. AN ALL-RECURSIVE LINEAR SOLUTION APPROACH FOR GENERAL SERIAL MANIPULATORS

In this approach, the CPC parameters of each link, the orientation and the translation parameters, are solved recursively by releasing the robot joints one at a time.

A. PROBLEM REFORMULATION

In order to apply the method shown in this section, there is a need to slightly modify the CPC kinematic model. The motion matrix of the CPC model is given as follows:

$$Q_i = \{ \begin{array}{ll} Rot(z, q_i), & \text{for revolute joint,} \\ \\ Trans(z, q_i), & \text{for prismatic joint,} \end{array} \qquad (6.4.1)$$

where

$$q_i = s_i q'_i, \; s_i \in \{+1, -1\} \qquad (6.4.2)$$

and q'_i is the ith joint variable. Note that the CPC convention requires that any two consecutive joint axes have a nonnegative inner product, that is, $b_{i,z} \geq 0$. In general, this requirement can be met by changing the sign of one of the joint variables of consecutive joints. This is true because changing the sign of the joint value is equivalent to reversing the direction of joint axis for both revolute and prismatic joints. Therefore, the convention of the CPC model is slightly extended by including a sign parameter, s_i, as shown in (6.4.2).

We now consider a robot with n joints. Its world–to–end–effector transformation matrix ${}^wT_n$ can be expressed as

$$
{}^wT_n = {}^wT_0 \cdots {}^{n-1}T_n
$$

Without loss of generality, we assume that the kinematic parameters of the joints from the end–effector to the $(i+1)$th joint have been known, and that the unknowns to be estimated are the world–to–base transformation, ${}^wT_0$, and the kinematic parameters of joints $1, ..., i$. Also, we assume that ${}^wT_n$ can be measured. Similar to the calibration procedure described in Section III, when calibrating the ith joint, only those joints with known kinematic parameters, plus the ith joint itself, are permitted to be moved. By moving those joints (from the end–effector to the ith joint) to two different configurations and recording their corresponding world–to–end–effector transformation matrices, we have

$$
{}^wT_{n,1} = {}^wT_i \, Q_{i,1} \, V_i \, {}^{i+1}T_{n,1} \tag{6.4.3}
$$

$$
{}^wT_{n,2} = {}^wT_i \, Q_{i,2} \, V_i \, {}^{i+1}T_{n,2} \tag{6.4.4}
$$

where ${}^wT_{n,1}$ and ${}^wT_{n,2}$ are the measured world–to–end–effector transformation matrices; ${}^{i+1}T_{n,1}$ and ${}^{i+1}T_{n,2}$ can be computed from the kinematic model since their kinematic parameters have already been known; $Q_{i,1}$ and $Q_{i,2}$ are the ith motion matrix at two joint positions. The problem now decomposes into many kinematic parameter estimation sub–problems of a single joint axis.

By multiplying ${}^wT_{n,1}{}^{-1}$ and ${}^wT_{n,2}{}^{-1}$ in both sides of (6.4.3) and (6.4.4), respectively, we have the following equality

$$
{}^wT_i Q_{i,1} \, V_i \, {}^{i+1}T_{n,1} \, {}^wT_{n,1}{}^{-1} = {}^wT_i Q_{i,2} \, V_i \, {}^{i+1}T_{n,2} \, {}^wT_{n,2}{}^{-1}
$$

or

$$
Q_{i,1} \, V_i \, {}^{i+1}T_{n,1} \, {}^wT_{n,1}{}^{-1} = Q_{i,2} \, V_i \, {}^{i+1}T_{n,2} \, {}^wT_{n,2}{}^{-1} \tag{6.4.5}
$$

Rearranging equation (6.4.5), results in

$$
\Delta Q_i V_i = V_i \, \Delta T_i \tag{6.4.6}
$$

where $\Delta Q_i = Q_{i,1}{}^{-1} T_{i,2}$, $\Delta T_i = {}^{i+1}T_{n,1}{}^{-1} \, {}^w T_{n,1}{}^{-1} \, {}^w T_{n,2} \, {}^{i+1}T_{n,2} \, {}^w T_{n,2}{}^{-1}$ and V_i is the unknown homogeneous transformation matrix to be estimated. Equation (6.4.6) in this form of representation is very similar to the equation for the hand/eye calibration problem. Unfortunately, the solution obtained in the hand/eye calibration problem cannot be directly applied to this problem. This is because the hand/eye calibration techniques given later in Chapter 8 all require that the axes of rotation be neither parallel nor anti–parallel. However in the single joint calibration problem, it is obvious that there is at most one effective rotation axis, i.e., the rotation axis of ΔQ_i. There is no rotation axis if the joint is prismatic. Fortunately, due to the special structure of the CPC kinematic model, a linear solution can still be obtained.

Equation (6.4.6) can be separated into two equations, one is the rotation matrix equation and the other is the translation vector equation. That is,

$$R_{\Delta Q_i} R_{V_i} = R_{V_i} R_{\Delta T_i} \tag{6.4.7}$$

$$R_{\Delta Q_i} p_{V_i} + p_{\Delta Q_i} = R_{V_i} p_{\Delta T_i} + p_{V_i} \tag{6.4.8}$$

where

$$R_{V_i} = R_i Rot(z, \beta_i) \tag{6.4.9}$$

$$p_{V_i} = R_{V_i} l_i \tag{6.4.10}$$

Furthermore, $R_{\Delta Q_i}$, $R_{\Delta T_i}$ and R_{V_i} are 3x3 rotation matrices, and $p_{\Delta Q_i}$, $p_{\Delta T_i}$ and p_{V_i} are 3x1 translation vectors of ΔQ_i, ΔT_i and V_i, respectively. In the following sections, we shall show how to solve the kinematic parameters of the prismatic and revolute joints from the above equations.

B. CALIBRATION OF A PRISMATIC JOINT

The redundant and the unknown parameters associated with a prismatic joint are first listed below for better clarity of our derivation:

1. The four given redundant parameters are β_i and l_i. These are typically set to zero if not used.
2. The unknowns are R_i and the sign parameters s_i.

From (6.4.1), $R_{\Delta Q_i} = I$ and $p_{\Delta Q_i} = [0 \; 0 \; \Delta d_i]^T$. By substituting $R_{\Delta Q_i}$ and $p_{\Delta Q_i}$ into (6.4.8), we have

$$p_{\Delta Q_i} = R_{V_i} p_{\Delta T} \tag{6.4.11}$$

Substituting (6.4.9) into (6.4.11), yields

$$R_i^T p_{\Delta Q i} = p'_{\Delta T i}$$

where $p'_{\Delta T i} = Rot(z, \beta_i) p_{\Delta T}$, and β_i is the given redundant parameter. Since $p_{\Delta Q i} = [0\ 0\ \Delta d_i]^T$,

$$b'_i \Delta q_i = p'_{\Delta T i} \qquad (6.4.12)$$

where $b'_i = [-b_{i,x}\ -b_{i,y}\ b_{i,z}]^T$ is the third row vector of the rotation matrix R_i.

Assuming that s observations are available; i.e., $\Delta d_{i,j}$ and $\Delta T_{i,j}$ for $j = 1$, 2, ..., s. Then b'_i can be determined by minimizing the following error function using a least-square method.

$$e = \sum_{j=1}^{m} \|b'_i \Delta q_{i,j} - p'_{\Delta T i,j}\|^2$$

where $b'^T_i b'_i = 1$. To solve the above equation, we first form the Lagrangian

$$e_L = \sum_{j=1}^{m} \left(b'_i \Delta q_{i,j} - p'_{\Delta T i,j}\right)^T \left(b'_i \Delta q_{i,j} - p'_{\Delta T i,j}\right) + \lambda\left(1 - b'^T_i b'_i\right) \quad (6.4.13)$$

The gradient of e_L in equation (6.4.13) is

$$\nabla e_L = 2 \sum_{j=1}^{m} \left(b'_i \Delta q_{i,j}^2 - p'_{\Delta T i,j} \Delta q_{i,j}\right) + 2\lambda b'_i$$

By letting $\nabla e_L = 0$, we obtain

$$b'_i = \frac{\sum_{j=1}^{m} \left(p'_{\Delta T i,j} \Delta q_{i,j}\right)}{\sum_{j=1}^{m} \left(\Delta q_{i,j}^2\right) - \lambda}$$

where λ can be determined such that b'_i is a unit vector. Consequently, using a Euclidian norm:

$$b'_i = \frac{\sum\limits_{j=1}^{m} \left(p'_{\Delta Ti,j} \Delta q_{i,j} \right)}{\left| \sum\limits_{j=1}^{m} \left(p'_{\Delta Ti,j} \Delta q_{i,j} \right) \right|} \qquad (6.4.14)$$

Notice that if the third component of b'_i is negative, then in order to be consistent with the CPC convention, we should change the sign of b'_i and let $s_i = -1$; otherwise, we let $s_i = 1$. Once the unit vector b'_i is obtained, the rotation matrix R_i can be computed in terms of the direction vector. The remaining CPC parameters can then be determined by

$$l_i = R_{Vi}{}^T p_{Vi}.$$

Remarks:
1. Equation (6.4.14) means that the least–square solution of the direction vector b'_i is just the weighted average of the translation vectors $p'_{\Delta Ti,j}$.
2. Intuitively, if the difference of the joint values $\Delta d_{i,j}$ can be made larger, more accurate results are to be expected.

C.   CALIBRATION OF A REVOLUTE JOINT
The redundant and the unknown parameters associated with a revolute joint are first listed below for clarity of the derivation:

1. Two given redundant parameters β_i and the last components of l_i. These are typically set to zero if not used.
2. The unknowns are R_i, the first two components of l_i and the sign parameters s_i.

For a revolute joint,

$$R_{\Delta Qi} = Rot(z, \Delta \theta_i) \qquad (6.4.15)$$

$$p_{\Delta Qi} = 0. \qquad (6.4.16)$$

Substituting (6.4.15) and (6.4.16) into (6.4.7) and (6.4.8), yields

$$Rot(z, \Delta\theta_i)R_{Vi} = R_{Vi} R_{\Delta Ti} \qquad (6.4.17)$$

$$Rot(z, \Delta\theta_i)p_{Vi} = R_{Vi}p_{\Delta Ti} + p_{Vi} \qquad (6.4.18)$$

By taking the transpose of both sides of equation (6.4.17), and then multiplying the unit vector $z = [0\ 0\ 1]^T$ on the right, we have

$$R_{Vi}{}^T z = R_{\Delta Ti}{}^T R_{Vi}{}^T z$$

By use of equation (6.4.9), we obtain

$$b'_i = D_i b'_i \qquad (6.4.19)$$

where b'_i is the same vector as in equation (6.4.12), and

$$D_i = Rot(z, \beta_i)R_{\Delta Ti}{}^T Rot(z, -\beta_i) \qquad (6.4.20)$$

Since in (6.4.19), D_i is a rotation matrix, b'_i can be found using the same method given in Section III. Next an alternative approach is provided.

If we have s observations, then we will have the following homogeneous equation (which is over-determined if $m > 1$)

$$\begin{bmatrix} D_1^T - I_{3\times 3} \\ \vdots \\ D_s^T - I_{3\times 3} \end{bmatrix} b'_i \equiv E b'_i = \varepsilon$$

where ε is the error vector due to measurement noise. The parameter vector b'_i can be estimated by minimizing $\|\varepsilon\|^2$ subject to the constant $\|b'_i\|^2 = 1$. It can be shown that the solution for b'_i is the unit eigenvector of $E^T E$ corresponding to the smallest eigenvalue.

The sign parameters s_i can be determined as follows. From (6.4.2), we can compute R_i from b'_i. Substituting the estimated R_i into (6.4.17), we have

$$Rot(z, s_i\Delta\theta_i) = R_{Vi} R_{\Delta Ti} R_{Vi}{}^T$$

Thus s_i can be obtained by checking the consistency of the above s equations.

Again in order to estimate b'_i more accurately, the joint angle difference, $\Delta\theta_i$ should be made larger. Otherwise, the D_i matrix, in equation (6.4.20), will approach a unit matrix when $\Delta\theta_i$ is very small. This will cause the estimation of the rotation axis to be very sensitive to noise.

After b'_i is obtained, we can compute the rotation matrix R_i and then post multiply it by $Rot(z, \beta_i)$ to obtain R_{v_i}. Next, by substituting R_{v_i} into (6.4.1) and rearranging it, we have

$$(Rot(z, \Delta\theta_i) - I)p_{v_i} = R_{v_i} p_{\Delta Ti}$$

Or

$$(Rot(z, \Delta\theta_i) - I)R_{v_i} l = R_{v_i} p_{\Delta Ti}$$

It is very easy to show that the rank of the upper-left 2x2 submatrix (denoted by E_i) of $(Rot(z, \Delta\theta_i) - I)R_{v_i}$ is 2 if $\Delta\theta_i$ is nonzero. Note that on both sides, the third row is exactly zero. Thus the first two components of l can be solved from the above equation, since the last entry of l is given.

D. DETERMINATION OF THE WORLD–TO–BASE (BASE) TRANSFORMATION MATRIX

After all the parameters associated with revolute and prismatic joints are calibrated, we are able to then compute the world–to–base transformation matrix $^WT_0$. Suppose that we have m observations, i.e., $j = 1, 2, ..., m$. By separating $^WT_0$ into the rotation and translation parts, we have

$$^WR_{nj} = {^WR_0}\,{^0R_{nj}} \qquad (6.4.21)$$

$$^Wp_{nj} = {^WR_0}\,{^0p_{nj}} + {^Wp_0} \qquad (6.4.22)$$

where $^0R_{nj}$ and $^0p_{nj}$ are respectively the 3x3 rotation matrix and 3x1 translation vector of the transformation matrix $^0T_{nj}$. The unknown matrix $^WT_0$ can be solved in two stages. In the first stage, $^WR_0$ is solved. The solved $^WR_0$ is then used to determine $^Wp_0$. From (6.4.21) we have the following matrix equations

$$A = {^WR_0}\,B$$

where $A = [{^WR_{n1}}\ ...\ {^WR_{nm}}]$ and $B = [{^0R_{n1}}\ ...\ {^0R_{nm}}]$. By solving the following fusion of rotation matrices problem:

$$\text{minimize } \|A^T - B^T\,{^WR_0}^T\| \qquad \text{subject to } {^WR_0}^T\,{^WR_0} = I,$$

we have the following procedure for finding the closed–form solution for $^wR_0$:

Step 1. Computation of the matrix $C = BA^T$.

Step 2. Computation of the singular value decomposition $C = U\Sigma V^T$, where the matrix Σ consists of all singular values of C.

Step 3: Compute $^wR_0 = U V^T$.

After the rotation matrix $^wR_0$ is obtained, the substitution of $^wR_0$ into (6.4.22) yields

$$^wp_0 = \frac{1}{m}\sum_{j=1}^{m} \left(^wp_{nj} - {^wR_0}{^0p_{nj}} \right)$$

which completes the calibration procedure.

V. EXTENSION OF THE HYBRID LINEAR SOLUTION TO GENERAL SERIAL MANIPULATORS

The technique discussed in Section III applies only to all-revolute robots such as the PUMA 560. In this section, the method is extended to kinematic parameter estimation of a robot featuring both revolute and prismatic joints.

A. SOLUTION FOR ORIENTATION PARAMETERS

The CPC orientation parameters associated with revolute joints and prismatic joints can be determined by the method given in Section IV. Note that when the procedure given in Section IV is used for determining the orientation of the prismatic joints, all redundant parameters can be either set to zero or equations (6.2.1) and (6.2.2) have to be modified to include these redundant parameters.

B. SOLUTION   FOR TRANSLATION PARAMETERS

Equation (6.2.2) can be rewritten as

$$^wp_n + \sum_{\substack{j:\ \text{for all} \\ \text{prismatic joints}}} \left[\ ^wr_{(j-1)z}\ d_j \right] = \sum_{\substack{i:\ \text{for all} \\ \text{revolute joints}}} \left[\ ^wr_{ix}\ l_{i,x} + {^wr_{iy}}\ l_{i,y} \right] + {^wR_0}l_0 \quad (6.5.2)$$

The left-hand side of (6.5.1) is known.   If the orientation parameters (equivalently, R_1, R_2, $\cdots$, R_n and $^WR_0$) are solved, the coefficient vectors in the right-hand side of (6.5.1) become known.   The vector l_i can thus be solved linearly from (6.5.1).   Let y_i be the left-hand side of (6.5.1) evaluated at the ith robot configuration, let $y \equiv [\, y_1{}^T,\; y_2{}^T,\; ...,\; y_m{}^T]^T$, and denote the $2(n-p)+3$ vector of unknown translation parameters by x, where m is the number of measurements, and p is the number of prismatic joints. The m measurement equations of the type (6.5.1) can be written in a matrix form

$$y = G\,x \tag{6.5.2}$$

where G consists of the coefficients the translation parameters $l_{i,x}$, $l_{i,y}$ and l_0 given in (6.5.2).   Similar to the proof of Theorem 6.3.2, matrix G is nonsingular if all joints are moved and if a sufficient number of measurements is taken.

VI.   NUMERICAL STUDIES

To investigate the robustness of the above procedures in the presence of measurement noise, the PUMA robot geometry was chosen (Figure 6.6.1) along with its CPC nominal parameters listed in Table 6.6.1.

The following issues were investigated:

1. The effect of pose measurement errors on the accuracy of kinematic parameter identification taking into account static errors of the measurement system.
2. The effect of random noise on pose measurement errors.
3. The effect of propagation errors in the orientation parameter estimation due to the recursive nature of the algorithm.
4. Comparison between an error model based identification technique and the linear solution method.

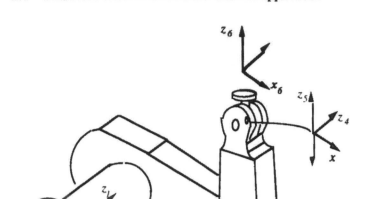

Figure 6.6.1 Geometry of the PUMA arm

Table 6.6.1 The nominal CPC parameters of the PUMA 560
(units: *mm* for linear and *rad* for angular parameters)

i	$z_{i,x}$	$z_{i,y}$	$z_{i,z}$	$l_{i,x}$	$l_{i,y}$	$l_{i,z}$	β_i
0	0	0	1	1000.0	1000.0	0	
1	0	1	0	0	-440.0	0	
2	0	0	1	431.82	0	0	
3	0	-1	0	-20.31	149.09	0	
4	0	1	0	0	-433.05	0	
5	0	-1	0	0	0	0	
6	0	0	1	0	0	94.85	0

Table 6.6.2 The " actual" CPC parameters of the PUMA 560
(units: *mm* for linear and *rad* for angular parameters)

i	$z_{i,x}$	$z_{i,y}$	$z_{i,z}$	$l_{i,x}$	$l_{i,y}$
0	-0.0137481	-0.0129580	0.9998215	996.79688	998.35938
1	-0.0004721	0.9999568	0.0092851	4.76562	-408.51562
2	0.0159926	-0.0188148	0.9996951	432.60124	-3.35938
3	0.0007844	-0.9999713	-0.0075304	-18.74750	148.69938
4	-0.0156100	0.9998355	-0.0092385	8.59376	-442.73750
5	0.0175819	-0.9998366	0.0042010	-4.60938	-4.60938
6	0.0102745	0.0020549	0.9999451	0.15624	-4.21876

$l_{6,z} = 96.10000$; $\beta_6 = -0.02325$.

Different levels of uniformly distributed random noise were injected to the nominal CPC parameters of the PUMA arm, to simulate an "actual" robot; Table 6.6.2 lists one example. It should be pointed out that the parameter deviations given in the table are slightly exaggerated.

In all test results presented in this section, 30 pose measurements uniformly covering the PUMA workspace were employed for verification. Orientation errors and position errors are defined in terms of the Euclidian norms of the pose position and orientation deviations, respectively.

A. THE EFFECT OF STATIC ERRORS IN THE MEASURING SYSTEM ON POSE MEASUREMENT ERRORS

Robot calibration accuracy depends critically on the accuracy of the calibration equipment used for end-effector pose measurements. For instance, a coordinate measuring machine, whose three linear axes are slightly misaligned due to manufacturing tolerances, may exhibit a deterministic error in the position reading which varies spatially within the machine workspace. In addition, the machine will have a random position error reading due to encoder resolution, compliance and deadzone effects. The relative significance of pose measuring error sources varies with the type of measurement techniques. In coordinate measuring done by stereo vision, the random error component due to pixel resolution usually dominates the effects of deterministic errors due to imperfect calibration of the camera perspective transformation matrix. On the other hand, in a laser tracking system, fixed errors in the tracking mirror gimbals will cause the distance measuring accuracy to be by several orders of magnitude worse than the resolution provided by the laser interferometers. As a general rule, pose measuring devices must be calibrated prior to their use for robot calibration, and this calibration is sometimes done using other devices which have greater accuracy. Some endpoint sensing devices are often equipped with self-calibration hardware. Pre-calibration of the pose measuring device improves its accuracy. Realistic calibration simulations need to account for both types of pose measuring errors. Specifically, the actual (unknown) kinematic error parameters of the measuring device may be included. The simplest case of static errors in the measurement devices is a constant bias in the position and orientation measurements.

In the simulation study presented in this section, two cases were investigated. In the first case, bias values varying from 0.001 mm to 0.128 mm were added to the z component of all pose measurements. The result is shown in Figure 6.6.2. In the second case, bias values from 0.0001 radians to 0.0128 radians were set for the three Z-Y-Z Euler angles of a rotation matrix which post-multiplied all rotation measurements of the robot pose. The result is shown in Fig. 6.6.3.

As can be seen from Figures 6.6.2 and 6.6.3, unknown biases in pose measurements do affect calibration accuracy. Any bias in robot position measurements almost linearly influences the position accuracy of the calibrated robot, but it does not affect the robot orientation calibration accuracy. On the other hand, bias in the orientation measurements affects both the position and orientation accuracy of the calibrated robot.

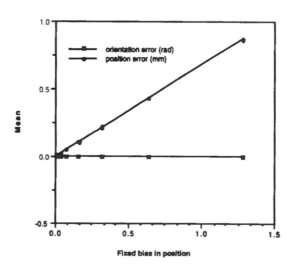

Figure 6.6.2 Pose errors against bias in position measurement

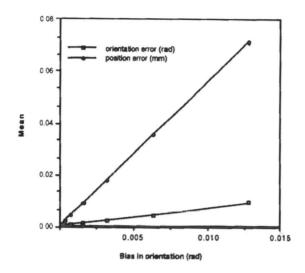

Figure 6.6.3 Pose errors against bias in orientation measurement

B. THE EFFECT OF RANDOM ERRORS IN THE MEASURING SYSTEM ON POSE MEASUREMENT ERRORS

Robot pose measurement errors caused by random errors in the measuring device are commonly modeled by either zero mean Gaussian noise or by zero mean uniformly distributed noise. In this simulation study, three intensity levels of a uniformly distributed random noise (listed in Table 6.6.3) were added to the Z-Y-Z Euler angles and positions of the tool to model robot pose measurement noise. Noise level 1 models a relatively accurate pose measuring device and noise level 3 represents a crude measuring system. Noise level 2 models a system with relatively high position measurement accuracy and low orientation accuracy.

Figure 6.6.4 depicts the mean orientation errors and the corresponding standard deviations vs. the number of pose measurements used for the parameter identification of each individual joint. Figure 6.6.5 shows the mean position errors and their standard deviations vs. the number of measurements used for the identification of translation parameters. When 24 and 30 measurements were used, no additional random measurements were taken. When 40 and 50 measurements were used, 10 and 20 random measurements were added respectively to the sequential measurements.

As seen from Figure 6.6.4, orientation errors are reduced when the number of pose measurements is increased, up to a point where further reduction is insignificant. However, by adding a small number of random pose measurements to the sequential measurements, position errors can be further reduced, as seen from Figure 6.6.5.

Table 6.6.3 Noise injected to "pose measurements"

	Noise level 1	Noise level 2	Noise level 3
Orientation noise (deg)	U[-0.01,0.01]	U[-0.05,0.05]	U[-0.05,0.05]
Position noise (mm)	U[-0.1,0.1]	U[-0.1,0.1]	U[-0.5,0.5]

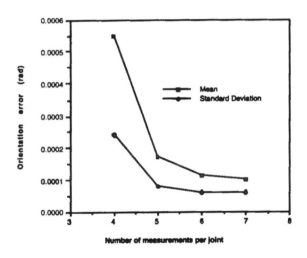

Figure 6.6.4 Orientation errors against number of pose measurements used for each joint identification

C. THE EFFECT OF PROPAGATION ERRORS IN THE ORIENTATION PARAMETER ESTIMATION DUE TO THE RECURSIVE NATURE OF THE ALGORITHM

Figure 6.6.6 shows the joint axis direction errors defined as $\|b_i - b_i^{id}\|$, where b_i and b_i^{id} are respectively the nominal and identified i^{th} joint axis direction vectors. In this test, 30 pose measurements, five for each joint were used. That is, each joint was moved to five different locations while other joints were kept fixed.

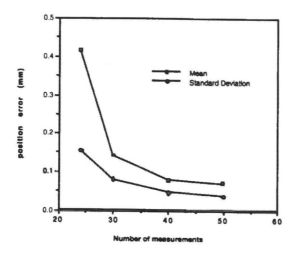

Figure 6.6.5 Position errors against number of pose measurements

From Figure 6.6.6, it can be seen that the propagation of orientation parameter errors was insignificant. This is due to the fact that the matrix C_i represents a relative rotation of the manipulator, induced by a relative rotation of the ith axis motion. Propagation errors from previous joint axis motions tend to eliminate one another.

D. COMPARISON BETWEEN AN ERROR MODEL BASED IDENTIFICATION TECHNIQUE AND THE LINEAR SOLUTION METHOD

Figures 6.6.7 and 6.6.8 compare two identification methods for different noise levels. For this simulation, the CPC error model was used as a representative of the error models. The CPC error model has performed well due to the completeness and parametrical continuity properties of the model. In this test, 30 pose measurements, five for each joint, were used for the linear solution method, whereas 15 random pose measurements were used for the error model based method. Increasing the number of pose measurements does not improve by much the accuracy of kinematic compensation for the error model based method.

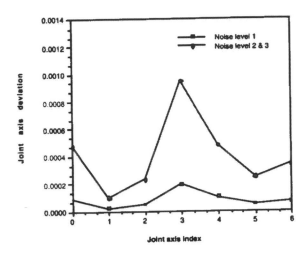

Figure 6.6.6 Joint axis deviations

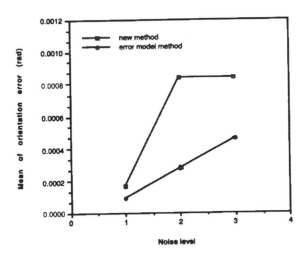

Figure 6.6.7 Orientation errors from two methods for
different noise levels listed in Table 6.6.3 with robot parameters
listed in Table 6.6.2

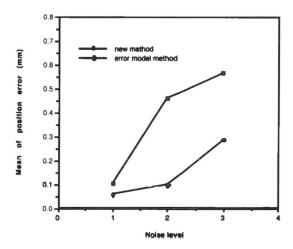

Figure 6.6.8 Position errors from two methods against
different noise levels listed in Table 6.6.3 with robot parameters
listed in Table 6.6.2

Figures 6.6.7 and 6.6.8 clearly show that error model based techniques
consistently outperform the linear solution method whenever parameter
deviations are reasonably small and only random pose measurement errors are
present. This is attributed to the staged solution of the linear method. While
the identified parameters are optimal in each stage, global optimality for the
entire set of kinematic parameters is not assured. On the other hand, under
certain conditions error model based algorithms always find the global optimal
solution.

Simulations were conducted to assess the performance of the two methods
in the presence of bias measurement errors. Both the position accuracy and
orientation accuracy achieved by the linear solution method are slightly better
than those of the error model method. This is because the linear solution
method employs relative orientation measurements for the estimation of most
orientation parameters of the robot.

The performance of the linear solution method is of course independent of
the size of kinematic parameter deviations. This method can be applied to
kinematic identification problems that have large parameter deviations. On
the other hand, as obtained in the simulation study, joint axis misalignment
that exceeds 15 degrees often causes a failure of the error model based
algorithm.

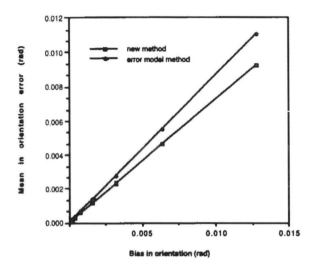

Figure 6.6.9 Orientation errors from two methods for
bias in orientation measurement

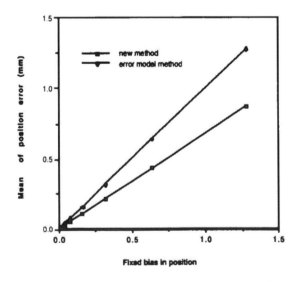

Figure 6.6.10 Position errors from two methods against
bias in position measurement

VII. SUMMARY AND REFERENCES

The idea of freezing the robot and moving one joint at a time, for the sake of identifying the pose of individual joint axis lines has been the topic of much research. For a detailed discussion of the use of regression techniques for line and circle fitting, the readers are referred to the books by Stone (1987) and by Mooring, Roth and Driels (1991). Originally, a vector analysis approach was introduced by Barker (1983) and later developed fully by Sklar (1988). The shape matrix approach of Broderick and Cipra (1988) also employed a measurement strategy by which each robot joint is individually and successively moved. A similar measurement strategy was employed by Lenz and Tsai (1989) for calibration of a Cartesian robot.

In this chapter, it is shown constructively that the CPC parameters of an n degrees of freedom robot can be linearly solved from the CPC kinematic model given $n+1$ measurements, provided that each revolute joint is released successively from the first joint to the last joint, and both the orientations and positions of the robot end-effector are measured (Zhuang and Roth (1993)). An attractive feature of the linear solution method is that no prior detailed knowledge of the nominal geometry of the robot is required other than its joint types. The method is applicable for identification of large kinematic deviations.

While the method presented in Section III eliminates propagation errors in the estimation of translation parameters, it is not presented in the most general form to allow its use for any arbitrary robot geometries. A revised scheme, outlined in Section IV, was proposed by Shih, Hung and Lin (1995), in which kinematic parameters are determined one axis at a time. The recursive method of Zhuang and Roth (1993) was further extended by Shih, Hung and Lin to estimate kinematic parameters of a binocular head using point measurements only. Simulation and experimental results demonstrated that the method of using point measurements can achieve higher accuracy than that of using pose measurements. One problem with recursive methods is that the ability to simultaneously solve for all robot translation parameters is lost, implying that more end-effector pose or point measurements have to be collected for an effective robot calibration.

Inspired by the ideas of Shih, Hung and Lin (1995), a modification to the linear hybrid approach of Section III presented in Section V appeared in Zhuang and Roth (1995). The linear kinematic identification method is now general enough to be applied to calibrating any serial manipulators under a unified framework. Furthermore, the method preserves an important advantage of the technique of Section III; i.e., the translation kinematic parameters of the manipulator can be simultaneously solved.

As mentioned, to obtain a linear identification solution, the robot kinematic parameters are to be computed in multiple sequential steps, which may cause propagation errors. As relative pose changes of the robot are utilized to identify the orientation parameters, propagation errors are shown to be insignificant in the identification of orientation parameters. Effects of certain static errors of the measuring devices are also reduced in this way. It is feasible to find the CPC translation parameters and orientation parameters related to prismatic joints by linearly solving a set of system equations even if the joint motions are random. This fact can be used to further improve the effectiveness of this technique. However, since absolute position measurements are used for the identification of the robot translation parameter vector, static errors will affect the estimated robot translation parameters. The computation of robot translation parameters using relative robot pose measurement is still an open research issue.

Using more than $n+1$ measurements to suppress measurement noise calls for a least square solution. The shown linear solution algorithms handle this case in a straightforward manner. This is another useful feature of the linear approaches.

Chapter 7

SIMULTANEOUS CALIBRATION OF A ROBOT AND A HAND-MOUNTED CAMERA

I. INTRODUCTION

In order to accurately measure the position and orientation of an object in the world coordinate system using the robot-camera system, various components of the system need to be calibrated. This includes the position and orientation of the robot base with respect to the robot world, the robot hand with respect to the robot base, the camera with respect to the robot hand, and the object with respect to the camera. These four tasks are respectively known as robot base calibration, manipulator calibration, hand/eye calibration and camera calibration. Camera calibration and manipulator calibration have already been discussed in much detail in the previous chapters. Issues regarding robot hand/eye calibration and base calibration will be investigated in Chapters 8 and 9, respectively.

In this chapter, a solution procedure is developed for the simultaneous calibration of a robot and a monocular camera rigidly mounted to the robot hand. The work was motivated by the following sequence of logical arguments: Recognizing that a by-product of a camera calibration phase is the full pose of the camera, with respect to a coordinate frame located on the camera calibration board, suggests a replacement of the stereo cameras with a single camera. Unlike in (Puskorius and Feldkamp (1987); Zhuang, Roth and Wang (1993)), where a camera calibration step was done once to pave the way for coordinate measuring using the calibrated stereo cameras, the monocular camera must be recalibrated upon each change of the robot configuration, as was done in Lenz and Tsai (1989). This first-step robot/camera pose measurements is to be followed by a robot kinematic identification step. The next logical reasoning is to merge these two steps using a generalized "camera model" that contains the robot kinematics.

With such a simultaneous calibration algorithm, calibration at different levels of complexity can be done under a unified framework. The chapter outlines three levels of calibration. The simplest level is to identify the camera model together with the hand/eye transformation, assuming that the relative pose of the robot hand with respect to the robot base and the relative pose of the robot base with respect to the robot world are accurately known. The second level of calibration is to identify the camera model, the hand/eye transformation and the base/world transformation, assuming that the internal robot geometry is accurately known. The third level, which is the most complex, is to calibrate the camera and the entire robot simultaneously.

It is assumed in all three calibration levels, that the ratio of scale factors of the camera-imaging system is given *a priori*. If this is not the case, then lens distortion information has to be provided. The ratio of scale factors for a given camera system can be obtained prior to robot calibration using the techniques developed in (Lenz and Tsai (1987); Zhuang, Roth, Xu and Wang (1993)), which were explained in detail in Chapter 2. Once the ratio of scale factors is obtained, it remains unchanged for the particular camera-imaging system.

The chapter is organized in the following manner. In Section II, the kinematic model of the camera-robot system is developed, the cost function for the kinematic identification is constructed, and the iterative solution algorithm is outlined. The Jacobian matrix that relates measurement errors to system parameter errors is derived in Section III. Section IV addresses a number of implementation issues. Section V extends the method to the stereo camera case. Discussion and references are given in Section VI.

II. KINEMATIC MODEL, COST FUNCTION AND SOLUTION STRATEGY

The basic geometry of the system is shown in Figure 7.2.1, which is similar to the one shown in Figure 4.2.4. Let $\{x_w, y_w, z_w\}$ denote the world coordinate system, normally placed at a convenient location outside the robot and camera. Let $\{x_b, y_b, z_b\}$ denote the base coordinate system of the robot, physically located within the structure of the robot base. Let $\{x, y, z\}$ denote the camera coordinate system, whose origin is at the optical center point O, and whose z axis coincides with the optical axis. Let $\{X, Y\}$ denote the image coordinate system (not shown in Figure 7.2.1, but one can refer to Figure 2.2.1) centered at O_I (the intersection of the optical axis z and the image plane). The axes $\{X, Y\}$ lie on a plane parallel to the x and y axes.

Recall that the 4x4 homogeneous transformation T_n relating the end-effector pose to the world coordinates is

$$T_n = A_0 A_1 \ldots A_{n-1} A_n$$

where A_0 is the 4x4 homogeneous transformation from the world coordinate system to the base coordinate system of the robot, A_i is the transformation from the $(i-1)$th to the ith link coordinate systems of the robot, and A_n is the transformation from the nth link coordinate system of the robot to the camera coordinate system.

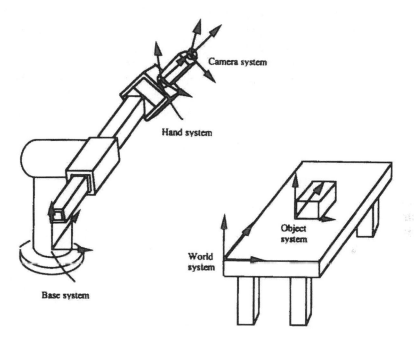

Figure 7.2.1   A robotic system and its coordinate system assignment

The transformations A_i can be represented in terms of any proper kinematic modeling convention. Let $\rho = [\rho_1 \; \rho_2 \; ... \; \rho_p]^T$ be a vector consisting of all link parameters of the robot, where p is the number of independent kinematic parameters in the robot. T_n is then a matrix function of ρ.

Recall also that the camera model, relating the world coordinate system $\{x_w, y_w, z_w\}$ to the image coordinate system $\{X, Y\}$ can be written as (refer to Chapter 2)

$$X\left(1 + kf_{xy}\right) = f_x \frac{r_1 x_w + r_2 y_w + r_3 z_w + t_x}{r_7 x_w + r_8 y_w + r_9 z_w + t_z} \tag{7.2.1}$$

$$Y\left(1 + kf_{xy}\right) = f_y \frac{r_4 y_w + r_5 y_w + r_6 z_w + t_y}{r_7 x_w + r_8 y_w + r_9 z_w + t_z} \tag{7.2.2}$$

where f_x and f_y are scale factors and k is radial distortion coefficient. Furthermore, f_{xy} is the same as r^2 defined in (2.2.18).

As seen in (7.2.1) and (7.2.2), the idea is to absorb all the kinematic parameters of the robot into the extrinsic parameters $(r_i, \; i = 1, \; 2, \; ..., \; 9)$ of the camera.

If the ratio of scale factors μ is known, the problem of simultaneous calibration of robot and camera can be stated as follows: Given a number of calibration points whose world coordinates are known and whose image coordinates are measured, estimate the robot kinematic parameter vector ρ and the camera parameters f_x, f_y, and k. The case in which μ is unknown is discussed in Section IV.

In order to apply optimization techniques to solve the calibration problem, a cost function needs to be constructed.

Define a 16x1 vector ϕ in the following manner:

$$\phi(\rho, f_x, f_y, k) \equiv [f_x r_1 \ f_y r_4 \ r_7 \ k r_7 \ f_x r_2 \ f_y r_5 \ r_8 \ k r_8 \ f_x r_3 \ f_y r_6 \ r_9 \\ k r_9 \ f_x t_x \ f_y t_y \ t_z \ k t_z]^T \qquad (7.2.3)$$

and a 2x16 matrix C as follows:

$$C = \begin{bmatrix} x_w & 0 & -x_w X & -f_{xy} x_w X & y_w & 0 & -y_w X & -f_{xy} y_w X & z_w & 0 & -z_w X & -f_{xy} z_w X & 1 & 0 & -X & -f_{xy} X \\ 0 & x_w & -x_w Y & -f_{xy} x_w Y & 0 & y_w & -y_w Y & -f_{xy} y_w Y & 0 & z_w & -z_w Y & -f_{xy} z_w Y & 0 & 1 & -Y & -f_{xy} Y \end{bmatrix} \qquad (7.2.4)$$

Equations (7.2.1)-(7.2.2) can then be rewritten in the following compact form

$$C \, \phi(\rho, f_x, k) = 0 \qquad (7.2.5)$$

Note that the dependence of ϕ on f_y is omitted as f_y can be recovered from f_x using (2.2.17).

In (7.2.5), C is a known coefficient matrix and ϕ is a vector function of the unknown parameters. Given a particular calibration point whose world coordinates $\{x_w, y_w, z_w\}$ and image coordinates (X, Y) are known, matrix C can be computed from (7.2.4). According to (7.2.5), a single calibration point can only provide two scalar equations. To estimate all unknown parameters, a sufficient number of calibration points have to be used. Let C_i, whose structure is given in (7.2.4), denote the computed coefficient matrix using the ith calibration point. Let m be the number of calibration points used for parameter estimation. The problem is then reduced to determining ρ, f_x and k that minimize the cost function E in a least squares sense, where

$$E = \sum_{i=1}^{s} \left[C_i \, \phi(\rho, f_x, k) \right]^T \left[C_i \, \phi(\rho, f_x, k) \right] \qquad (7.2.6)$$

where s is the number of pose measurements. With no loss of generality, we can now assume that the robot changes its configuration at each measurement.

There are two cases: 1) there is only one calibration point ($m = s$), and 2) there are l points on the calibration fixture ($m = ls$). In both cases, the analysis is the same.

An iterative procedure is needed to obtain the optimal solution. To properly apply a nonlinear least squares algorithm to the problem, the following issues have to be addressed:

1. Choice of an initial condition, and
2. The structure of the Jacobian matrix.

It is assumed for practicality that p^0, f_x^0 and k^0, the nominal values of p, f_x and k are known. In the case of robot-camera calibration, the nominal robot kinematic parameter vector p^0 is usually provided from the data sheets of the robot, except for those parameters associated with A_0 and A_n, which need to be roughly determined by a proper gauging device. The initial values of the scale factors s_x and s_y can be set from the camera specifications. The nominal value of the distortion coefficient k is set to zero. Finally the nominal value of the focal length may be read from the lens itself.

It is assumed that the nominal parameters are in the neighborhood of the actual ones. With this assumption and from (7.2.5),

$$C_i(\phi^0 + d\phi) = 0, \qquad i = 1, 2, ..., s. \qquad (7.2.7)$$

where the superscript 0 denotes a nominal quantity. It is further assumed that the function ϕ is differentiable at the nominal parameters. Then the following relationship holds up to a first order approximation,

$$C_i \phi^0 \cong - C_i \left(\frac{\partial \phi}{\partial p} dp + \frac{\partial \phi}{\partial k} dk + \left(\frac{\partial \phi}{\partial f_x} + \mu \frac{\partial \phi}{\partial f_y} \right) df_x \right)_{p^0, f_x^0, f_y^0, k^0}$$

$$i = 1, 2, ..., s.$$

The coefficient matrix in the right hand side of the above equation is a Jacobian, which can either be computed explicitly as shown in the next section, or be approximated using finite differences. The error parameters dp, df_x and dk may be solved from the equation using a weighted least squares algorithm, and can then be used to update p, f_x and k. The process is iterated until the amount of adjustment on each of the parameters is below a prescribed threshold.

III.   THE IDENTIFICATION JACOBIAN

In this case, the Identification Jacobian is a Jacobian matrix relating measurement residuals to parameter errors $d\rho$, df_x and dk.   The Identification Jacobian plays a central role not only in the identification of the unknown parameter vector, but also in the study of optimal selection of measurement configurations and accuracy compensation. It is shown in this section that the original robot identification Jacobian is used as one of the building blocks of the robot/camera Identification Jacobian.

It will be helpful at this point to introduce some more compact notation. A function $vec(X)$ will denote a vector valued function of a matrix X, resulting from stacking the matrix columns one on top of the other, first column on top, second column right under and so on. Thus, ϕ in Equation (7.2.3) can be compactly rewritten as

$$\phi = vec(FT_n) \tag{7.3.1}$$

where

$$F = \begin{bmatrix} f_x & 0 & 0 & 0 \\ 0 & f_y & 0 & 0 \\ 0 & 0 & 1 & 0 \\ 0 & 0 & k & 0 \end{bmatrix}$$

Assume initially that the nominal vector ϕ^0 deviates from ϕ^*, the optimal solution by a small amount. Thus,

$$\phi^* \cong \phi^0 + d\phi \tag{7.3.2}$$

where $d\phi$ is a differential change of ϕ.   To simplify the notation, the superscript 0 from now on may be omitted if no confusion arises. By definition of F,

$$d\phi \; : \; vec(dF\,T_n + F\,dT_n) = vec(dF\,T_n) + vec(F\,dT_n) \tag{7.3.3}$$

Define the vectors

$$J_x = [r_1 \; 0 \; 0 \; 0 \; r_2 \; 0 \; 0 \; 0 \; r_3 \; 0 \; 0 \; 0 \; t_x \; 0 \; 0 \; 0]^T$$
$$J_y = [0 \; r_4 \; 0 \; 0 \; 0 \; r_5 \; 0 \; 0 \; 0 \; r_6 \; 0 \; 0 \; 0 \; t_y \; 0 \; 0]^T$$
$$J_k = [0 \; 0 \; 0 \; r_7 \; 0 \; 0 \; 0 \; r_8 \; 0 \; 0 \; 0 \; r_9 \; 0 \; 0 \; 0 \; t_z]^T$$

With simple algebraic manipulations, the first term in the right hand side of (7.3.3), representing the contribution to the error model by the camera intrinsic error parameters, can then be written as

$$vec(dF\ T_n) = (J_x + \mu J_y)df_x + J_k dk \tag{7.3.4}$$

More involved algebraic manipulations need to be performed to expand the second term in the right hand side of (7.3.3), representing the contribution to the error model by the robot kinematic error parameters. A main objective is to allow robot calibration researchers and practitioners to retain the original robot error model, hence a large portion of the kinematic identification software. Define

$$\Delta \equiv T_n^{-1} dT_n \tag{7.3.5}$$

Then, the nontrivial elements of Δ, written as $\delta = [\delta_x,\ \delta_y,\ \delta_z]^T$ and $d = [d_x, d_y, d_z]^T$ are respectively the 3x1 rotational and positional error vectors of T_n. Two steps are now needed to relate $vec(FdT_n)$ to the robot kinematic parameter error vector dp. First, $vec(F\ dT_n)$ is related to the pose error vector $[\delta^T, d^T]^T$ by a linear transformation. The vector $vec(F\ dT_n)$ is then related to dp by an additional linear transformation mapping dp to $[\delta^T, d^T]^T$.
Multiplying both sides of (7.3.5) by FT_n yields

$$FdT_n = FT_n\Delta. \tag{7.3.6}$$

F can be decomposed into

$$F = GH$$

where

$$G = \begin{bmatrix} 1 & 0 & 0 & 0 \\ 0 & 1 & 0 & 0 \\ 0 & 0 & 1 & 0 \\ 0 & 0 & k & 0 \end{bmatrix}$$

$$H = \begin{bmatrix} f_x & 0 & 0 & 0 \\ 0 & f_y & 0 & 0 \\ 0 & 0 & 1 & 0 \\ 0 & 0 & 0 & 0 \end{bmatrix}$$

Then from (7.3.6),

$$vec(FdT_n) = vec(FT_n\Delta) = vec(GHT_n\Delta).$$

A crucial step in the derivation of the robot/camera error model is to invoke the Kronecker product of matrices. That is,

$$vec(GHT_n\Delta)=(I\otimes G)vec(HT_n\Delta)=((I\otimes G))\begin{bmatrix} vec\begin{pmatrix} H_{3:3}R\Omega(\delta) \\ 0_{1\times3} \\ H_{3:3}Rd \\ 0 \end{pmatrix} \end{bmatrix}$$

(7.3.7)

where Ω is defined in (5.2.5), I is a 3x3 identity matrix, $H_{3:3}$ is the upper-left 3x3 submatrix of H, and R is the rotation matrix of T_n. The symbol $\otimes$ denotes a Kronecker product of two matrices (Chen (1970)); that is,

$$A\otimes B \equiv \begin{bmatrix} a_{11}B & \cdots & a_{1n}B \\ \vdots & & \vdots \\ a_{n1}B & \cdots & a_{nn}B \end{bmatrix}$$

where a_{ij} is the ijth element of A.

In the error-model based robot calibration literature, the transformation from dp to $[\rho^T, d^T]^T$ is available. More specifically, one has

$$\delta = \sum_{j=1}^{p} J_{\delta j}\, dp_j \qquad (6.3.8)$$

$$d = \sum_{j=1}^{p} J_{dj}\, dp_j \qquad (6.3.9)$$

where $J_{\delta i}$ and J_{di} are 3x1 vectors. The details of J_δ and J_{di} in terms of a specific kinem tic modeling convention can be found in Chapter 5. After substituting (7.3.8)-(7.3.9) into (7.3.6) and with some algebraic manipulations, one obtains

$$vec(FdT_n)=(I\otimes G)\sum_{j=1}^{p}\begin{bmatrix} vec\begin{pmatrix} H_{3:3}R\Omega(J_{\delta j}) \\ 0_{1x3} \\ H_{3:3}RJ_{dj} \\ 0 \end{pmatrix} \end{bmatrix} dp_j \qquad (6.3.10)$$

where p is the number of independent kinematic parameters in the robot. Substituting (7.3.10) and (7.3.4) into (7.3.3) yields

$$C_i \phi \cong -C_i J_k dk - C_i(J_x + \mu J_y)df_x - C_i(I \otimes G)\sum_{j=1}^{p} \left[\begin{array}{c} vec\left(\begin{array}{c} H_{3:3}R\Omega(J_{dj}) \\ 0_{1\times 3} \\ H_{3:3}RJ_{dj} \end{array} \right) \\ 0 \end{array} \right] dp_j$$

$$i = 1, 2, ..., s. \quad (7.3.11)$$

To write the result in a matrix form, let

$$J_i = \left[\begin{array}{ccccc} -C_i J_k & -C_i(J_x + \mu J_y) & C_i(I \otimes G)\left[\begin{array}{c} vec\left(\begin{array}{c} H_{3:3}R\Omega(J_{d1}) \\ 0_{1\times 3} \\ H_{3:3}RJ_{d1} \end{array} \right) \\ 0 \end{array} \right] & ... C_i(I \otimes G)\left[\begin{array}{c} vec\left(\begin{array}{c} H_{3:3}R\Omega(J_{dp}) \\ 0_{1\times 3} \\ H_{3:3}RJ_{dp} \end{array} \right) \\ 0 \end{array} \right] \end{array} \right]$$

$$(7.3.12)$$

J_i is a $2\times(p+2)$ matrix; and also let

$$dp^{aug} \equiv [dk \quad df_x \quad dp^T]^T. \quad (7.3.13)$$

dp^{aug} is a $(p+2)\times 1$ vector. Equation (7.3.11) can then be rewritten as

$$C_i \phi = J_i dp^{aug} \quad i = 1, 2, ..., m.$$

Let

$$C \equiv \left[\begin{array}{c} C_1 \\ C_2 \\ \vdots \\ C_m \end{array} \right]$$

and

$$J \equiv \left[\begin{array}{c} J_1 \\ J_2 \\ \vdots \\ J_m \end{array} \right]$$

Equation (7.3.11) can then be written in a single matrix form,

$$C\phi = J\,d\rho^{aus} \tag{7.3.14}$$

The above equation provides the relationship between the measurement residual error vector $C\phi$ and the augmented parameter error vector $d\rho^{aus}$. The matrix J is termed the *Identification Jacobian* for robot-camera calibration.

IV. IMPLEMENTATION ISSUES

A. CAMERA PARAMETERS

The distortion parameter k needs not be computed each time the robot and the camera systems are calibrated. If k is known, the error model derived in the last section can be simplified. In this case, ϕ is a 12x1 vector,

$$\phi = [f_x r_1 \; f_y r_4 \; r_7 \; f_x r_2 \; f_y r_5 \; r_8 \; f_x r_3 \; f_y r_6 \; r_9 \; f_x t_x \; f_y t_y \; t_z]^T$$

C_i is a 2x12 matrix,

$$C = \begin{bmatrix} x_w & 0 & -x_w X & y_w & 0 & -y_w X & z_w & 0 & -z_w X & 1 & 0 & -X \\ 0 & x_w & -x_w Y & 0 & y_w & -y_w Y & 0 & z_w & -z_w Y & 0 & 1 & -Y \end{bmatrix}$$

where the subscript i is omitted. The 12x1 vectors J_x and J_y are,

$$J_x \equiv [r_1 \; 0 \; 0 \; r_2 \; 0 \; 0 \; r_3 \; 0 \; 0 \; t_x \; 0 \; 0]^T$$
$$J_y \equiv [0 \; r_4 \; 0 \; 0 \; r_5 \; 0 \; 0 \; r_6 \; 0 \; 0 \; t_y \; 0]^T$$

J_i is a 2x$(p+2)$ matrix,

$$J_i = \left[-C_i J_x \;\; -C_i J_y \;\; C_i \begin{bmatrix} (I \otimes (H_{3:3} R)) vec(\Omega (J_{\delta 1})) \\ H_{3:3} R J_{d1} \end{bmatrix} \cdots C_i \begin{bmatrix} (I \otimes (H_{3:3} R)) vec(\Omega (J_{\delta p})) \\ H_{3:3} R J_{dp} \end{bmatrix} \right]$$

and $d\rho^{aus}$ is a $(p+2)$x1 vector,

$$d\rho^{aus} \equiv [df_x, \; df_y, \; d\rho^T]^T.$$

The above new set of equations should be used to replace the corresponding equations in Section III for parameter identification if k is known.

The scale factors s_x and s_y of the camera can be determined using camera calibration techniques such as that given in (Lenz and Tsai (1987)) or be identified together with the robot parameters and the focal length by the method described in this chapter. These parameters need only to be determined once for a particular camera system. For the system described in this section, the scale factors do not need to be known in advance as these are identified in combinations with other parameters.

B. ROBOT PARAMETERS

A sufficient number of independent link parameters need to be used to express any variation of the actual robot structure away from the nominal design. This number, for a serial manipulator consisting of rigid links connected by low pair joints, is $4N - 2P + 6$, where N is the number of degrees of freedom and P is the number of prismatic joints.

Not all parameters in ρ need to be calibrated at each time. It is natural to accommodate different complexities of calibration with the solution method described above. Let us discuss three such levels of calibration.

The simplest level of calibration is to identify the camera parameters k, f_x and f_y together with the parameters that specify the hand/eye transformation A_n, assuming that the transformation from the world coordinate system to the robot hand coordinate system is known. This type of calibration is necessary whenever the relative pose of the camera with respect to the robot is changed. We use six link parameters to specify the transformation A_n. Together with the two camera parameters, there is a total of eight parameters to be estimated. That is, the dimension of $d\rho^{aug}$ given in (7.3.13) is 8.

The second level of calibration is to identify the camera parameters together with the parameters that specify the hand/eye transformation A_n and the base/world transformation A_0, assuming that the robot geometry is accurately known. This type of calibration is necessary whenever the camera changes its location with respect to the robot hand and the robot also changes its location with respect to an external reference object. Since 4 additional parameters for the transformation A_0 (one can treat $N = 1$ in this case), the dimension of ρ^{aug} is 12.

The third level, which is the most complex level of calibration, is to calibrate the entire robot-camera system. In this case, the dimension of ρ^{aug} is $4N - 2P + 8$, among which two are camera parameters.

C. CHANGE OF REFERENCE COORDINATE SYSTEM

The pose error vector of T_n, $[\delta^T, d^T]^T$, is given in the world coordinate system because T_n is defined as the transformation from the world coordinate system to the camera coordinate system. The robot error model given in (7.3.7)-(7.3.8) should be consistent with this convention. If the pose error vector is represented in the camera coordinate system, as has been the case in some robot kinematic error models, the expression given in (7.3.7) has to be modified to accommodate the change of the reference coordinate system.

The transformation from the camera coordinate system to the world coordinate system is T_n^{-1}. Let δ' and d' be respectively the rotational and positional error vectors of T_n^{-1}. Also let

$$T_n^{-1} \equiv \begin{bmatrix} R' & t' \\ 0_{1\times 3} & 1 \end{bmatrix}$$

Then

$$R = (R')^T \qquad (7.4.1)$$
$$t = -(R')^T t' \qquad (7.4.2)$$

Perturbing both sides of (7.4.1) and (7.4.2) yields

$$dR = (dR')^T \qquad (7.4.3)$$
$$d = -(dR')^T t' - (R')^T d' \qquad (7.4.4)$$

where by definition $d = dt$ and $d' = dt'$. From Equation (7.4.3), one has

$$\Omega(\delta) = -R^T \Omega(\delta')R$$

which, after simple vector and matrix manipulations, reduces to

$$\delta = -R^T \delta' \qquad (7.4.5)$$

From (7.4.4),

$$d = -dRt' - Rd' = R\,\Omega(d')t' - Rd'$$

Or

$$d = -R\,\Omega(t')d' - Rd' \qquad (7.4.6)$$

The error vectors δ and d in (7.3.7) must be substituted according to (7.4.5) and (7.4.6) whenever the robot error model (i.e. $J_{\delta i}$ and J_{di}) is given in terms of the camera coordinate frame.   Equations (7.3.8)-(7.3.13) must also be modified accordingly.

D. OBSERVABILITY OF THE UNKNOWN PARAMETERS

In the robot calibration literature, the observability of the kinematic error parameter vector dp is defined in terms of the Identification Jacobian. If the Identification Jacobian is full rank, the error parameter vector is said to be observable.

It is difficult to obtain analytical observability results for the robot/camera calibration since the structure of the Jacobian matrix in this case is very complex (refer to 7.3.13)). However, the Identification Jacobian shown in this chapter can be used for optimal off-line search of robot measurement configurations, which can significantly improve calibration quality. The extension of the techniques to robot-camera calibration is straightforward after the identification Jacobian matrix is made available analytically.

E. VERIFICATION OF THE CALIBRATION RESULTS

It is difficult to obtain a highly accurate reference against which the accuracy performance of a robot-camera calibration task is evaluated. In the absence of an accurate external device, one may use the following two approaches to assess the accuracy performance.

Approach I: Verification on the image plane.

Assume that f_x, f_y, k and ρ have been identified. A set of robot configurations, which are different from those used for identification, is also given. In addition, a set of world coordinates of the calibration points in each robot configuration is given. The coordinates of the corresponding image points are measured. Using the identified camera and robot parameters together with the robot joint variables at each configuration, and the world coordinates of the calibration points, one can compute the predicted image coordinates of each image point (refer to (7.2.1)-(7.2.2)). The Euclidean norm of the difference between the measured and computed image coordinates at each image point can be defined as a *2D calibration error*. After 2D calibration errors are computed for all image points at all robot configurations, the mean and standard deviation of the 2D calibration errors can be obtained.

Approach II: Verification in the world coordinate system.

One may compute 3D world coordinates of each calibration point using the calibrated robot and camera parameters. This is possible by using more than one view, that is more than one robot configuration, for the same calibration point. Two views of a point are sufficient to compute its world coordinates using stereo triangulation. Using more than two views calls for a least squares fitting. The Euclidean norm of the difference between the computed world coordinates of the calibration point and its given world coordinates can be defined as a *3D calibration error*. One can compute the mean and standard deviation of the 3D calibration errors by repeating this procedure for all calibration points.

V.   EXTENSION TO STEREO-CAMERA CASE

When two cameras are used, a pair of equations of the form (7.2.5) can be obtained for each object point measurement; that is,

$$C_j \, \phi(\rho_j, f_{x,j}, k_j) = 0, \quad j = R, L \qquad (7.5.1)$$

where the subscript R denotes those quantities related to the right camera and L to the left camera. Among the parameter vectors ρ_R and ρ_L, most elements are identical as the transformation from the world coordinate frame to the $(n-1)$th robot link coordinate frame is shared by the two cameras. Let ρ and ρ_c represent the kinematic parameter vectors from the world to the right camera frames and from the right to the left camera frames, respectively. The dimension of ρ is again $4N - 2P + 6$, and that of ρ_c is 6. Equation (7.5.1) can thus be rewritten as

$$C_R \, \phi(\rho_c, f_{x,R}, k_R) = 0$$
$$C_L \, \phi(\rho, \rho_c, f_{x,L}, k_L) = 0$$

In most applications, the two cameras have very similar optical characteristics. The distortion coefficients k_R and k_L can thus be treated identically. Denote these by k. One then obtains the following pair of equations,

$$C_R \, \phi(\rho_c, f_{x,R}, k) = 0 \qquad (7.5.2)$$
$$C_L \, \phi(\rho, \rho_c, f_{x,L}, k) = 0 \qquad (7.5.3)$$

There is a total of $4N - 2P + 15$ parameters, among which three are camera parameters. According to (7.5.2)-(7.5.3), a single calibration point can provide four scalar equations. To estimate all unknown parameters, a sufficient number of calibration points have to be used. Let the subscript i denote the quantities corresponding to the ith calibration point. One has,

$$C_{R,i} \, \phi(\rho_c, f_{x,R}, k_R) = 0$$
$$C_{L,i} \, \phi(\rho, \rho_c, f_{x,L}, k_L) = 0$$

for $i = 1, 2, ..., s$. The identification procedure outlined in the last section can be applied to find the unknown robot and camera parameter vectors. Due to the use of stereo cameras, the number of parameters to be estimated is increased by seven, which means that the computational cost in this case is higher than that of the monocular-camera case.

However, the stereo-camera setup allows the identification of the tool-camera transformation without the need of external sensing devices.

An important feature of the one-stage approach with stereo cameras is that stereo matching is not needed, because the computation of the C_R and C_L matrices does not require matched image points.

VI.   DISCUSSION AND REFERENCES

One may solve a system calibration problem in a multi-stage process. It may start by calibrating the camera, follow by calibrating the hand/camera, and then use the camera and hand/eye models to calibrate the manipulator. After all system components are individually calibrated, the pose of an object in the robot world system can be determined.

Such a multistage approach has pros and cons. Its main advantages are:

1. Since system calibration is performed by calibrating its components or subsystems separately, each component calibration task is relatively simple.
2. If some of the system components have changed their location or parameters, calibration needs only to be repeated for these system components. For example, if the camera changed its focal length, only the camera needs to be recalibrated.

The drawbacks of the multi-stage approach are:

1. Parameter estimation errors in early stages propagate to the later stages. For example, any errors in the estimation of camera parameters may degrade the estimation quality of the robot link parameters.
2. The validity of the hand/eye calibration stage is questionable in certain cases. More specifically, most literature sources on hand/eye calibration make the assumption that the relative motions of the robot and the sensor are accurately known (Shiu and Ahmad (1989)).   While the relative motion of the sensor is measured by an external device, the relative motion of the robot is computed by the use of the robot kinematic model, which may not be sufficiently accurate prior to stringent robot calibration.

The concept of *autonomous calibration* through the creation of closed kinematic chains and use of internal sensory information only was developed by Bennett and Hollerbach (1991). More specifically, their system set-up consisted of a stereo camera system mounted on the robot hand. Each camera was able to perform a single rotation about a selected axis. The internal sensory information included the robot joint positions, the 2D image sensor reading and two additional joint angle readings coming from the motorized cameras. That paper proposed a simultaneous calibration strategy for both the robot and the stereo cameras using a unified mathematical model for the entire system.

In this chapter, we concentrated on passive camera systems. In this method, unlike the technique presented by Bennett and Hollerbach (1991), which used two instrumented cameras, a single passive camera is used. The method was originally presented in Zhuang, Wang and Roth (1993, 1995). A single camera system has a much larger field of view, compared to a stereo cameras system, consequently enlarging the measurable workspace of the robot manipulator. In addition, the robot-camera model developed in this chapter includes a radial camera lens distortion parameter.

One may argue that since the estimation problem is nonlinear and the dimension of the unknown parameter vector is large, such an algorithm may be very time-consuming and may require a very good initial guess. It is shown in this chapter that simultaneous calibration of a camera-robot system requires the addition of only two camera parameters to the parameter set that is normally used in robot calibration. Furthermore, the method developed for the construction of error models allows robot calibration practitioners to use their own previously developed robot error models as a compatible part of the camera-robot kinematic identification software. The identification problem can thus be solved using standard nonlinear least squares procedures such as the Levenberg-Marquardt algorithm (Marquardt (1963)). A good initial guess of the unknown parameters of the system is usually not hard to obtain since the nominal geometry of the robot is known in most cases and additional parameters can be roughly measured by proper gauging devices.

Chapter 8

ROBOTIC HAND/EYE CALIBRATION

I. INTRODUCTION

Robotic hand/eye calibration is the process of identifying the fixed yet unknown position and orientation of the robot sensor (typically, one or more cameras mounted on the robot hand) with respect to the robot hand coordinate frame. The robot hand frame, also known as the end-effector frame, is the 3D coordinate frame that is often used within the robot control software. The robot controller is designed to allow the computation of the 3D position and orientation of the hand frame, using the forward kinematics equations and the encoder reading of each joint position. In this chapter, we assume that the robot forward kinematics is known precisely.

For hand/eye calibration it is sometimes irrelevant to find (or characterize) the sensor frame with respect to a world reference frame. Because robot motion commands are specified relative to the robot hand frame and object positions are measured relative to the sensor frame, it is necessary to determine the *relative* pose between the coordinate frames of the robot hand and that of the sensor mounted on it. This relative pose cannot be measured directly as both the robot hand frame and the sensor frame are usually located inside the robot and the sensor. The problem of estimating this pose is the topic of this chapter.

A solution approach is to move the robot at least twice, each by a known amount, and to observe the resulting sensor motion induced by the robot motion; refer to Figure 8.1.1 (Shiu and Ahmad (1989)). Let T_6 be the transformation describing the geometry of the robot and OBJ be the transformation representing the object in the camera frame; refer to Figure 8.1.2. If the robot is moved from position T_{61} to T_{62}, and the position of the fixed object relative to the camera frame is found to be OBJ_1 and OBJ_2, respectively, then the following equation is obtained:

$$T_{61}X \ OBJ_1 = T_{62}X \ OBJ_2$$

The above equation can be rewritten as

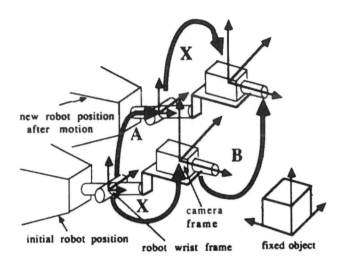

Figure 8.1.1 The hand-eye calibration problem
(From Shiu, Y. C. and S. Ahmad (1989). "Calibration of Wrist-Mounted
Robotic Sensors by Solving Homogeneous Transform Equations of the Form
$AX = XB$," *IEEE Trans. Robotics & Automation*, Vol. 5, No. 1, pp. 16-27.
With permission)

$$T_{62}^{-1}T_{61}X = X \; OBJ_2 OBJ_1^{-1}$$

By defining $A \equiv T_{62}^{-1}T_{61}$, the relative motion made by the robot, and $B \equiv OBJ_2 OBJ_1^{-1}$, the relative motion of the sensor, one obtains the following homogeneous transformation equation,

$$AX = XB \qquad\qquad (8.1.1)$$

where X is the 4x4 transformation from the robot tool coordinate frame to the sensor coordinate frame, A is the 4x4 transformation of the robot gripper frame from its ith to the $(i+1)$th position, and B is the 4x4 transformation of the sensor frame, also from its ith to the $(i+1)$th position. Under the assumption that the robot is well-calibrated, the transformation A can be readily computed from the given joint reading at each joint configuration. A typical scenario for the construction of the matrix B is, for instance, a hand-mounted camera viewing a calibration board. If the camera is recalibrated at each robot position, the B transformation becomes readily available. There are other scenarios that lead to the same equation (8.1.1).

Because such an equation has six independent unknowns, one needs to make two or more robot displacements to obtain under certain conditions a unique solution. Mathematically, this is equivalent to solving a system of equations of the form

$$A_i X = X B_i \quad i = 1, 2 \tag{8.1.2}$$

where i indicates the ith motion. Equation (8.1.2) can be broken into the following orientation and position equations

$$R_{A,i} R_X = R_X R_{B,i} \quad i = 1, 2 \tag{8.1.3}$$

and

$$R_{A,i} p_X + p_{A,i} = R_X p_{B,i} + p_X \quad i = 1, 2 \tag{8.1.4}$$

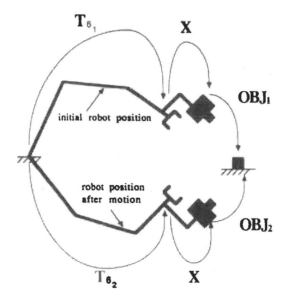

Figure 8.1.2 The formulation of the hand-eye calibration problem
(From Shiu, Y. C. and S. Ahmad (1989). "Calibration of Wrist-Mounted Robotic Sensors by Solving Homogeneous Transform Equations of the Form $AX = XB$," *IEEE Trans. Robotics & Automation*, Vol. 5, No. 1, pp. 16-27. With permission)

where R_X and p_X are the 3x3 rotational and 3x1 position parts of X, respectively. Similar definitions are used for $R_{A,i}$, $R_{B,i}$, $p_{A,i}$ and $p_{B,i}$. If R_X is known, p_X may then be solved linearly from the system of position equations of the type (8.1.4). Therefore, the more challenging part of the solution is to find R_X from the system of rotation equations (8.1.3).

In this chapter, two techniques for hand/eye calibration are discussed. The first method is a two-stage approach, in which the rotation part of the unknown transformation is obtained from the rotation equations (8.1.3), and the translation part is then determined using the known rotation matrix from the position equations (8.1.4). The second method is basically a one-stage nonlinear iterative algorithm. The calibration is performed using either both rotation and position equations (8.1.3) and (8.1.4), or only the position equation (8.1.4), depending on whether or not the orientation part of B_i is given or not.

The chapter is organized in the following manner. An introduction to quaternion algebra is given in Section II. In Section III, a linear solution method based on algebra of quaternions is given. Section IV presents a nonlinear solution algorithm along with a discussion about the observability of the hand/eye parameters. Representative simulation results are presented in Section V. Notes and references are provided at the end of this chapter.

II.   REVIEW OF QUATERNION ALGEBRA

A.   QUATERNIONS

A quaternion is an extension of the concept of a complex number. Let a be a *quaternion* written in the form,

$$a = a_0 + a_1 i + a_2 j + a_3 k \qquad (8.2.1)$$

where a_i, $i = 0, 1, 2, 3$, are real numbers, and i, j and k are quaternion units. Note that i, j and k are orthonormal 3D vectors; that is $i \times j = k, j \times k = i$, and $k \times i = j$. Denote

$$a = [a_0, \ a^T]^T$$

where a_0 and a are referred to as *scalar quaternion* and *vector quaternion*, respectively. A *Hamiltonian conjugate* of a quaternion a, denoted by a^*, is defined as:

$$a^* \equiv [a_0, \ -a^T]^T$$

Clearly, a quaternion can be treated as a 4D vector.

B. QUATERNION ALGEBRA
a. Addition and Subtraction

Let $a = [a_0, a^T]^T$ and $b = [b_0, b^T]^T$ be two quaternions. Then addition and subtraction, $\pm$, of two quaternions are defined as follows

$$a \pm b = [a_0 \pm b_0, a^T \pm b^T]^T$$

Both operations obey the associative and commutative laws.

b. Multiplication

Multiplication of two quaternions, denoted by $\circ$, creates a new quaternion,

$$a \circ b = [a_0, a^T]^T \circ [b_0, b^T]^T$$
$$= a_0 b_0 + b_0 a + + a_0 b + a \circ b$$
$$= a_0 b_0 + b_0 a + a_0 b - a \circ b + a \times b$$

which can written as

$$a \circ b = [a_0 b_0 - a \circ b, (a_0 b + b_0 a + a \times b)^T]^T \qquad (8.2.2)$$

quaternion multiplication is associative and distributive, but not necessarily commutative.

c. Norm, Division and Inversion

The norm of a, denoted by $N(a)$, is defined as

$$N(a) = a^T a = \|a\|^2$$

The inverse of a nonzero a is given by

$$a^{-1} = a^*/N(a)$$

d. Unit Quaternions and Euler Parameters

Let q be a quaternion. If

$$N(q) = 1$$

then q is referred to as a *unit quaternion*. For a unit quaternion

$$q^{-1} = q^*$$

A unit quaternion can be expressed as

$$q = \cos(\theta/2) + \sin(\theta/2)k \qquad (8.2.3)$$

where θ is a rotation angle and k is a rotation axis. Let $q = [q_0, q_1, q_2, q_3]^T$; q_i are called *Euler parameters*. That is, for an arbitrary rotation in 3D space, characterized by angle of rotation θ about an axis k, the corresponding quaternion $q = [q_0, q^T]^T$ is

$$q = \sin(\theta/2)k, \qquad q_0 = \cos(\theta/2)$$

e. Unit Quaternions and Rotation Matrices

Whenever $y_0 = 0$, the quaternion y is referred to as a *pure quaternion*. If the vector part y of a pure quaternion y is obtained from a 3D vector x through a rotation R_q,

$$y = R_q x \qquad (8.2.4)$$

then the pure quaternion y can be shown to be written as

$$y = q \cdot x \cdot q^* \qquad (8.2.5)$$

where the vector of the Euler parameters q is related to the rotation matrix R_q through the following equation,

$$R_q = \begin{bmatrix} q_0^2 + q_1^2 - q_2^2 - q_3^2 & 2(q_1 q_2 - q_0 q_3) & 2(q_1 q_3 + q_0 q_2) \\ 2(q_1 q_2 + q_0 q_3) & q_0^2 - q_1^2 + q_2^2 - q_3^2 & 2(q_2 q_3 - q_0 q_1) \\ 2(q_1 q_3 - q_0 q_2) & 2(q_2 q_3 + q_0 q_1) & q_0^2 - q_1^2 - q_2^2 + q_3^2 \end{bmatrix} \qquad (8.2.6)$$

Note that $\pm q$ both correspond to the same rotation matrix R_q, or in other words, quaternions which differ only in terms of their signs represent the same rotation.

III. A LINEAR SOLUTION

The linear solution method of the hand/eye calibration problem involves two steps. First, the rotation part of the hand/eye transformation matrix X is solved from (8.1.3), and the translation part of X is then determined from (8.1.4).

A. SOLUTION FOR THE ROTATION MATRIX

The solution is based on quaternion algebra. The following two lemmas, due to Shiu and Ahmad (1989), are important for the derivation of the linear solution.

Lemma **8.3.1**: The eigenvalues of a general rotation matrix $Rot(k, \theta)$ of the type (3.2.3) are $\{1, e^{j\theta}, e^{-j\theta}\}$.

Proof: It can be shown that the general rotation matrix can be written in the following form (Shiu and Ahmad (1989)),

$$Rot(k, \theta) = E \begin{bmatrix} 1 & 0 & 0 \\ 0 & e^{j\theta} & 0 \\ 0 & 0 & e^{-j\theta} \end{bmatrix} E^{-1}$$

where E is a nonsingular matrix consisting of the eigenvectors of $Rot(k, \theta)$. A similarity transformation preserves the eigenvalues.   []

Lemma **8.3.2**: If R_A and R_B are rotation matrices such that $R_A R = R R_B$ for any rotation matrix R, then R_A and R_B must correspond to rotations that have the same angle of rotation.

Proof: A rotation matrix is always invertible. Thus R_A and R_B are similar since $R_A R_X = R_X R_B$. Since similar matrices have the same eigenvalues, so are R_A and R_B. By Lemma 8.3.1, R_A and R_B must have the same angle of rotation.   []

Denote $R_A = Rot(k_A, \theta)$, $R_B = Rot(k_B, \theta)$, $R_X = Rot(k_X, \omega)$. Then,

Lemma **8.3.3**: $R_A R_X = R_X R_B$ is equivalent to the following vector equation

$$\sin(\theta/2)\sin(\omega/2)(k_A + k_B) \times k_X = \sin(\theta/2)\cos(\omega/2)(k_B - k_A) \quad (8.3.1)$$

Proof: Let $a \equiv [a, a^T]^T$, $b \equiv [b, b^T]^T$ and $x \equiv [x, x^T]^T$ be quaternions that correspond to the rotation matrices R_A, R_B and R_X, where a, b, and x are scalars and a, b and x are 3x1 vectors. Let u_i and b_i be the ith column 3D vector of the identity matrix and R_B, respectively. The 3D vectors u_i and b_i induce (zero scalar) pure quaternions. The equation $R_A R_X = R_X R_B$ can be written as

$$R_A R_X u_i = R_X b_i \qquad i = 1, 2, 3 \qquad (8.3.2)$$

However according to (8.2.5),

$$R_X b_i = \mathbf{x} \circ b_i \circ \mathbf{x}^* \qquad i = 1, 2, 3 \qquad (8.3.3)$$

where $R_X b_i$ has been treated as a pure quaternion. Furthermore,

$$R_A R_X u_i = \mathbf{a} \circ \mathbf{x} \circ u_i \circ \mathbf{x}^* \circ \mathbf{a}^* \qquad i = 1, 2, 3 \qquad (8.3.4)$$

where $R_X b_i$ has been also treated as a pure quaternion. By (8.3.2),

$$\mathbf{a} \circ \mathbf{x} \circ u_i \circ \mathbf{x}^* \circ \mathbf{a}^* = \mathbf{x} \circ b_i \circ \mathbf{x}^* \qquad i = 1, 2, 3 \qquad (8.3.5)$$

Because **a**, **b** and **x** are all unit quaternions, (8.3.5) may be rearranged into

$$u_i \circ \mathbf{y} = \mathbf{y} \circ b_i \qquad i = 1, 2, 3 \qquad (8.3.6)$$

where

$$\mathbf{y} \equiv [y, \, y^T]^T \equiv \mathbf{x}^* \circ \mathbf{a}^* \circ \mathbf{x} \qquad (8.3.7)$$

From (8.3.6),

$$\mathbf{y}^* \circ u_i \circ \mathbf{y} = b_i \qquad i = 1, 2, 3$$

However

$$b_i = \mathbf{b} \circ u_i \circ \mathbf{b}^* \qquad i = 1, 2, 3$$

The above equations yield a unique solution,

$$\mathbf{y} = \mathbf{b}^* \qquad (8.3.8)$$

By substituting (8.3.8) into (8.3.7), the following equation for **x** is obtained,

$$\mathbf{x} \circ \mathbf{b}^* = \mathbf{a}^* \circ \mathbf{x} \qquad (8.3.9)$$

Expanding (8.3.9), according to (8.2.2), provides

$$[b \bullet x, (- xb + bXx)^T]^T = [a \bullet x, (- xa - aXx)^T]^T$$

The above equation is equivalent to

$$(b + a)Xx = x(b - a) \tag{8.3.10}$$

because the scalar part $(b - a) \bullet x = 0$ is implied by (8.3.10). However for R_A, R_B and R_X being rotation matrices,

$$a = \sin(\theta/2)k_A$$
$$b = \sin(\theta/2)k_B$$
$$x = \sin(\omega/2)k_X$$
$$x = \cos(\omega/2)$$

Substituting the above relationships into (8.3.10) yields (8.3.1).   []

If $\theta \neq 0$, then from (8.3.1)

$$\tan(\omega/2)(k_A + k_B)Xk_X = k_B - k_A \tag{8.3.11}$$

Recall that $\Omega(v)$ denotes a skew symmetric matrix that corresponds to $v \equiv [v_x, v_y, v_z]^T$. Define

$$G_1 \equiv \Omega(k_A + k_B)$$
$$h_1 \equiv k_B - k_A$$

and

$$z \equiv \tan(\omega/2)k_X \tag{8.3.12}$$

Equation (8.3.11) can then be equivalently written as

$$G_1 z = h_1 \tag{8.3.13}$$

As $Rank(G_1) = 2$, one needs two such matrix equations to obtain a unique solution. Let $R_{Ai}R_X = R_X R_{Bi}$, $i = 1$, 2, be two given matrix equations. $R_{Ai} = Rot(k_{Ai}, \theta_i)$ and $R_{Bi} = Rot(k_{Bi}, \theta_i)$. Then if $\theta_i \neq 0$, one can stack the equations of the form (8.3.13) into

$$Gz = h \tag{8.3.14}$$

where

$$G = \begin{bmatrix} \Omega(k_{A1} + k_{B1}) \\ \Omega(k_{A2} + k_{B2}) \end{bmatrix}$$

$$h = \begin{bmatrix} k_{B1} - k_{A1} \\ k_{B2} - k_{A2} \end{bmatrix}$$

and z is given in (8.3.12).

Lemma 8.3.4: If $\omega \neq \pi$, then the necessary and sufficient condition for (8.3.14) to have a unique solution z is

$$k_{A1} \times k_{A2} \neq 0 \qquad (8.3.15)$$

Proof: A necessary and sufficient condition for z to be a unique solution is that rank$(G) = 3$ because (8.3.14) is a consistent system of linear equations. By elementary matrix transformations on G, an equivalent condition is

$$(k_{A1} + k_{B1}) \times (k_{A2} + k_{B2}) \neq 0 \qquad (8.3.16)$$

According to Shiu and Ahmad (1989), the rotation matrix R_X satisfies the following equation,

$$k_{Ai} = R_X k_{Bi} \qquad (8.3.17)$$

Substituting (8.3.17) into (8.3.16) yields

$$((I + R_X)k_{A1}) \times ((I + R_X)k_{A2}) \neq 0 \qquad (8.3.18)$$

The eigenvalues of $I + R_X$ are $\{2,\ 1 + \cos\omega + j\sin\omega,\ 1 + \cos\omega - j\sin\omega\}$. Therefore if $\omega \neq \pi$, Equation (8.3.18) is equivalent to (8.3.16). Since k_{A1} and k_{A2} undergo the same scaling, rotation and translation, these vectors will not become parallel after the scaling, rotation and translation if they were not parallel prior to these operations. Equation (8.3.18) is thus equivalent to (8.3.15). □

Equation (8.3.14) may be extended to accommodate more measurement equations to allow a least squares solution of z. Finally if $\|z\| \neq 0$, then k_X and ω can be obtained from (8.3.12) by

$$k_X = z/\|z\| \tag{8.3.19}$$

$$\omega = 2\mathrm{atan}(\|z\|) \tag{8.3.20}$$

If $\|z\| \equiv 0$, then from (8.3.12) $\omega = 0$ and k_X is arbitrarily chosen as $[0, 0, 1]^T$.

In summary, we have the following fact:

Theorem 8.3.1:   A consistent system of two rotation matrix equations $R_{Ai}R_X = R_X R_{Bi}$, $i = 1, 2$, has a unique solution if the axes of rotation for R_{A1} and R_{A2} are neither parallel nor anti-parallel to one another and the rotation angles of R_{A1} and R_{A2} are neither zero nor π.

Remark: The sign of the rotation axis of R_A is non-unique whenever its rotation angle equals π. However one can pick up an arbitrary sign for the rotation axis of R_A since both choices correspond to the same physical system. On the other hand, for a chosen rotation axis k_A, there are two rotation axes $\pm k_B$, both satisfying the equation $R_A R_X = R_X R_B$. This is where the non-uniqueness of the solution comes from. Suppose that both rotation angles of R_{A1} and R_{A2} equal π, then k_{A1} and k_{A2} correspond to $\pm k_{B1}$ and $\pm k_{B2}$ respectively. We can in this case substitute each one of the sets $\{k_{A1}, k_{A2}, k_{B1}, k_{B2}\}$, $\{k_{A1}, k_{A2}, k_{B1}, -k_{B2}\}$, $\{k_{A1}, k_{A2}, -k_{B1}, k_{B2}\}$, and $\{k_{A1}, k_{A2}, -k_{B1}, -k_{B2}\}$ into (8.3.14) to obtain four different values of z, hence four different values of R_X, out of which only one is the correct physical answer. Similar analysis can be done for the case in which only one of the rotation angles of R_{A1} and R_{A2} equals π. Thus, the following statement can be made:

If both rotation angles of R_{A1} and R_{A2} equal π, then four solutions, out of which one is the true solution, can be obtained by solving (8.3.14). If only one of the rotation angles of R_{A1} and R_{A2} equals π, then two solutions, out of which one is the true solution, can be obtained. In cases of multiple solutions, one can in general pick up the correct solution based on constraints of the physical system.

B. SOLUTION FOR THE TRANSLATION VECTOR

After R_X is determined, it is straightforward to compute the translation vector p_X from (8.1.4). Rearranging these equations provides

$$(R_{A,i} - I)p_X = R_X p_{B,i} - p_{A,i} \qquad i = 1, 2 \tag{8.3.21}$$

The following lemmas, due to Shiu and Ahmad (1989), are needed in the proof of the conditions for a unique solution of the translation vector.

Lemma 8.3.5: If R is a 3x3 rotation matrix and its corresponding rotation angle is neither 0 nor π, any row of $(R - I)$ is a linear combination of the transposes of the two eigenvectors corresponding to the two non-unity eigenvalues of R.

Proof: Recall that rotation matrix R whose angle of rotation is neither 0 nor π has the eigenvalues $\{1,\ e^{j\theta}\ \text{and}\ e^{-j\theta}\}$. The eigenvalues of $R - I$ are thus 0, $e^{j\theta} - 1$ and $e^{-j\theta} - 1$. Denote the eigenvectors of R as e_1, e_2 and e_3 and let $e_i = [e_{i,x},\ e_{i,y},\ e_{i,z}]^T$. Then the first row of R can be written as

$$(e^{j\theta} - 1)e_{2,x}\, e_2^T + (e^{-j\theta} - 1)e_{3,x}\, e_3^T$$

Similar expressions can be derived for the second and third rows of R.   □

Lemma 8.3.6: For the two rotation matrices R_1 and R_2 whose axes of rotation are neither parallel nor anti-parallel to one another and whose angles of rotation are neither 0 nor π, it is impossible that the sets of vectors $\{e_2, e_3, f_2\}$ and $\{e_2, e_3, f_3\}$ are both linearly dependent, where e_2 and e_3 are the eigenvectors corresponding to the two non-unity eigenvalues of R_1, and f_2 and f_3 are those corresponding to the two non-unity eigenvalues of R_2.

Proof: To prove that $\{e_2, e_3, f_2\}$ is linearly independent, we must prove that $k_1 = k_2 = k_3 = 0$ if

$$k_1 e_2 + k_2 e_3 + k_3 f_2 = 0$$

Multiplying both sides of the above equation by e_1 yields

$$k_3 e_1^T f_2 = 0$$

because $e_1^T e_2 = e_1^T e_3 = 0$. If $e_1^T f_2 \neq 0$, then $k_3 = 0$ and

$$k_1 e_2 + k_2 e_3 = 0.$$

Since e_2, e_3 are linearly independent, then $k_1 = k_2 = 0$. Thus $\{e_2, e_3, f_2\}$ are linearly independent if $e_1^T f_2 \neq 0$. Similarly one can show that if $e_1^T f_3 \neq 0$, then $\{e_2, e_3, f_3\}$ are linearly independent. However since the axes of rotation for the two rotations are neither parallel nor anti-parallel, $e_1{}^Tf_2$ and $e_1{}^Tf_3$ cannot be simultaneously zero. This proves the lemma.   []

The following theorem provides the conditions for a unique solution of the translation vector.

Theorem **8.3.2**: A consistent system of two translation equations of the type (8.3.21) has a unique solution if the axes of rotation for R_{A1} and R_{A2} are neither parallel nor anti-parallel to one another and the rotation angles of R_{A1} and R_{A2} are neither zero nor π.

Proof: Equation (8.3.21) can be written in a matrix equation form,

$$\begin{bmatrix} R_{A1} - I \\ R_{A2} - I \end{bmatrix} p_X = \begin{bmatrix} R_X\, p_{B,1} - p_{A,1} \\ R_X\, p_{B,2} - p_{A,2} \end{bmatrix} \tag{8.3.22}$$

Since the rank of $(R_{A1} - I)$ is two, we can pick up two independent equations from it.   According to Lemma 8.3.5, both of which are linear combinations of the transposes of e_2 and e_3. We can now pick one equation from $(R_{A2} - I)$, which is a linear combination of the transposes of f_2 and f_3. Thus the rank of the coefficient matrix of p_X is three, which means that a unique solution for p_X can be determined by solving the matrix equation.   []

IV. A NONLINEAR ITERATIVE SOLUTION

To obtain a nonlinear iterative solution, the hand/eye calibration problem needs to be reformulated. In this section, the new formulation is presented first, followed by the derivation of the Identification Jacobian based on this formulation. Some analysis of the observability of the unknown hand/eye parameters is also given in this section.

A. AN ALTERNATIVE MATHEMATICAL FORMULATION OF THE HAND/EYE CALIBRATION PROBLEM

Again, the solution is derived by using quaternion algebra. In addition to the representation of the rotation part of the hand-eye transformation given in (8.3.11), the following lemma gives the representation for its translation part. Recall that $R_A = Rot(k_A, \theta)$, $R_B = Rot(k_B, \theta)$, $R_X = Rot(k_X, \omega)$.

Lemma **8.4.1**:   If $\theta \neq 0$, the equation $R_A p_X + p_A = R_X p_B + p_X$ is equivalent to the following equation

$$\Omega(u + p_B)z = p_B - u \tag{8.4.1}$$

Or

$$- \Omega(z)(u + p_B) = p_B - u \tag{8.4.2}$$

where $u \equiv (R_A - I)p_X + p_A$.

Proof: Lemma 8.4.1 can be proved in a similar way to Lemma 8.3.3. Let $x \equiv (x, x)$ be a quaternion corresponding to the rotation matrix R_X, where x is a scalar and x is a 3x1 vector. Then

$$R_X p_B = u \tag{8.4.3}$$

However,

$$R_X p_B = x \circ p_B \circ x^*$$

Thus

$$u = x \circ p_B \circ x^*$$

where $R_X p_B$ has been treated as pure quaternions. Rearranging the above equation into

$$x \bullet p_B = u \circ x$$

and expanding it provides

$$(- p_B \bullet x, \, xp_B + x \times p_B) = (- u \bullet x, \, xu + u \times x) \tag{8.4.4}$$

Equation (8.4.4) is equivalent to

$$(p_B + u) \times x = x(p_B - u) \tag{8.4.5}$$

since the scalar part $(p_B - u) \bullet x = 0$ is implied by (8.4.5). One obtains (8.4.1) by using $z \equiv x/x$, and

$$x = \sin(\omega/2)k_X$$

$$x = \cos(\omega/2)$$

Equation (8.4.2) is just another form of (8.4.5), in which the unknown position vector p_X appears linearly. []

As a minimum, two sets of equations (8.3.11) (or (8.3.13)) and (8.4.2) are needed for a unique solution of z and p_X. After z is computed, k_X and ω can be obtained from (8.3.19)-(8.3.20). Equations (8.3.11) and (8.4.2) provide an alternative formulation of the hand/eye calibration problem. Although this formulation is derived using quaternions, the application of the final results does not require any knowledge of quaternions.

B.   THE COST FUNCTION

Let

$$f(z) \equiv \Omega(k_A + k_B)z - k_B + k_A = 0 \tag{8.4.6}$$

$$g(z, p_X) \equiv \Omega(u + p_B)z - p_B + u = 0 \tag{8.4.7}$$

where $u \equiv (R_A - I)p_X + p_A$. Note that (8.4.6) and (8.4.7) are derived from (8.3.11) and (8.4.2), respectively. The 6x1 vector, $[f(z)^T, g(z, p_X)^T]^T$, is referred to as the *measurement residual vector of X*.   The vectors $f(z)$ and $g(z, p_X)$ are further called the rotation part and the position part of the measurement vector, respectively. The vectors $f(z)$ and $g(z, p_X)$ will be both zero if there is no error in A and B.

Once z and p_X are estimated, the unknown hand/eye transformation matrix can be readily constructed.

Let the hand/eye parameter vector be $\rho = [z^T, p_X^T]^T$. The problem is now to choose ρ that minimizes the cost function E, where

$$E = \sum_{i=1}^{s} f_i(\rho)^T f_i(\rho) + w \sum_{i=1}^{s} g_i(\rho)^T g_i(\rho) \tag{8.4.8}$$

where i denotes the ith measurement, s is the total number of measurements, and w is a weighting factor that balances the contributions of the position and orientation error cost factors to the total cost.   The cost function is a nonlinear function of ρ. Therefore, an iterative procedure is needed to obtain its solution.   To properly apply a gradient optimization technique such as the Gauss-Newton algorithm to the problem at hand, the following issues have to be addressed:

1.   The initial condition,
2.   The structure of the Jacobian (which will be explained shortly), and
3.   The updating rule.

It is assumed and is practically justifiable that X^0, the nominal value of X,   is known.   In the case of hand/eye calibration, the camera usually provides the full pose information; therefore, the nominal solution can be provided by a non-iterative procedure such as those discussed in Section II, or simply by roughly measuring the hand/eye transformation using a gauging device. In the case of calibrating a tool mounted on a manipulator, the tool full pose information may not be available. Engineering drawings that specify nominal dimensions of the tool are usually available, from which the initial hand/tool transformation can be computed. These nominal solutions can be used as initial conditions of the Gauss-Newton algorithm.

C.   THE IDENTIFICATION JACOBIAN

In this section, it is assumed that the weighting factor w in (8.4.8) has been properly chosen. This can be achieved by scaling the position equation (8.4.7). In the remaining part of this section, w will not be mentioned explicitly.

Assume for the moment that the initial solutions z^0 and p_x^0 deviate by a small amount from z^* and p_x^*, the optimal solutions. To a first order approximation,

$$f(z^0 + dz) \approx f(z^0) + \frac{\partial f}{\partial z} dz \qquad (8.4.9)$$

$$g(z^0 + dz, p_x^0 + dp_x) \approx g(z^0, p_x^0) + \frac{\partial g}{\partial z} dz + \frac{\partial g}{\partial p_x} dp_x \qquad (8.4.10)$$

where

$$\frac{\partial f}{\partial z} = \Omega(k_A + k_B)$$

$$\frac{\partial g}{\partial z} = \Omega(p_B + p_A + (R_A - I)p_x^0)$$

$$\frac{\partial g}{\partial p_x} = (I - \Omega(z^0))(R_A - I)$$

Equations (8.4.9) and (8.4.10) can be rewritten into

$$\begin{bmatrix} \frac{\partial f}{\partial z} & 0 \\ \frac{\partial g}{\partial z} & \frac{\partial g}{\partial p_x} \end{bmatrix} \begin{bmatrix} dz \\ dp_x \end{bmatrix} = - \begin{bmatrix} f(z^0) \\ g(z^0, p_x^0) \end{bmatrix} \qquad (8.4.11)$$

It can be shown that the coefficient matrix in (8.4.11) is singular because the rank of the first block diagonal matrix is two. Therefore, at least two sets of equations must be used to solve for dz and dp_x. Let

$$h \equiv - \begin{bmatrix} f_1(z) \\ f_2(z) \\ g_1(z,\ p_x) \\ g_2(z,\ p_x) \end{bmatrix} \qquad (8.4.12)$$

$$J \equiv \begin{bmatrix} \dfrac{\partial f_1}{\partial z} & 0 \\[2mm] \dfrac{\partial f_2}{\partial z} & 0 \\[2mm] \dfrac{\partial g_1}{\partial z} & \dfrac{\partial g_1}{\partial p_x} \\[2mm] \dfrac{\partial g_2}{\partial z} & \dfrac{\partial g_2}{\partial p_x} \end{bmatrix} \qquad (8.4.13)$$

where, as in Section II, the subscripts "1" and "2" in (8.4.12)-(8.4.13) denote the quantities evaluated using the first and second measurements, and the superscript "0" is dropped. In the above equations, h is a *measurement residual vector*, and J is an *Identification Jacobian for the hand/eye calibration*. The dimensions of ρ, h, and J are respectively 6x1, 12x1, and 12x6. Let $d\rho$ denote the *pose error vector of* X. With this notation, we have

$$J\ d\rho = h \qquad (8.4.14)$$

Equation (8.4.14) provides the relationship between the measurement residual error vector h and the parameter error vector $d\rho$ of X.

D. OBSERVABILITY ISSUES

The observability of the pose error vector $d\rho$ can be investigated through the study of the Identification Jacobian. If the Identification Jacobian is full rank, the error parameters are observable. The next theorem reveals the observability conditions of $d\rho$.

Theorem 8.4.1:   Assume that (8.4.14) is used for the estimation of the pose error parameter vector $d\rho$. The vector $d\rho$ is observable if

1. The rotation axes of R_{A1} and R_{A2} are neither parallel nor anti-parallel to one another,
2. Both rotation angles of R_{A1} and R_{A2} equal neither zero nor π, and
3. The rotation angle of R_X does not equal π.

The vectors and matrices are defined in (8.1.3).

Proof: From the definition of observability, dp is observable if the Jacobian matrix is full rank. The Jacobian matrix in this case is

$$J \equiv \begin{bmatrix} \Omega(k_{A1} + k_{B1}) & 0 \\ \Omega(k_{A2} + k_{B2}) & 0 \\ \Omega(p_{b1} + u_1) & (I - \Omega(z))(R_{A1} - I) \\ \Omega(p_{b2} + u_2) & (I - \Omega(z))(R_{A2} - I) \end{bmatrix}$$

J is full rank if each of its two block diagonal matrices is full rank. It is shown in Lemma 8.3.4 that an equivalent condition for the upper left block diagonal matrix to be full rank is

$$((I + R_X)k_{A1}) \times ((I + R_X)k_{A2}) \neq 0.$$

The eigenvalues of $I + R_X$ are $\{2, 1 + \cos\omega + j \sin\omega, 1 + \cos\omega - j \sin\omega\}$. Therefore if $\omega \neq \pi$, then the above equation is equivalent to

$$k_{A1} \times k_{A2} \neq 0.$$

which is the first condition of the theorem.   The lower right diagonal matrix of J can be written as

$$\begin{bmatrix} (I - \Omega(z))(R_{A1} - I) \\ (I - \Omega(z))(R_{A2} - I) \end{bmatrix} = \begin{bmatrix} I - \Omega(z) & 0 \\ 0 & I - \Omega(z) \end{bmatrix} \begin{bmatrix} R_{A1} - I \\ R_{A2} - I \end{bmatrix}$$

By using the symbolic manipulation software package MACSYMA®, it was shown that the determinant of $I - \Omega(z)$ is $(z_3^2 + 1)$, where z_3 is the third element of z. Thus $I - \Omega(z)$ is full rank. In addition recall, from Theorem 8.3.2, that

$$Rank\left(\begin{bmatrix} R_{A1} - I \\ R_{A2} - I \end{bmatrix}\right) = 3$$

if the first two conditions are satisfied. Consequently,

$$Rank\left(\begin{bmatrix} (I - \Omega(z))(R_{A1} - I) \\ (I - \Omega(z))(R_{A2} - I) \end{bmatrix}\right) = 3$$

This completes the proof of the theorem.   []

Remarks:
1. The conditions provided by Theorem 8.4.1 are sufficient.
2. More than two measurements may be used to reduce the effect of measurement noise.

It can be seen that using only the position measurements of B, one may still be able to obtain the unknown hand/eye transformation. In such a case, (8.4.7) becomes the only governing equation, and the Identification Jacobian using two measurements becomes

$$J \equiv \begin{bmatrix} \Omega(p_{b1} + u_1) & (I - \Omega(z))(R_{A1} - I) \\ \Omega(p_{b2} + u_2) & (I - \Omega(z))(R_{A2} - I) \end{bmatrix}$$

which is a 6x6 matrix. By (8.4.3), the above equation can be rewritten as

$$J \equiv \begin{bmatrix} \Omega((R_x + I)p_{b1}) & (I - \Omega(z))(R_{A1} - I) \\ \Omega((R_x + I)p_{b2}) & (I - \Omega(z))(R_{A2} - I) \end{bmatrix} \qquad (8.4.14)$$

Theorem 8.4.2: Assume that the above Identification Jacobian is adopted for kinematic identification; that is, only the position vectors of B_i, $i = 1, 2$, are used for the estimation of the hand/eye parameter error vector dp. The vector dp is observable only if

1. The rotation axes of R_{A1} and R_{A2} are neither parallel nor anti-parallel to one another,
2. Both rotation angles of R_{A1} and R_{A2} are nonzero, and
3. p_{B1} and p_{B2} are nonzero and are neither parallel nor anti-parallel.

Proof: The Jacobian given in (8.4.14) is full rank only if its columns are linearly independent. Based on the proof of Theorem 8.4.1, the second three columns are linearly independent only if the first two conditions are satisfied. Similar to the proof of Theorem 8.4.1, an equivalent condition for the first three columns to be linearly independent is

$$((I + R_X)p_{B1})\times((I + R_X)p_{B2}) \neq 0.$$

Therefore if $\omega \neq \pi$, the above equation is equivalent to

$$p_{B1} \times p_{B2} \neq 0.$$

which is the third condition. This completes the proof of the theorem.   []

Again, more than two measurements may be used to reduce measurement noise effect.

V. SIMULATION RESULTS

The algorithm was tested by generating various arbitrary samples of X and A_i from which the matrices B_i could be computed. We chose 80 different values uniformly covering the parameter space of X. The norms of p_X, the position part of X, were kept at 800 mm. For each X, up to 15 sets of $\{A_i, B_i\}$ were computed. The homogeneous transformations A_i were also randomly selected under the constraint that the norms of p_A were in a range of [500mm, 1000mm]. The homogeneous transformations B_i were solved from $B_i = X^{-1}A_iX$.

Three algorithms were extensively tested in the simulation study:
1. The two-stage method discussed in Section II,
2. The iterative one-stage method with full pose measurements of B_i, and
3. The iterative one-stage method with position measurements of B_i only.

Whenever no measurement noise was injected into the system, all three algorithms listed before produced correct results for all simulation runs.

Let us now consider an actual system. A_i is computed by using a forward kinematic model of the robot and the set of joint readings; thus, it is orthonormal, although it may not be accurate. Adding angular noise preserves the orthonormality of R_{Ai}. On the other hand, B_i is obtained in general by an external sensing device; therefore, R_{Bi} may not be orthonormal. Orthonormalization of R_{Bi} before plugging it into either the Jacobian matrix or (8.1.2) is unnecessary since both the two-stage and one-stage algorithms can handle the case that R_{Bi} is not orthonormal. To simulate a real system, measurement uncertainties were modeled as uniformly distributed random numbers added to the three Euler angles of R_{Ai}, the position vector p_{Ai}, and the nontrivial elements of B_i.

Different types of noise intensities were tested. The Euclidian norms of d_i and δ_i were used to define the *position* and *rotation errors of X*, that is, the deviation of an estimate from the true displacement. Means and standard deviations of $\|d_i\|$ and $\|\delta_i\|$ were computed for each set of measurements from all 80 samples of X, providing the first two moments of the ensemble statistics.

Table 8.5.1 lists three different types of noise used in the simulations. U[a, b] denotes the uniformly distributed random noise in a range of [a, b]. Type I noise models a moderate noise level for both position and orientation measurements. Types II and III have higher noise levels except that in type III, the noise level for the sensor orientation measurement is zero. This case was designed to study the effect of sensor orientation measurement errors on the performance of the algorithms.

Table 8.5.1 Noise intensity levels used in simulation

	Positioning noise of p_A and p_B (mm)	Noise of angles of R_A (rad)	Noise of elements of R_B (rad)
Type I	U[-0.25, 0.25]	U[-0.0005, 0.0005]	U[-0.0005, 0.0005]
Type II	U[-1.0, 1.0]	U[-0.0025, 0.0025]	U[-0.0025, 0.0025]
Type III	U[-1.0, 1.0]	U[-0.0025, 0.0025]	None

The performance of the algorithms is illustrated by Figures 8.5.1-8.5.3. Simulations consistently showed that the one-stage iterative algorithm using either full poses of B_i or only its position vectors produced identical results. Gaining full understanding of this phenomenon is still an open research issue. Only one curve, that is the one-stage method, was then used in these figures to assess both algorithms.

The horizontal axes in Figures 8.5.1-8.5.3 represent the number of homogeneous transformation equations for the estimation problem. The vertical axes of Figures 8.5.1-8.5.3 show the mean and standard deviation of the rotation and position estimation errors.

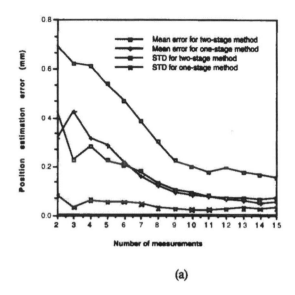

(a)

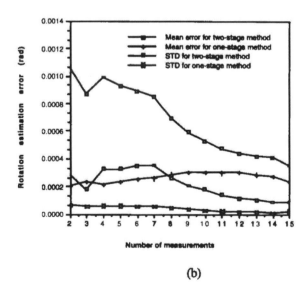

(b)

Figure 8.5.1 Estimation errors for type I noise

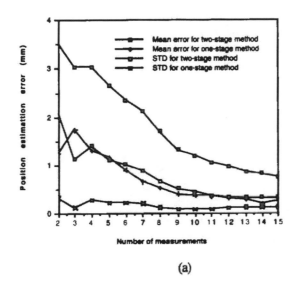

(a)

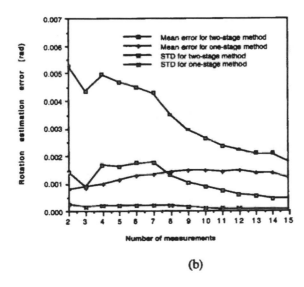

(b)

Figure 8.5.2 Estimation errors for type II noise

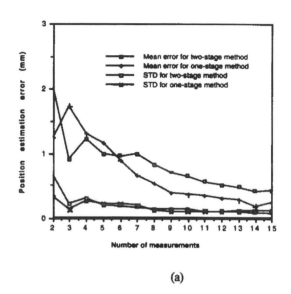

(a)

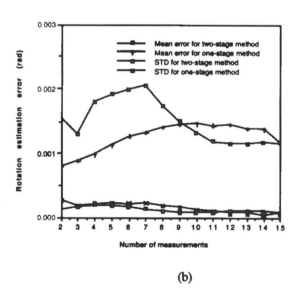

(b)

Figure 8.5.3 Estimation errors for type III noise

By studying these simulation results, the following remarks can be made:

1. In general, the accuracy performance of the one-stage algorithm is better than that of the two-stage algorithm.
2. When the noise intensity in the rotation measurement is significant, as modeled by types I and II noise processes, the advantage of the one-stage algorithm is clear, since large sensor orientation noise causes significant propagation errors to the estimation of the position vector of X in the two-stage method.
3. Whenever sensor orientation measurements are noise-free, as represented by type III noise, the performance of the two-stage algorithm is improved (Figure 8.5.3). This is due to relatively smaller rotation errors propagated to position errors. On the other hand, the performance of the one-stage algorithm remains the same, as sensor orientation measurements are not used in the one-stage algorithm.
4. As expected, the estimation error increases almost proportionally to the noise intensity (comparing Figure 8.5.1 with Figure 8.5.3).
5. The iterative algorithm usually converges after a few iterations if the initial conditions are within a ball of radius 20 (mm) for position parameters of X and a ball of radius 0.01 (rad) for its rotation parameters; both are centered at their optimal parameters.
6. On the whole, by using additional measurements, estimation errors decrease gradually. This trend slows down significantly after the number of measurements becomes greater than 8.
7. A weighting matrix may be used in the cost function (8.4.8). By selecting different values of the elements in the weighting matrix, one can emphasize the position estimation accuracy over the orientation estimation accuracy or vice versa. It should also be mentioned that in the simulations, metric units were actually used for translation parameters and position measurements. Using different units will produce different estimation results.

VI.   DISCUSSION AND REFERENCES

A number of approaches have been proposed in the literature for hand/eye calibration. Shiu and Ahmad (1989) proposed moving the robot twice and observing the resulting sensor motion. That is, the robot is moved by a known amount and the resulting motion is measured. They proposed taking two or more robot motions to obtain a unique solution of the hand/eye transformation under certain conditions. The size of the coefficient matrix in their formulation is, however, unneccesarily large.

Tsai and Lenz (1989) solved the same problem independently using a more efficient linear algorithm. The number of unknowns in their method stays the same no matter how many measurements are used simultaneously. Simulation and experiments were reported and analyzed to show the efficiency and accuracy of the algorithm. Their paper is highly insightful.

Chou and Kamel (1988) presented a solution based on quaternions. In this approach, a system of nonlinear equations is iteratively solved using the Newton-Raphson procedure. They later improved the approach by transforming equation (8.1.3) into a system of linear equations and the derivation of a closed-form solution to the system of linear equations using the generalized inverse method with singular-value decomposition analysis.

Zhuang and Roth (1991b) solved the rotational equations using a linear approach that was also based on quaternions. Although quaternions were applied to obtain a unique solution of R_x, the actual results were presented without referring to quaternions, resulting in an algorithm that is easy to understand and implement. This has been the basis for the material described in Section III.

The above mentioned methods have two common characteristics: (1) Their algorithms all consist of two stages. The two-stage algorithms may be optimal in each stage. However, these may not be globally optimal due to the coupling of the rotation and position equations. (2) These algorithms all require the complete knowledge of A_i and B_i. While A_i may be obtained with ease if the robot forward kinematics is relatively accurate, it is sometimes difficult to obtain complete information about B_i since it may not be practical to use a sensor measurement device which is capable of measuring both relative rotations and translations of the robot sensor with sufficient accuracy.

An old version of the one-stage iterative algorithm given in Section IV was first reported by Zhuang and Shiu (1993). Not only is this algorithm usually less sensitive to noise compared to the two-stage algorithms, it can handle the cases in which the robot sensor orientation information is not available. The calibration is performed using both rotation and position parts of equation (8.1.2) or using only the position part of the equations, depending on whether the orientation part of B_i is given or not. Unlike other calibration approaches, the one-stage algorithm obtains all kinematic parameters of X simultaneously, thus eliminating possible propagation errors and improving noise sensitivities. With this algorithm, the parameters of X are computable even without sensor orientation measurements. Comparative simulation studies show that the accuracy performance of the iterative algorithm is, in general, better than that of non-iterative two-stage algorithms, except when noise in sensor orientation measurements is negligible compared with that in sensor positional measurements. This version of the iterative algorithm was used in the simulation study.

The key component of the iterative algorithm is an Identification Jacobian relating measurement residuals to pose error parameters of the unknown hand/eye transformation. The Jacobian matrix given by Zhuang and Shiu (1993) is, however, not well-structured in the sense that the dimension of the Jacobian matrix is unnecessarily large. For instance, that Jacobian is a 24x6 matrix when two pose measurements are used. Due to the complexity of the Jacobian, conditions for the observability of the error parameters, defined in terms of the rank of the Jacobian matrix, are difficult to derive.

A more compact Identification Jacobian was presented by Zhuang and Qu (1994). The derivation is straightforward and simple, owing to the alternative mathematical formulation of the hand/eye calibration problem given in Section III.A. The new Jacobian had a much lower dimension. The Jacobian is a 12x6 matrix when two pose measurements are used. Observability conditions of the pose error parameters in the unknown hand/eye transformation are also easy to prove using this new Jacobian. The material given in Section IV is mainly based on Zhuang and Qu (1994).

As a by-product, the Identification Jacobian given in Section III.B can be used for optimal selection of robot measurement configurations. This issue will be discussed in detail in Chapter 11.

The problem of determining position and orientation of a tool mounted on a robot manipulator is similar to that of hand/eye calibration. Normally, the tool is mounted on the flange of the robot arm. The flange coordinate frame can be treated as the robot hand frame. The transformation from the robot base frame to the flange frame is given by the robot forward kinematic model. The transformation from the flange frame to the tool frame, which is established at a convenient location on the tool, may be obtained from the designer's engineering drawing. Due to mechanical tolerance, this nominal transformation may not be sufficiently accurate. Tool calibration is required in such a case. Identification of the tool location in the robot flange frame is similar to robot hand/eye calibration, except that a tool is usually passive. An external measuring device is needed to acquire poses of the tool in different robot measurement configurations. Commonly used measuring systems such as coordinate measuring machines, theodolites and laser tracking systems may only provide position information of the tool. The approach proposed in this chapter is thus a viable technique for robot tool calibration.

Chapter 9

ROBOTIC BASE CALIBRATION

I. INTRODUCTION

The transformations of a robotic system that most frequently need to be calibrated are the base and tool transformations. This chapter deals with the calibration of base transformation.

Techniques for base calibration can be classified into those using pose measurements and those using only position measurements. Readers are referred to Section 6.D for the calibration algorithm of the base transformation when full poses of the end-effector are measured. Metrology practitioners generally agree that orientation measurements are expensive to obtain. This chapter presents a number of methods that can effectively compute the base transformation whenever only end-effector positions are available.

II. PROBLEM STATEMENT

Let $\{W\}$, $\{B\}$ and $\{T\}$ be the robot world, base and tool coordinate frames, respectively. The world coordinate frame is a reference frame assigned at any convenient location. For example, when a Coordinate Measuring Machine (CMM) is used to measure the pose of the robot, the CMM axes can be used to define the world coordinates. The base frame is assigned somewhere inside the base of the robot. For instance, for the PUMA 560, the base frame origin is usually located at the intersection of the first and second joint axes. The tool frame is normally attached to the end-effector.

Figure 9.2.1 illustrates an example of such coordinate frame assignments. It is assumed that the homogeneous transformation ^bT_t from the base frame to the tool frame is known with sufficient accuracy. This implies that, before the robot base is localized in the world coordinate fram, the accuracy of the robot itself is much higher than that of the robot with respect to its environment.

It is assumed that only point measurements ^wp_t of the end-effector are provided. Let ^wT_b and ^wT_t be the homogeneous transformations relating the base to the world frames and the tool to the world frames, respectively. Let ^wR_t and ^wp_t be the rotation matrix and the position vector of ^wT_t, respectively. Similarly for ^wR_b, ^wp_b, ^bR_t and ^bp_t. The following relationship then holds:

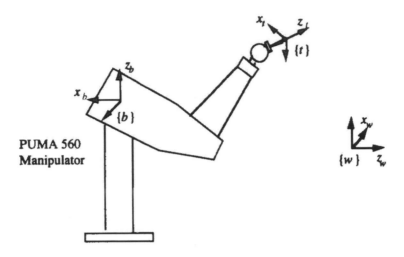

Fig. 9.2.1 An example of link frame assignment

$$^{w}p_{t} = {}^{w}R_{b}{}^{b}p_{t} + {}^{w}p_{b}$$

The vector $^{b}p_{t}$ is a function of the robot joint variables and link parameters. What is needed is to do is to estimate $^{w}R_{b}$ and $^{w}p_{b}$ provided that $^{b}p_{t}$ and $^{w}p_{t}$ are given at a number of robot configurations. The position vector $^{w}p_{t}$ is measured by an end-effector position measuring device and $^{b}p_{t}$ is computed by using the robot forward kinematic solver. Since the robot link parameters are part of in the forward kinematics, the robot internal geometry must be known accurately.

More generally, we can define the following calibration problem: Let the given 3D data points $^{1}p_{i}$ and $^{2}p_{i}$, $i = 1, 2, ..., m$, be represented in two coordinate frames. Estimate the rotation R and translation t that transform $^{1}p_{i}$ to $^{2}p_{i}$; i.e.,

$$^{2}p_{i} = R\,^{1}p_{i} + t \quad i = 1, 2, ..., m. \tag{9.2.1}$$

In the robotics literature, this problem is also referred to as the *robot localization problem*.

In the linear approaches to be discussed in this chapter, the unknown rotation is determined first from the data points, and is then used together with the data points to determine the unknown translation.

III. ESTIMATION OF THE BASE ORIENTATION

There are a few established techniques for solving this problem. For a comprehensive treatment of this subject, readers are referred to (Kanatani (1994)). In this section, we provide a detailed description of a quaternion-based method and summarize briefly two other methods.

A. QUATERNION-BASED ALGORITHMS
From (9.2.1),

$$R \, {}^1p = {}^2p - t \tag{9.3.1}$$

where the subscript i has been temporarily dropped.
   Recall that (refer to (8.2.5)),

$$R \, {}^1p = q \circ {}^1p \circ q^* \tag{9.3.2}$$

where $R \, {}^1p$ is treated as a pure quaternion. Substituting (9.3.2) into (9.3.1) yields

$$q \circ {}^1p - ({}^2p - t) \circ q = 0 \tag{9.3.3}$$

where $\circ$ denotes a quaternion product. Expanding (9.3.3) provides

$$(q \bullet {}^1p, \; q \, {}^1p - {}^1p \times q) = (q \bullet ({}^2p - t), \; q \, ({}^2p - t) - q \times ({}^2p - t)) \tag{9.3.4}$$

Equation (9.3.4) is equivalent to

$$({}^1p + ({}^2p - t)) \times q = q \, ({}^1p - ({}^2p - t)) \tag{9.3.5}$$

because the scalar part $({}^1p - ({}^2p - t)) \bullet q = 0$ is implied by (9.3.5). However for R to be a rotation matrix, $q = \sin(\theta/2)k$ and $q = \cos(\theta/2)$, where k and θ are the rotation axis and angle of R, respectively. Substituting these relationships into (9.3.5) yields

$$\tan(\theta/2)({}^2p - t + {}^1p) \times k = {}^1p - {}^2p + t \tag{9.3.6}$$

Let $\Omega(v)$ denote a skew symmetric matrix which corresponds to $v \equiv [\, v_x, \, v_y, \, v_z]^T$,

$$\Omega(v) \equiv \begin{bmatrix} 0 & -v_z & v_y \\ v_z & 0 & -v_x \\ -v_y & v_x & 0 \end{bmatrix}$$

Equation (9.3.6) can be rewritten as

$$\Omega(^2p + {}^1p)x - \Omega(t)x - t = {}^1p - {}^2p \qquad (9.3.7)$$

where

$$x = \tan(\theta/2)\,k \qquad (9.3.8)$$

$$y = -\Omega(t)x - t . \qquad (9.3.9)$$

Equation (9.3.7) can then be written as

$$\Omega(^2p + {}^1p)x + y = {}^1p - {}^2p \qquad (9.3.10)$$

Let us assume that m measurements are available,

$$\Omega(^2p_i + {}^1p_i)x + y = {}^1p_i - {}^2p_i \quad i = 1, 2,..., m \qquad (9.3.11)$$

Eliminating y from the above equations yields

$$\Omega(a_i + b_i)x = a_i - b_i \quad i = 1, 2,..., m\text{-}1 \qquad (9.3.12)$$

where $a_i \equiv {}^1p_i - {}^1p_{i+1}$, and $b_i \equiv {}^2p_i - {}^2p_{i+1}$. A minimum of three measurements is needed to solve for x from a system of equations of the form (9.3.12). Define

$$A = \begin{bmatrix} \Omega(a_1 + b_1) \\ \Omega(a_2 + b_2) \end{bmatrix} \qquad (9.3.13)$$

and

$$b = \begin{bmatrix} a_1 - b_1 \\ a_2 - b_2 \end{bmatrix} \qquad (9.3.14)$$

where A is a 6x3 matrix, and b is a 6x1 vector.   Two equations of the form (9.3.12) can then be written as

$$Ax = b \qquad\qquad (9.3.15)$$

Remark: A weight matrix W may be introduced in this stage to account for measurement noise based on prior knowledge of the measurement system. That is, we may define a cost function E, where

$$E = (Ax = b)^T W (Ax = b)$$

and x may be solved by a weighted least squares procedure or a best linear unbiased estimate algorithm if W is the noise covariance matrix. The above formulation is readily amenable to an efficient recursive procedure. For simplicity, however, we assume that $W = I$ in the following derivation.

Lemma 9.3.1: Assume that $\theta \neq \pi$. The equivalent condition for

$$a_1 \times a_2 \neq 0 \qquad\qquad (9.3.16)$$

is that the rank of the following matrix is three,

$$\begin{bmatrix} \Omega(a_1 + b_1) \\ \Omega(a_2 + b_2) \end{bmatrix}$$

Proof: The above matrix being full rank is equivalent to the following condition,

$$(a_1 + b_1) \times (a_2 + b_2) \neq 0 \qquad\qquad (9.3.17)$$

However, from (9.2.1)

$$Ra_i = b_i \quad i = 1, 2, \dots, m\text{-}1 \qquad\qquad (9.3.18)$$

Substituting (9.3.18) into (9.3.17) yields

$$((I + R)a_1) \times ((I + R)a_2) \neq 0 \qquad\qquad (9.3.19)$$

The eigenvalues of $I + R$ are $\{2, 1 + \cos\theta + j \sin\theta, 1 + \cos\theta - j \sin\theta\}$. Therefore if $\theta \neq \pi$, Equation (9.3.19) is equivalent to (9.3.16). This completes the proof of the lemma. $\qquad\qquad\qquad$ []

Remark: Equation (9.3.16) can be easily extended to accommodate the case in which $m > 3$.

After x is solved from Equation (9.3.15) by a least squares procedure, k and θ can be obtained by

$$k = x/\|x\| \tag{9.3.20a}$$
$$\theta = 2\operatorname{atan}(x_{max}/k_{max}) \tag{9.3.20b}$$

Another algorithm for rotation estimation that utilizes quaternions is as follows:

Let us first compute the 3x3 correlation matrix

$$K \equiv \sum_{i=1}^{n-1} w_i a_i b_i^T \tag{9.3.21}$$

where w_i is a positive weight for the ith datum.

Using the elements of K, we define a 4x4 symmetric matrix L as

$$L = \begin{bmatrix} K_{11} + K_{22} + K_{33} & K_{32} - K_{23} & K_{13} - K_{31} & K_{21} - K_{12} \\ K_{32} - K_{23} & K_{11} - K_{22} - K_{33} & K_{12} + K_{21} & K_{31} + K_{13} \\ K_{13} - K_{31} & K_{12} + K_{21} & -K_{11} + K_{22} - K_{33} & K_{23} + K_{32} \\ K_{21} - K_{12} & K_{31} + K_{13} & K_{23} + K_{32} & -K_{11} - K_{22} + K_{33} \end{bmatrix}$$

Then the unit quaternion $q = (q, q^T)^T$ that corresponds to the unknown base rotation matrix is the eigenvector of L that corresponds to the largest eigenvalue. Furthermore,

$$q = \sin(\theta/2)k \tag{9.3.22a}$$
$$q = \cos(\theta/2) \tag{9.3.22b}$$

where again $k \equiv [k_x, k_y, k_z]^T$ and θ are the rotation axis and rotation angle of R, respectively. Once k and θ are solved from (9.3.22), the rotation matrix R is then computed from the following form,

$$R = \begin{bmatrix} k_x^2 v\theta + c\theta & k_x k_y v\theta - k_z s\theta & k_x k_z v\theta + k_y s\theta \\ k_x k_y v\theta + k_z s\theta & k_y^2 v\theta + c\theta & k_y k_z v\theta - k_x s\theta \\ k_x k_z v\theta - k_y s\theta & k_y k_z v\theta + k_x s\theta & k_z^2 v\theta + c\theta \end{bmatrix} \qquad (9.3.23)$$

where $v\theta \equiv 1 - cos(\theta)$, $c\theta \equiv cos(\theta)$, and $s\theta \equiv sin(\theta)$.

B. SVD-BASED METHOD

If UDV^T is a singular value decomposition of K given in (9.3.21), then a least-squares fit of the rotation matrix R can be shown to be

$$R = U \, \text{diag}(1, 1, \det(UV^T)) \, V^T \qquad (9.3.24)$$

The solution is unique if $\text{rank}(K) > 1$ and $\det(UV^T) = 1$, or if $\text{rank}(K) > 1$ and the minimum singular value is a simple root.

IV. ESTIMATION OF THE BASE POSITION

After the rotation matrix R is estimated using one of the methods given in Section III, the translation vector t can be solved from (9.2.1) as follows,

$$t = {}^2p_i - R \, {}^1p_i \qquad i = 1, 2, ..., m.$$

The least squares solution for t is then

$$t = \frac{1}{m} \sum_{i=1}^{m} w_i \left[{}^2p_i - R \, {}^1p_i \right] \qquad (9.4.1)$$

where w_i is a positive weight for the ith datum. Note that the weights in (9.4.1) and in (9.3.21) need not be the same.

Another way of computing the translation vector t is to use (9.3.9) without reuse of the measurement data. More specifically, (9.3.9) can be rewritten as

$$(\Omega(x) - I)t = y \qquad (9.4.2)$$

Lemma 9.4.1: The solution of (9.4.2) exists and is unique.

Proof: By using the symbolic manipulation package $MACSYMA^{®}$ it can be shown that the determinant of $I - \Omega(x)$ is $(x_3^2 + 1)$, where x_3 is the third element of x. Thus $I - \Omega(x)$ is full rank. Consequently, t can be uniquely determined from (9.4.2). This proves the lemma.   []

V. EXPERIMENTAL RESULTS

The objective of the experiment was to compare the PUMA 560 base calibration results obtained from the above algorithms. Even though the data was collected by a coordinate measuring machine (CMM) many observations and conclusion are still applicable for vision-aided calibration experimentation.

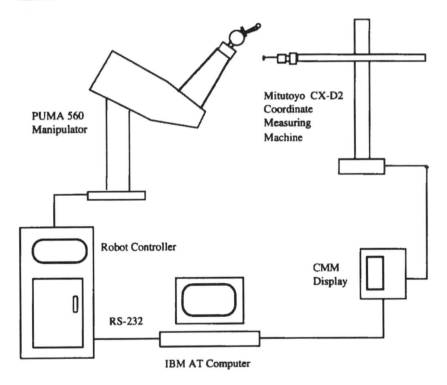

Figure 9.5.1 Schematic of equipment set-up for PUMA 560 base calibration experiments

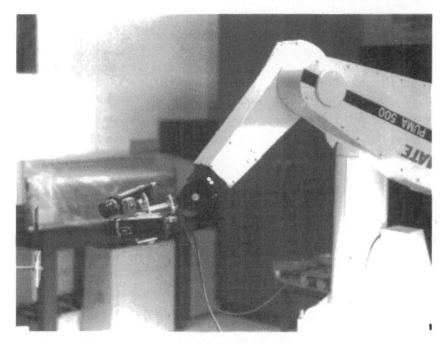

Figure 9.5.2 Actual experiment set-up for PUMA 560 base localization

The experimental system (Figure 9.5.1) consisted of a Puma 560 robot, a Mitutoyo Model CX-D2 CMM and an IBM PC-AT computer. The CMM has a rated repeatability of 0.01 *mm* and a work volume of 400x500x800 *mm*^{3}; refer to Figure 9.5.2 for the hardware setup.

The robot base coordinate frame was placed at the intersection of the first and second joint axes of the PUMA arm. The tool coordinate frame was placed at the tooling ball center of the end-effector. The CMM position indicators were set to zero after driving each axis to its travel limit. This established the coordinate frame for the CMM, which was then defined as the world coordinate frame for the measurement process.

The schematic geometry of the PUMA arm is shown in Figure 9.2.1. The nominal D-H link parameters of A_1, A_2, ..., A_5, and A_6, were obtained from the engineering drawings of the PUMA arm.

A set of 100 position measurement points within the intersection of work volumes of the PUMA and the CMM, from both left- and right-arm configurations of the PUMA robot, was collected in the experiment, among which 50 measurements were used for identification purposes and the other 50 positions were used for calibration verification. The algorithms given in Sections III and IV were used to estimate the parameters of the system. All algorithms produced identical results. Table 9.5.1 present the verification results. Listed are the mean, maximum and standard deviation of $\|p - p^p\|$ against the number of measurements, where p and p^p are the measured and the predicted end-effector positions, respectively. Only those results that use up to 10 measurements are listed as no significant improvement was observed by using more measurements.

It can be seen that by carefully identifying the parameters in the transformation ^wT_b alone, the mean positioning errors can be reduced to about 1.0 mm. Any further accuracy improvement requires the calibration of the internal robot structure.

Table 9.5.1 The position errors using the identified ^wT_b
together with the nominal robot link parameters (unit: *mm*)

# measurement	Mean Error	Maximum Error	STD
3	0.932	1.919	0.555
4	0.763	1.556	0.385
5	0.693	1.449	0.327
6	0.668	1.388	0.290
7	0.680	1.180	0.312
8	0.679	1.023	0.290
9	0.673	0.964	0.298
10	0.675	0.962	0.287

VI.   SUMMARY AND REFERENCES

This chapter reviewed a number of linear approaches for calibration of the transformation from robot world coordinate frame to the robot base coordinate frame.

In the literature, a linear least squares solution method for solving the robot localization problem by using Clifford algebra was proposed in Ravani and Ge (1991). Other linear solutions of the absolute orientation problem were proposed in computer vision literature (Horn (1987) and (Horn, Hilden and Negahdaripour (1988)). The quaternion method was proposed in Horn (1987) and was also discussed in (Umeyama (1991), Kanatani (1993) and Zhuang, Roth and Sudhakar (1991)). The singular value decomposition method is discussed in (Arun, Huang and Blostein (1987) and Kanatani (1993)). The D-H parameters of PUMA 560 were cited from Fu, Gonzalez and Lee (1987).

Chapter 10

SIMULTANEOUS CALIBRATION OF ROBOTIC BASE AND TOOL

I.   INTRODUCTION

Many robotic applications require accurate knowledge of the geometric inter-relationships among the robot, its sensors and a reference frame. Figure 10.1.1 illustrates schematically the typical geometry of a robotic cell. The world coordinate frame is an external reference frame, which is usually defined by the robot calibration measurement setup. The base coordinate frame is normally defined inside the robot structure. Its default location is often provided by the robot manufacturer; however users are allowed to redefine it. The flange coordinate frame is defined on the mounting surface of the robot end-effector. The tool frame is assigned at a convenient location within the end-effector. Formal definitions of the homogeneous transformations A, B, X and Y (shown in Figure 10.1.1) will be given in Section II.

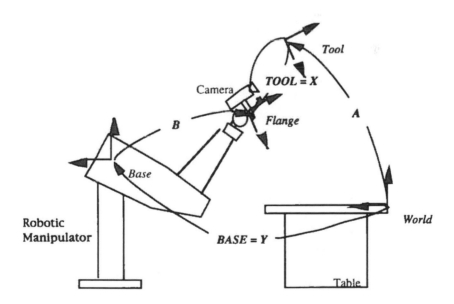

Figure 10.1.1 Geometry of a robotic system

In most practical situations the robot base to flange geometry needs not be calibrated as often as the other two transformations, namely the $BASE$ and $TOOL$ transformations.   That is, a complete manipulator calibration needs to be done relatively infrequently, for instance after a mechanical repair or scheduled maintenance. For a well calibrated manipulator, one may occasionally need to perform partial calibration whenever a tool is changed or whenever the robot world frame is redefined.   The most frequently used methods for such partial calibration actions are based on error models, which are special cases of the complete error models defined in Chapter 5.

The chapter presents a linear solution for the simultaneous identification of both the $BASE$ and $TOOL$ transformations. This problem is formulated as solving a system of homogeneous transformation equations of the form $A_i X = Y B_i$, $i = 1, 2, ..., m$, where m is the number of pose measurements.

This chapter is organized into the following sections.   In Section II, the identification problem of the robot-to-world and tool-to-flange transformations is formulated.   The linear solution to this problem is derived in Section III. Performance assessment of the method through simulations is presented in Section IV. The chapter ends with references and discussion.

II.   PROBLEM STATEMENT

For better readability we shall allow ourselves to be somewhat repetitious about coordinate frame definitions.   Let T_n be a 4x4 homogeneous transformation relating the position and orientation of the robot end-effector to the world coordinate frame.

$$T_n = A_0 A_1 .. A_{n-1} A_n \qquad (10.2.1)$$

Each joint is assumed to be ideal, either revolute or prismatic.   Frame $\{i\}$, a Cartesian coordinate frame $\{x_i, y_i, z_i\}$, $i = -1, 0,..., n$, is established for each link.   The world, base and tool frames are denoted as the -1th, 0th and nth link frames, respectively.

$BASE$, the transformation from the world to the base frames, is fixed and defined as $BASE \equiv A_0$. $FLANGE$, the transformation from the $(n-1)$th to the flange frames, contains the joint variable q_n while $TOOL$, the transformation from the flange to the tool frames, is fixed.   Thus $FLANGE \cdot TOOL \equiv A_n$. The robot kinematic model Equation (10.2.1) can thus be rewritten as follows:

$$T_n = BASE \cdot A_1 .. A_{n-1} \cdot FLANGE \cdot TOOL \qquad (10.2.2)$$

The nominal values of transformations A_1, ..., A_{n-1} and $FLANGE$ can be obtained from the design specifications of the robot. Their actual values can be estimated using a complete robot calibration process. It is assumed in this chapter that these transformations are known with sufficient precision. The $BASE$ and $TOOL$ transformations are application-dependent, and should be recalibrated more frequently. Since a homogeneous transformation is always invertible, let (referring to Figure 10.1.1)

$$A \equiv T_n \tag{10.2.3}$$

$$B \equiv A_1..A_{n-1} \bullet FLANGE \tag{10.2.4}$$

$$X \equiv TOOL^{-1} \tag{10.2.5}$$

$$Y \equiv BASE \tag{10.2.6}$$

Then by Equation (10.2.2), the following homogeneous transformation equation is obtained:

$$AX = YB \tag{10.2.7}$$

The homogeneous transformation A is known from end-effector pose measurements, B is computed using the calibrated manipulator internal-link forward kinematics, X is the inverse of the unknown $TOOL$ transformation, and Y is the unknown $BASE$ transformation.

The problem is now reduced to that of solving a system of matrix equations of the type of Equation (10.2.7). That is, X and Y are to be determined uniquely from a set of equations of the type (10.2.7), given A_i and B_i, which are obtained at different measurement configurations.

III. A LINEAR SOLUTION

Readers who folowed closely the derivations in Chapters 8 and 9 surely begin by now to see the general pattern of the quaternion-based method. This chapter continues the same path.

Let R_A, R_X, R_Y and R_B be the respective 3×3 upper-left block rotation sub-matrices of A, X, Y and B, and let p_A, p_X, p_Y and p_B be the respective 3×1 translational vectors obtained from the columns of these matrices. Equation (10.2.7) can then be decomposed into a rotation equation (10.3.1) and a position equation (10.3.2),

$$R_A R_X = R_Y R_B \tag{10.3.1}$$

$$R_A p_X + p_A = R_Y p_B + p_Y \tag{10.3.2}$$

Equation (10.3.2) is a linear vector equation in p_X and p_Y if R_Y is known. Upon obtaining R_X and R_Y, p_X and p_Y may be solved linearly from a set of equations of the type Equation (10.3.2). We thus first focus on the linear solution of the rotation equation (10.3.1).

A. SOLUTION FOR ROTATION MATRICES

Readers are referred to Section 8.II for background materials on Quaternion Algebra.

Let $a \equiv (a_0, a)$, $b \equiv (b_0, b)$, $x \equiv (x_0, x)$ and $y \equiv (y_0, y)$ be unit quaternions that correspond to the rotation matrices R_A, R_B, R_X and R_Y. Note that a_0 and a are the scalar and 3×1 vector parts of a. Similarly for the other unit quaternions. Let u_j and b_j be the jth column vector of the identity matrix and the B matrix, respectively. The vectors correspond to the respective pure quaternions u_j and b_j. Let $\circ$ denote a quaternion product. Finally, for any rotation matrix R, let $R = Rot(k, \theta)$, where k and θ are respectively the rotation axis and rotation angle.

Lemma 10.3.1:   The rotation equation $R_A R_X = R_Y R_B$ is equivalent to the following quaternions equation

$$a \circ x = y \circ b \qquad (10.3.3)$$

This fact is a straightforward extension of some of the results given in Chapter 8.

Define a 6×1 vector w and a 3×1 vector c as follows

$$w \equiv [x'/y_0, \ y'/y_0]^t \qquad (10.3.4)$$

$$c \equiv b - (b_0/a_0)a \qquad (10.3.5)$$

and a 3×6 matrix G as

$$G \equiv [\ a_0 I + \Omega(a) + aa^t/a_0, \ -b_0 I + \Omega(b) - ab^t/a_0] \qquad (10.3.6)$$

Lemma 10.3.2:   If $\theta_A \neq \pi$ and $\theta_Y \neq \pi$, then $R_A R_X = R_Y R_B$ is equivalent to the vector equation

$$Gw = c \qquad (10.3.7)$$

Proof: Expanding Equation (10.3.3) using quaternion products yields

$$(a_0 x_0 - a.x, \ a_0 x + x_0 b + a \times x) = (b_0 y_0 - b.y, \ b_0 y + y_0 b - b \times y) \quad (10.3.8)$$

If $\theta_A \neq \pi$, then $a_0 \neq 0$. x_0 can be solved from the scalar part of Equation (10.3.8),

$$x_0 = (a/a_0).x + (b_0/a_0)y - (b/a_0).y \quad (10.3.9)$$

Substituting Equation (10.3.9) into the vector part of Equation (10.3.8) yields

$$[(b_0/a_0)a - b]y_0 + a_0 x + a(a/a_0).x - a(b/a_0).y - b_0 y + a \times x + b \times y = 0 \quad (10.3.10)$$

If $\theta_Y \neq \pi$, then $y_0 \neq 0$. Dividing both sides of Equation (10.3.10) by y_0 yields

$$(a_0 I + \Omega(a) + aa^t/a_0)x/y_0 + (-b_0 I + \Omega(b) - ab^t/a_0)y/y_0 = b - (b_0/a_0)a \quad (10.3.11)$$

where use has been made of the relationship $v \times x = \Omega(v)x$ for $v = a$ or b. Finally, by the definitions of G, w and c, Equation (10.3.7) is obtained.   []

We now examine under what conditions $\theta_A = \pi$ occurs. Let

$$R_A \equiv \begin{bmatrix} n_x & o_x & a_x \\ n_y & o_y & a_y \\ n_z & o_z & a_z \end{bmatrix}$$

According to (Paul, 1981),

$$2\cos\theta_A = n_x + o_y + a_z - 1 \quad (10.3.12a)$$

and

$$4\sin^2\theta_A = (o_z - a_y)^2 + (a_x - n_z)^2 + (n_y - o_x)^2 \quad (10.3.12b)$$

Therefore, $\theta_A = \pi$ is equivalent to the following simultaneous set of conditions,

$$n_x + o_y + a_z = -1 \quad (10.3.13a)$$
$$o_z - a_y = 0 \quad (10.3.13b)$$
$$a_x - n_z = 0 \quad (10.3.13c)$$
$$n_y - o_x = 0 \quad (10.3.13d)$$

These conditions can be checked for each measurement configuration, and if satisfied this pose shall be avoided.

A similar analysis can be done to identify the conditions under which $\theta_Y = \pi$ occurs. To avoid these singular conditions, care has to be taken when placing either the base frame or the world frame.

Example: A rotation R_A that satisfies the conditions given in Equation (10.3.13) is shown below:

$$R_A \equiv \begin{bmatrix} -1 & 0 & 0 \\ 0 & 0 & 1 \\ 0 & 1 & 0 \end{bmatrix}$$

In this example, the tool frame can be obtained by rotating the world frame 180 degrees about the axis $[0, 0.707, 0.707]^T$.

Equation (10.3.7) consists of 3 scalar equations with 6 unknowns. A unique solution of w therefore requires multiple measurements. Let

$$G w = C \qquad (10.3.14)$$

where G (a (3m)x6 matrix) and C (a (3m)×1 vector) are defined as

$$G \equiv \begin{bmatrix} G_1 \\ G_2 \\ \vdots \\ G_m \end{bmatrix} \qquad (10.3.15a)$$

$$C \equiv \begin{bmatrix} c_1 \\ c_2 \\ \vdots \\ c_m \end{bmatrix} \qquad (10.3.15b)$$

In Equation (10.3.15), G_i and c_i have, respectively, the same structure as G and c in Equations (10.3.6) and (10.3.5). The SVD algorithm can be applied to solve for w. After $w \equiv [w_1 \ w_2 \ w_3 \ w_4 \ w_5 \ w_6]^t$ is obtained, x and y can be determined from Equation (10.3.4) using the constraints $y\, y^* = x\, x^* = 1$ and Equation (10.3.9). More specifically,

$$y_0 = \pm \{1 + w_4^2 + w_5^2 + w_6^2\}^{-1/2} \qquad\qquad (10.3.16a)$$

$$y = y_0[w_4 \ w_5 \ w_6]^t \qquad\qquad (10.3.16b)$$

$$x = y_0[w_1 \ w_2 \ w_3]^t \qquad\qquad (10.3.16c)$$

$$x_0 = \pm \{1 - x_1^2 - x_2^2 - x_3^2\}^{-1/2}. \qquad\qquad (10.3.16d)$$

Since $\pm q$, where q is an arbitrary quaternion, corresponds to the same rotation matrix R_q, one needs only to determine the sign of x_0 by using Equation (10.3.9).

Before proving that at least three measurements are required to obtain a unique solution, the following simplifications are made:

Assume that there are three rotation equations,

$$R_{Ai}R_X = R_Y R_{Bi} \qquad i = 1, 2, 3 \qquad\qquad (10.3.17)$$

Let $R_X' = R_{A1}R_X$, and $R_Y' = R_Y R_{B1}$. Then Equation (10.3.18) given next is equivalent to Equation (10.3.17) since once R_X' and R_Y' are obtained, R_X and R_Y can be determined uniquely.

$$R_X' = R_Y' \qquad\qquad (10.3.18a)$$

$$R_{Ai}'R_X' = R_Y'R_{Bi}' \qquad i = 2, 3 \qquad\qquad (10.3.18b)$$

where $R_{Ai}' \equiv R_{Ai}R_{A1}^t$, and $R_{Bi}' \equiv R_{Bi}'R_{B1}^t$. By Equation (10.3.18a), R_{A2}' and R_{B2}' in Equation (10.3.18b) must have the same rotation angle (Lemma 8.2.1, Chapter 8 (?)).

Define

$$R_{Ai}'R_{A1} = Rot(k_{Ai}', \beta_i) \qquad i = 2, 3 \qquad\qquad (10.3.19a)$$

$$R_{Bi}'R_{Bi} = Rot(k_{Bi}', \beta_i) \cdot \qquad i = 2, 3 \qquad\qquad (10.3.19b)$$

β_i, termed the ith *relative rotation angle*, plays a role in determining the uniqueness of the solution for both the rotation matrices and the position vectors.

Theorem 10.3.1:   A necessary condition for unique identification of R_X and R_Y from Equation (10.3.14) is that $m \geq 3$, where m is the number of pose measurements.

Proof: Let $m = 2$. By Equation (10.3.18), the 6x6 matrix G in Equation (10.3.15a) is reduced to

$$G \equiv \begin{bmatrix} I_{3 \times 3} & -I_{3 \times 3} \\ G_{A2} & G_{B2} \end{bmatrix}$$ (10.3.20)

where $[G_{A2} \; G_{B2}] \equiv G_2$. By elementary matrix transformations, G in Equation (10.3.20) can be reduced to

$$G \equiv \begin{bmatrix} I_{3 \times 3} & 0_{3 \times 3} \\ 0_{3 \times 3} & G_{A2} + G_{B2} \end{bmatrix}$$ (10.3.21)

A necessary and sufficient condition for Equation (10.3.12) to have a unique solution is that $Rank(G) = 6$, or equivalently, $Rank(G_{A2} + G_{B2}) = 3$. By using the software package MACSYMA®, $G_{A2} + G_{B2}$ can be triangularized and simplified. The resulting three diagonal elements after simplification are $\{- k_{B2,1}(k_{B2,1} - k_{A2,1})sin^2(\beta_2), - (k_{B2,3} + k_{A2,3})((k_{A2,1}k_{B2,2} - k_{A2,2}k_{B2,1})sin(\beta_2) - (k_{A2,3} + k_{B2,3})cos(\beta_2), 0\}$, where $[k_{A2,1} \; k_{A2,2} \; k_{A2,3}]^t \equiv k_{A2}'$ and $[k_{B2,1} \; k_{B2,2} \; k_{B2,3}]^t \equiv k_{B2}'$. Since the third diagonal element of the reduced matrix is zero, $Rank(G_{A2} + G_{B2}) < 3$. Consequently for a unique solution of w it is necessary that $m > 2$. ∎

Three pose measurements are not always sufficient for obtaining a unique solution of Equation (10.3.14), as can be seen in the following theorem.

Theorem 10.3.2: Let $m = 3$. If the axes of rotation for R_{A2}' and R_{A3}' are neither parallel nor antiparallel one to another and the relative rotation angles β_2 and β_3 are both neither zero nor π, then there exists a unique solution of R_X and R_Y.

Proof: Substituting Equation (10.3.18a) into Equation (10.3.18b) yields

$$R_{Ai}'R_X' = R_X'R_{Bi}'$$ $i = 2, 3$

The problem is now reduced to that of a robot hand/eye calibration, which has been solved in Chapter 8. After R_X' is obtained, R_X and R_Y can be computed from $R_X' \neq R_{A1}R_X$ and $R_X' \neq R_Y R_{B1}$, respectively. ∎

As seen in the above theorem, three pose measurements yield only two relative measurement equations.

Theorem 10.3.2 also implies that the orientation of the robot end-effector has to be changed from one measurement configuration to another to guarantee the existence of solution for the unknown rotation matrices. The conditions that the relative angle $\beta_i = \pi$ occurs are similar to those given in Equation (10.3.13), with the modification that the rotation R_A is replaced by the relative rotation $R_{Ai}^t R_{A1}$.

B. SOLUTION FOR POSITION VECTORS

Following the solution of R_x and R_y, p_X and p_Y become readily computable. Equation (10.3.2) can be rewritten as

$$F \begin{bmatrix} p_X \\ p_Y \end{bmatrix} = d \qquad (10.3.23)$$

where F (a 3x6 matrix) and d (a 3x1 vector) are defined as

$$F \equiv [R_A, \quad -I_{3x3}] \qquad (10.3.24a)$$

$$d \equiv R_Y p_B - p_A \qquad (10.3.24b)$$

If more than one measurement is provided, let

$$F \begin{bmatrix} p_X \\ p_Y \end{bmatrix} = D \qquad (10.3.25)$$

where F (a $(3m)$x6 matrix) and D (a $(3m)$x1 vector) are defined similarly to the definitions given in Equation (10.3.16). Again the Singular Value Decomposition algorithm can be applied to solve for p_Y and p_B.

It will be shown that at least 3 measurements have to be used to have a unique solution of Equation (10.3.25). Recall that R_{Ai} is the ith measurement of R_A, and

$$R_{Ai}^t R_{A1} = Rot(k_{Ai}', \beta_i) \qquad i = 2, 3$$

$Rot(k_{Ai}', \beta_i)$ can be decomposed into [2]

$$Rot(k_{Ai}', \beta_i) = E_i H_i E_i^{-1} \quad i = 2, 3 \qquad (10.3.26)$$

where

$$H_i \equiv \begin{bmatrix} 1 & 0 & 0 \\ 0 & e^{j\beta_i} & 0 \\ 0 & 0 & e^{-j\beta_i} \end{bmatrix} \qquad (10.3.27)$$

and E_i is an orthogonal matrix whose columns consist of the eigenvectors of $R(k_{Ai}', \beta_i)$. Furthermore,

$$E_i^{-1} = E_i' \quad i = 2, 3 \tag{10.3.28}$$

after the eigenvectors are normalized.

Theorem 10.3.3: Given R_X and R_Y, the necessary and sufficient conditions for a unique solution of p_X and p_Y from Equation (10.3.25) is that

(i) $m = 3$; and
(ii) $R_{Ai} \neq R_{Aj}$ for i, j $\in$ {1, 2, 3} and i $\neq$ j.

Proof: We note that since all rotations R_{Ai} must be different from one another, $\beta_i \neq 0$ and $\beta_i \neq \beta_j$ for i, j $\in$ {2, 3} and i $\neq$ j. This is because β_i is a relative rotation angle. If β_i equals zero, then the ith measurement is identical to the first measurement, and the number of measurements is reduced. Similarly for the case that $\beta_i = \beta_j$.
    Let $m = 3$,

$$F \equiv \begin{bmatrix} R_{A1} & -I_{3\times3} \\ R_{A2} & -I_{3\times3} \\ R_{A3} & -I_{3\times3} \end{bmatrix} \tag{10.3.29}$$

If rank$(F) = 6$, a unique solution can be obtained since Equation (10.3.24) is consistent in the absence of measurement noise. Following a sequence of elementary transformations, F can be written as

$$F = P F_1 Q \tag{10.3.30}$$

where P is a 9x9 nonsingular matrix and Q is a 6x6 nonsingular matrix, and

$$F_1 \equiv \begin{bmatrix} 0_{3\times3} & -I_{3\times3} \\ (Rot(k_{A2}', \beta_2) - I)R_{A1} & 0_{3\times3} \\ (Rot(k_{A3}', \beta_3) - I)R_{A1} & 0_{3\times3} \end{bmatrix} \tag{10.3.31}$$

where $R(k_{Ai}', \beta_i)$ is defined in Equation (10.3.26). Since both P and Q are full rank, the solution condition is reduced to rank$(F_1) = 6$. By Equation (10.3.26), F_1 can be further decomposed as follows

$$
F_1 = \begin{bmatrix} I_{3\times3} & 0_{3\times3} & 0_{3\times3} \\ 0_{3\times3} & E_2 & 0_{3\times3} \\ 0_{3\times3} & 0_{3\times3} & E_3 \end{bmatrix} \begin{bmatrix} 0_{3\times3} \\ (H_2 - I)E_2^t \\ (H_3 - I)E_3^t \end{bmatrix} \begin{bmatrix} -I_{3\times3} \\ 0_{3\times3} \\ 0_{3\times3} \end{bmatrix} \begin{bmatrix} I_{3\times3} & 0_{3\times3} \\ 0_{3\times3} & R_{A1} \end{bmatrix}
$$

$$
\equiv F_{11} F_{12} F_{13} \tag{10.3.32}
$$

$Rank(F_1) = Rank(F_{12})$ since F_{11} and F_{13} are full rank. Let $E_i \equiv [e_{i1}\ e_{i2}\ e_{i3}]$. Then

$$
(I - H_i)E_i^t = \begin{bmatrix} 0_{1\times3} \\ (1 - e^{j\beta_i})e_{i2}^t \\ (1 - e^{j\beta_i})e_{i3}^t \end{bmatrix} \tag{10.3.33}
$$

for $i = 2, 3$. e_{32} and e_{33} are not simultaneous linear combinations of e_{22} and e_{23} if $\beta_2 \neq \beta_3$ since e_{32} and e_{33} do not simultaneously lie on the plane spanned by e_{22} and e_{23}. The coefficients $1 - e^{j\beta_i} \neq 0$ and $1 - e^{j\beta_i} \neq 0$ for $i = 2, 3$ if $\beta_2 \neq 0$ and $\beta_3 \neq 0$. Thus there are three independent vectors among $(1 - e^{j\beta_2})e_{22}$, $(1 - e^{j\beta_2})e_{23}$, $(1 - e^{j\beta_3})e_{32}$, and $(1 - e^{j\beta_3})e_{33}$ if $\beta_2 \neq 0$, $\beta_3 \neq 0$, and $\beta_i \neq \beta_j$. Consequently $Rank(F_{12}) = 6$, and a unique solution can be determined by solving Equation (10.3.25).

The proof of the "necessary" part is trivial. If $m = 2$, then $Rank(F_{12}) \leq 5$ since $Rank((I - H_2)E_2^t) \leq 2$ by checking Equation (10.3.33). In this case, F is singular therefore there is no unique solution for p_X and p_Y. By the argument given in the proof of the sufficient condition, F is also singular if $R_{Ai} = R_{Aj}$ for any $i, j \in \{2, 3\}$ and $i \neq j$.   []

In summary, to obtain the homogeneous transformations X and Y from homogeneous transformation equations of the form $A X = Y B$, three or more different robot pose measurements are required. The rotation angles θ_{Ai} must not be equal to π when solving for rotation matrices R_X and R_Y from Equation (10.3.15). Furthermore, while changing the orientation of the robot end-effector, one should avoid moving the robot end-effector to certain poses specified by Equation (10.3.14).

The overall solution procedure is as follows:

1. Equation (10.3.15) is used to compute the vector w using measurement data.
2. Equation (10.3.17) is then used to compute x and y from w.
3. R_X and R_Y are recovered from x and y using Equation (8.2.6).
4. Equation (10.3.25) is used to compute p_X and p_Y using the measurement data as well as the computed R_X and R_Y.

IV.   SIMULATION STUDIES

Simulation studies were conducted to assess the performance of the solution method. When measurement noise was not present, it was found that the procedure discussed in Section II produces exact results if singular points were removed (that is, those points that correspond to $\theta_A \neq \pi$).

To investigate the robustness of the procedures in the presence of measurement noise, the nominal PUMA robot geometry was chosen for simulation using the Denavit-Hartenberg (D-H) parameters listed in Table 10.4.1.

Table 10.4.1 The nominal D-H parameters of the PUMA 560

i	θ_i (deg)	d_i (mm)	α_i (deg)	a_i (mm)
1	90.0	0	-90	0
2	0.0	149.09	0	431.8
3	90.0	0	90	-20.32
4	0.0	433.07	-90	0
5	0.0	0	90	0
FLANGE	0.0	56.25	0	0

Note that the listed values for θ_i in Table 10.4.1 are joint offsets at the arm home position.

Two arbitrarily chosen homogeneous transformations for simulating "actual" $BASE$ and $TOOL$ are given in Equation (10.4.1).

$$
BASE = \begin{bmatrix}
-0.99908 & -0.03266 & 0.02786 & 164.226 \\
0.02737 & 0.01553 & 0.99950 & 301.638 \\
-0.03308 & 0.99935 & -0.01462 & -962.841 \\
0.00000 & 0.00000 & 0.00000 & 1.00000
\end{bmatrix}
$$

$$(10.4.1a)$$

$$TOOL = \begin{bmatrix} -0.97651 & -0.09468 & -0.19356 & 9.190 \\ 0.06362 & -0.98493 & 0.16082 & 5.397 \\ -0.20587 & 0.14473 & 0.96782 & 62.628 \\ 0.00000 & 0.00000 & 0.00000 & 1.00000 \end{bmatrix}$$

(10.4.1b)

One of three uniformly distributed random noise processes (denoted by U[a, b], listed in Table 10.4.2) was added to the Z-Y-Z Euler angles and positions of the tool to model measurement noise.   Other measurement uncertainties were ignored. Noise level 1 represents the noise of a moderately accurate measurement instrument while noise level 3 simulates that of a relatively inaccurate device.

In all test results presented in this section, 30 pose measurements uniformly spaced in the PUMA workspace were employed for verification purposes. Orientation errors and position errors are defined in terms of the Euclidian norm of pose position and orientation deviations, respectively. Three types of tests were conducted to explore the following three issues.

Table 10.4.2 Noise injected to "pose measurements"

	Orientation noise (deg)	Position noise (mm)
Noise level 1	U[-0.01, 0.01]	U[-0.1, 0.1]
Noise level 2	U[-0.05, 0.05]	U[-0.1, 0.1]
Noise level 3	U[-0.05, 0.05]	U[-0.5, 0.5]

A. NUMBER OF POSE MEASUREMENTS REQUIRED FOR CALIBRATION

In this type of test, the nominal PUMA geometry listed in Table 10.4.1 and noise level 1 listed in Table 10.4.2 were adopted. Figures 10.4.1 and 10.4.2 depict position and orientation errors against the number of pose measurements. The position error is defined as the norm of the difference between the computed position using the identified $BASE$ and $TOOL$ and the actual position obtained using the given $BASE$ and $TOOL$. The orientation error is defined in a similar way.

It can be seen that pose errors were reduced when the number of pose measurements was increased, as seen from Figure 10.4.1. This trend continues until adding more measurements would not significantly reduce pose errors. The threshold value for the number of pose measurements is around 6. Adding measurements can sometimes degrade the estimation quality, as seen in Figure 10.4.2.

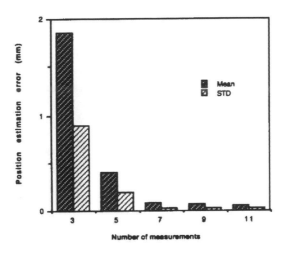

Figure 10.4.1 Position errors against number of pose measurements

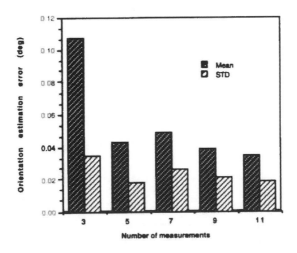

Figure 10.4.2 Orientation errors against number of pose measurements

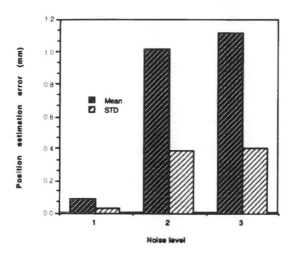

Figure 10.4.3 Position errors against different levels of measurement noise

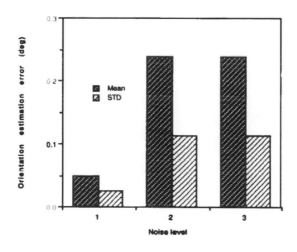

Figure 10.4.4 Orientation errors against different levels of measurement noise

B. CALIBRATION EFFECTIVENESS UNDER DIFFERENT MEASUREMENT NOISE LEVELS

Figures 10.4.3 and 10.4.4 show the pose errors under different noise levels with seven pose measurements being used. These figures suggest that pose errors using the identified *BASE* and *TOOL* transformations are roughly proportional to the intensity of the measurement noise. Furthermore, it was verified that pose orientation errors are independent of the accuracy of pose position measurements. However pose position errors depend on both orientation and position measurement accuracy.

C. CALIBRATION EFFECTIVENESS WHEN THE NOMINAL ROBOT GEOMETRY DEVIATES FROM ITS ACTUAL ONE AND JOINT READINGS ARE NOT PERFECT

To simulate an actual manipulator, uniformly distributed random noise processes were also injected to the PUMA joint variables and link parameters. Three noise models tested in the simulation are listed in Table 10.4.3.

Table 10.4.3 Noise injected to robot link parameters and joint variables

	Rotation parameters (*deg*)	Translation parameters (*mm*)	Joint variables (*deg*)
Model 1	U[0.0, 0.0]	U[0.0, 0.0]	U[0.0, 0.0]
Model 2	U[-0.01, 0.01]	U[-0.1, 0.1]	U[-0.01, 0.01]
Model 3	U[-0.05, 0.05]	U[-0.5, 0.5]	U[-0.05, 0.05]

Model 1 represents a perfect robot and Model 3 is a fair representation of an actual one. The position and orientation errors exhibited by these models are listed in Figures 10.4.5 and 10.4.6. The 1 *mm* figure for position errors contributed by Model 3 may be tolerable in some applications. If an application demands higher accuracy, users must consider a complete calibration of the robot. With complete calibration, PUMA arm position errors can be reduced to within 0.4 *mm*.

V. SUMMARY AND REFERENCES

The linear approach for robot simultaneous *BASE* and *TOOL* calibration by solving homogeneous transformation equations of the form $A X = Y B$ was first proposed by Zhuang, Roth and Sudhakar (1994). The method is applicable to most robots used in an off-line programming environment. When cameras are mounted on the robot hand, the eye-hand transformation can be treated as a special case of the tool-flange transformation, if the camera frame is considered as a "tool frame". This approach can thus be applied to robot/world and eye/hand calibration. The results may also be useful to other problems that can be formulated into homogeneous transformation equations of the form $A X = Y B$. Even though the derivations require the knowledge of quaternions algebra, the results do not. Users can easily implement the linear least squares procedure with minimal or no knowledge of advanced mathematical concepts.

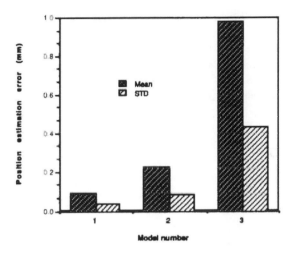

Figure 10.4.5 Position errors against different levels of model uncertainties

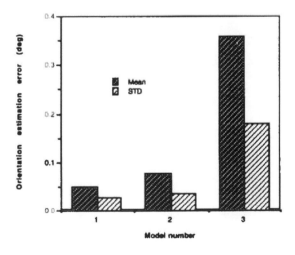

Figure 10.4.6 Orientation errors against different levels of model uncertainties

A drawback of this approach is that since the geometric parameters of the system are estimated in a two-stage process, estimation errors from the first stage propagate to the second stage. Moreover, complete robot pose measurements are needed. While complete poses are not hard to obtain when hand-mounted cameras are used, this is considered to be a difficult and expensive task if other measurement devices are employed. Some of the *BASE* and *TOOL* parameters become unobservable if one uses only partial pose information, refer to Section 5.4.D.

Users can also apply an iterative procedure to calibrate the *BASE* and *TOOL* transformations, which is a special case of the complete robot calibration. The error models suitable for this application have been discussed in Chapter 5.

Simulation has shown that for a PUMA-type robot, by calibrating base and tool its positioning error can be reduced to 1 mm. On the other hand, as demonstrated by Zhuang, Wang and Roth (1993) and by many other researchers, a complete calibration of the PUMA arm can reduce the position error to 0.3 mm.

As has been mentioned, one can either use an error-model based method or the linear method given in this chapter to calibrate simultaneously the base and tool transformations. The error-model based technique has the advantage of being easy to implement, but it needs initial conditions. On the other hand, the linear method does not need any initial guess of unknown robot kinematic parameters. It is up to the user to select a method to fit a particular application.

Chapter 11

ROBOT ACCURACY COMPENSATION

I. INTRODUCTION

Robot accuracy compensation is a process by which robot pose errors are compensated through corrections to the joint variables, based on the kinematic parameter errors that have been identified during the kinematic identification phase of a robot calibration procedure.

Techniques for robot accuracy compensation can be classified into two basic categories: model-based and interpolation methods. In the latter case, positioning errors of the robot end-effector at certain grid points of the robot workspace are measured. Interpolation techniques such as the bilinear interpolation procedure and polynomial fitting are then used to fit correct joint commands to positioning errors throughout the region of interest within the robot workspace.

In a model-based method, a robot kinematic model is used to describe robot pose errors, assuming that the dominant error sources are geometric errors such as joint axis misalignments and joint offsets. Calibration involves the fitting of robot model parameters to measured pose errors. The calibrated robot model can then be used to predict errors of the robot at various robot end-effector locations or to compensate for such errors.

Among model-based techniques, one of the simplest is the so-called Pose Redefinition method, utilizing the nominal inverse kinematic solution applied to properly redefined world coordinates of each task point. Most model-based techniques, however, are gradient-based, using the Newton-Raphson (or Inverse Jacobian) algorithm, the Damped Least-Squares (DLS) procedure, or the Linear Quadratic Regulator (LQR) algorithm.

This chapter is organized in the following manner. The workspace-mapping technique is discussed in Section II, followed by a presentation of the pose redefinition method in Section III. Gradient-based techniques are described in Section IV. Remarks and references are given in Section V.

II. WORKSPACE-MAPPING METHOD

As has been mentioned, all interpolation methods involve the measurement of positioning errors of the robot end-effector at specific grid points of the robot workspace. A suitable interpolation technique is then applied to fit joint command correction according to positioning errors throughout a region of interest within the robot workspace.

Out of several interpolation methods, we chose to describe in detail the bilinear interpolation technique, due to its simplicity and effectiveness. In this section, we restrict the discussion to two-dimensional calibration and compensation.

A. SYSTEM SETUPS

It is assumed that the system consists of a robot, a hand-mounted camera, and precision camera calibration board featuring a known grid pattern of dots. From a system's control view point, there are two options. The first is to have the camera as an integral part of the robotic system, by letting the controller supervise both the camera and the robot, as shown in Figure 11.2.1. In option 2, the camera and its vision processor are not part of the robot system (Figure 11.2.2). The camera is mounted on the robot end-effector for the purpose of calibration and is to be removed as soon as the calibration task is over.

The calibration board may be mounted on a precision motorized positioner stage to enlarge the measurable workspace of the robot; refer to Figure 11.2.3.

B. ACCURACY COMPENSATION SUB-TASKS

The following actions are essential in any accuracy compensation task:

1   *Repeatability assessment of the robotic system.* The robot is repeatedly moved to a predefined location. The end-effector positions of the robot relative to a fixed frame in the workspace (defined by a dot pattern on the calibration board, for instance) are recorded by the vision system. Mean and standard deviation (or variance) of the robot end-effector positions are computed. This process is repeated in a number of different predefined locations sampled throughout the region of interest within the robot workspace. The obtained repeatability sets the upper bound of the accuracy that the robot can achieve after the calibration experiment.

2   *Determination of the positioning errors of the robot system.* The region of interest of the robot workspace in this case is defined by the calibration board (or the precision stage together with the board). The dot patterns on the glass plate define the grid points to be measured by the robot. The robot is moved to various positions such that all these grid points are measured. The end-effector positions at these grid points are computed. The differences between the measured position values and the desired position values of the robot end-effector, defined as the positioning errors, can thus be calculated at each grid point.

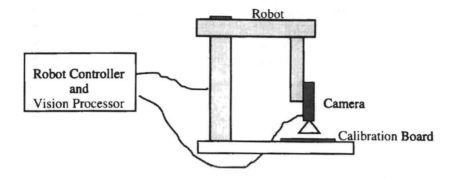

Figure 11.2.1 System configuration - Option 1

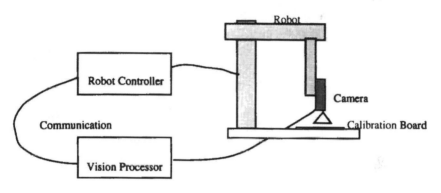

Figure 11.2.2 System configuration - Option 2

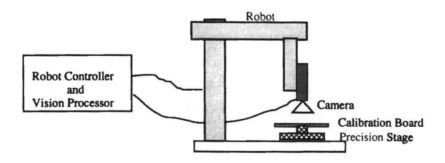

Figure 11.2.3 A precision positioner stage used to enlarge
the measurable workspace

3   *Compensation for the robot positioning errors.*   After entire robot positioning errors are identified at the grid points, errors at the region of interest of the robot workspace can be compensated by interpolation techniques such as the bilinear interpolation. Specifically, given the predicted errors at four neighboring grid points, it is possible to interpolate errors at any point within the rectangle defined by the four points.

C. THE BILINEAR INTERPOLATION

In this section, we chose to describe the bilinear interpolation, one of the simplest yet effective interpolation techniques for accuracy compensation in 2D robot calibrations.

Let point (u, v) represent the desired task-space coordinates of a component placed by the robot on a given plane. Point $(u - e_x, v - e_y)$ will then represent the actual task-space robot position command, taking into account the robot calibration correction actions, to assure bringing the component to the desired point.

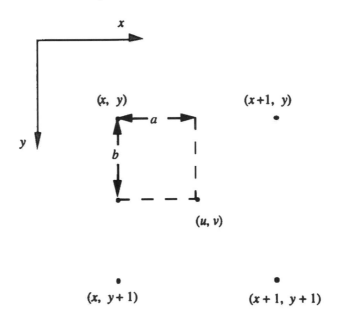

Figure 11.2.4 Bilinear interpolation

The calibration measurement involves sending the robot to four test points and performing precision measurements of the position errors $e_x(\bullet, \bullet)$, $e_y(\bullet, \bullet)$ at these points. The objective of the interpolation is to find the coordinate errors $e_x(u, v)$ and $e_y(u, v)$ along the x and y directions of this arbitrary task point (u, v), given the coordinate errors of its neighboring points. In the bilinear interpolation, it is assumed that at the four nearest neighbors of point (u, v), which are (x, y), $(x+1, y)$, $(x, y+1)$ and $(x+1, y+1)$, positioning errors are given. The geometric relationship of these neighboring points is shown in Figure 11.2.4. The bilinear interpolation assigns e_x and e_y, the errors along x and y direction at (u, v), to the point (u, v) via the following relation:

$$e_x(u, v) = (1 - a)(1 - b)\, e_x(x, y) + a(1 - b)\, e_x(x + 1, y) +$$
$$b(1 - a)\, e_x(x, y + 1) + a\, b\, e_x(x + 1, y + 1) \qquad (11.2.1a)$$
$$e_y(u, v) = (1 - a)(1 - b)\, e_y(x, y) + a(1 - b)\, e_y(x + 1, y) +$$
$$b(1 - a)\, e_y(x, y + 1) + a\, b\, e_y(x + 1, y + 1) \qquad (11.2.1b)$$

where $0 \leq a \leq 1$ and $0 \leq b \leq 1$ are the distances from (x, y) to (u, v) along the x and y axes, respectively; refer to Figure 11.2.4. Readers may verify that these two equations satisfy the constraints at the four grid points.

Note that the distance between two (horizontal or vertical) neighboring grid points are normalized to one. If in practice this is not the case, then a scale factor must be introduced for each direction.

This method is probably the simplest form of an interpolation-based pose redefinition technique for accuracy compensation.

III.   MODEL-BASED POSE-REDEFINITION ALGORITHM

Once robot parameter errors are identified, these can be used to predict robot end-effector pose errors at various robot joint configurations. If the pose error is sufficiently small, to justify the use of linearized accuracy error models at that joint configuration, this predicted pose error can be used to modify the desired pose, based on which a set of new values of the joint position vectors can be determined.

Let the desired robot pose be T^0. Also let T denote the pose computed by use of the estimated error parameter vector. The predicted pose error vector is $y \equiv [d^T, \delta^T]^T$, where δ and d are the orientation and position error vector, and the differential pose error satisfies the following relationship,

$$dT = T - T^0 = T^0 \, \delta T \qquad (11.3.1)$$

where

$$\delta T = \begin{bmatrix} \Omega(\delta) & d \\ 0_{1x3} & 0 \end{bmatrix}$$

From (11.3.1)

$$T = T^0(I + \delta T) \tag{11.3.2}$$

To a first order approximation (Paul, 1981),

$$I + \delta T = Trans(dx, dy, dz) Rot(x, \delta x) Rot(y, \delta y) Rot(z, \delta z) \tag{11.3.3}$$

Thus

$$T = T^0 Trans(dx, dy, dz) Rot(x, \delta x) Rot(y, \delta y) Rot(z, \delta z) \tag{11.3.4}$$

In order to compensate for pose errors caused by the errors in the robot kinematic parameter vector, a *redefined* pose, T', used for computing a corrected joint vector is constructed as follows:

$$T' = T^0 Trans(-dx, -dy, -dz) Rot(x, -\delta x) Rot(y, -\delta y) Rot(z, -\delta z) \tag{11.3.5}$$

T', solved from (11.3.5), is then used together with the robot nominal kinematic model to determine the required joint commands that drive the manipulator to the desired pose T^0.

Note that for a simpler notation, the dependence of the pose and its error vector on the joint variables and kinematic parameter vectors is not shown in this section.

IV.   GRADIENT-BASED ALGORITHMS

A.   SOLUTION STRATEGY

Following the kinematic identification phase, the identified kinematic error parameters can be used to predict robot end-effector pose errors at various robot joint configurations. If the parameter errors are sufficiently small, the predicted pose errors y can be related to joint corrections by a manipulator Jacobian,

$$y + J(\rho, q^0) dq \cong 0$$

or

$$y \cong -J(\rho, q^0) dq \tag{11.4.1}$$

where q^0 is the nominal joint vector, dq is the joint correction vector, and J is the *manipulator Jacobian*. The correction dq can then be uniquely solved from (11.4.1) if the Jacobian matrix is nonsingular, and in turn it can be used to modify the nominal joint command to reduce the pose error in this particular robot joint configuration, refer to Figure 11.4.1.

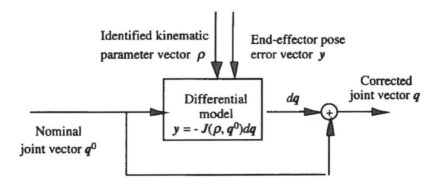

Figure 11.4.1. Gradient-based accuracy compensation scheme

If J is differentiable with respect to ρ, then

$$J(\rho^0, q^0) \cong J(\rho, q^0) = (\nabla J \, d\rho + J(\rho^0, q^0)) dq + \text{high order terms}$$

where ∇ is a gradient operator. Whenever the kinematic error parameter vector $d\rho = \rho - \rho^0$ is small, one can use the following approximation,

$$J(\rho^0, q^0) dq \cong J(\rho, q^0) dq$$

where $J(\rho^0, q^0)$ is the *nominal Jacobian*.

In this section we first derive the manipulator Jacobian as a key component in this accuracy compensation scheme. A number of gradient-based compensation algorithms are then presented.

B. DERIVATION OF THE MANIPULATOR JACOBIAN

For an n degrees of freedom manipulator, the dimension of the Jacobian matrix is $6 \times n$; the first three row vectors are associated with the end-effector position error vector d, whereas the last three correspond to its orientation error vector. Each column vector, on the other hand, represents the position and orientation error generated by the corresponding individual joint. Let us start by determining each column vector of the Jacobian matrix as a function of link parameters and joint variables. Let J_{pi} and J_{oi} denote the 3×1 column vectors of the Jacobian matrix associated with the position and orientation errors, respectively. That is, the Jacobian matrix is partitioned so that

$$J = \begin{bmatrix} J_{p1} J_{p2} \cdots J_{pn} \\ J_{o1} J_{o2} \cdots J_{on} \end{bmatrix}$$

(11.4.2)

The position error vector can then be written as a linear combination of the column vectors J_{pi}, as follows

$$d = J_{p1} dq_1 + J_{p2} dq_2 + \ldots + J_{pn} dq_n$$

(11.4.3)

If joint i is prismatic, it produces a translational error at the end-effector in the same direction as that of the joint axis. Let e_{i-1} be the unit vector pointing along the direction of joint axis i, as shown in Figure 11.4.2, and let dd_i be the scalar joint error in this direction, we then obtain,

$$J_{pi} dq_i = e_{i-1} dd_i$$

(11.4.4)

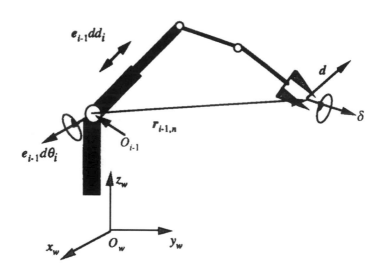

Figure 11.4.2 Position and orientation errors of the end-effector generated by the errors in the individual joint

As shown in Figure 11.4.2, if joint i is revolute, the composite of distal links is rotated from links i to n by a total orientation error do_i as follows:

$$do_i = e_{i-1} d\theta_i$$

(11.4.5)

The orientation error produces certain position error at the end-effector. Let $r_{i-1,n}$ be the position vector from O_{i-1} to the end-effector as shown in Figure 11.4.2. Then the position error due to the orientation error db_i is

$$J_{pi}dq_i = db_i \times r_{i-1,n} = (e_{i-1} \times r_{i-1,n})d\theta_i \qquad (11.4.6)$$

Thus the contribution of each joint to the end-effector position error is determined by either (11.4.4) or (11.4.6), depending on the type of joint.

Similarly, the orientation of the end-effector can be expressed as a linear combination of the column vectors J_{oi} as follows

$$\delta = J_{o1}dq_1 + J_{o2}dq_2 + \dots + J_{on}dq_n \qquad (11.4.7)$$

Whenever joint i is prismatic, it does not generate an orientation error at the end-effector, therefore

$$J_{oi}dq_i = 0 \qquad (11.4.8)$$

If, on the other hand, the joint is revolute, the orientation error is given by

$$J_{oi}dq_i = db_i = e_{i-1}d\theta_i \qquad (11.4.9)$$

In summary, for a prismatic joint,

$$\begin{bmatrix} J_{pi} \\ J_{oi} \end{bmatrix} = \begin{bmatrix} e_{i-1} \\ 0_{3\times1} \end{bmatrix} \qquad (11.4.10)$$

and for a revolute joint,

$$\begin{bmatrix} J_{pi} \\ J_{oi} \end{bmatrix} = \begin{bmatrix} e_{i-1} \times r_{i-1,n} \\ e_{i-1} \end{bmatrix} \qquad (11.4.11)$$

The vectors e_{i-1} and $r_{i-1,n}$ are functions of joint displacements. These vectors can be computed using the following method. The direction of joint axis $i-1$ is represented with respect to coordinate frame $i-1$ by $e = [0, 0, 1]^T$, as the joint axis is along the z_{i-1} axis. This vector can be transformed to a vector e_{i-1} defined in terms of the base frame using 3x3 rotation submatrices $^{j-1}R_j$ as follows

$$e_{i-1} = {}^wR_0{}^0R_1(q_1) \ ... \ {}^{i-2}R_{i-1}(q_{i-1})e \qquad (11.4.12)$$

The position vector $r_{i-1,n}$ can be computed through the link transformation matrices ${}^{j-1}A_j$. Let $\underline{r}_{i-1,n}$ be the 4x1 augmented vector[1] of $r_{i-1,n}$, and $\underline{r} = [0, 0, 0, 1]^T$ be the augmented position vector representing the origin of its coordinate frame, then $r_{i-1,n}$ is given by

$$\underline{r}_{i-1,n} = {}^wA_0{}^0A_1(q_1) \ ... \ {}^{n-1}A_n(q_n)\underline{r} \ - {}^wA_0{}^0A_1(q_1) \ ... \ {}^{i-2}A_{i-1}(q_{i-1})\underline{r}$$
$$(11.4.13)$$

where the first term accounts for the position vector from origin O_w to the end-effector and the second term is the vector from O_w to O_{i-1}.

C.   A NEWTON-RAPHSON COMPENSATION ALGORITHM
When the number of degrees of freedom of the robot is 6, one may use the Newton-Raphson algorithm to compute joint corrections given pose errors. If the Jacobian matrix is nonsingular, then by (11.4.1),

$$dq = - J \, (\rho^0, \, q^0)^{-1}y \qquad (11.4.14)$$

The algorithm breaks down whenever the robot is at its singular joint configurations; that is, at which the rank of the Jacobian matrix is less than six.

It is straightforward to recursively apply this algorithm to keep improving the compensation accuracy. In the optimization literature, this method is referred to as the Newton's algorithm.

D.   DLS AND LQR ALGORITHMS
To assure that a solution exists even when the manipulator is at its singularity zones, we treat the accuracy compensation as an optimization problem. The following cost function is adopted:

$$\Phi = \left(y + J \, dq\right)^T Q \left(y + Jdq\right) + dq^TGdq \qquad (11.4.15)$$

where Q is symmetric and nonnegative definite and G is symmetric positive definite. This modified cost function ensures the existence and uniqueness of the optimal solution at all robot configurations. The closed-form solution which minimizes Φ is

---

[1] An augmented vector here means that the vector is represented in homogeneous coordinates

$$dq = -(J^T Q J + G)^{-1} J^T Q y \qquad (11.4.16)$$

The above approach is named the Damped Least Squares (DLS) algorithm. The computation of dq is repeated at every task point.

The recursive version for the particular case of $G = \text{diag}(\gamma_1, \gamma_2, ..., \gamma_n)$ is the Linear Quadratic Regulator (LQR) algorithm.

Let J_i be the ith column vector of the Jacobian matrix J.

$$y = J \, dq \equiv [J_1, J_2, \cdots, J_n] dq \qquad (11.4.17)$$

Let $u \equiv [u_1, u_2, \cdots, u_n]^T \equiv dq$. Equation (11.4.17) can be rewritten as a discrete-time state equation in the following manner:

$$y_{k+1} = y_k + J_k u_k \quad k = 1, 2, ..., n \qquad (11.4.18a)$$
$$y_1 = y \qquad (11.4.18b)$$

Note that y_1 is the known robot pose error vector before compensation and y_{n+1} is the pose error vector after compensation.

The LQR problem is to minimize

$$\Phi(u, y_{n+1}) = y_{n+1}^T Q y_{n+1} + \sum_{k=1}^{n} \gamma_k u_k^2 \qquad (11.4.19)$$

subject to the constraint equations (11.4.18), where the coefficients γ_k are strictly positive.

The optimal solution u_k^* is given by the feedback control law

$$u_k^* = v_k y_k \qquad (11.4.20)$$

where the row vector of gains v_k is obtained from a recursive solution of a related Riccati difference equation

$$v_{n-k} = -\frac{J_{n-k}^T P_k}{J_{n-k}^T P_k J_{n-k} + \gamma_{n-k}} \qquad k = 0, 1, ..., n-1 \quad (11.4.21)$$

where

$$P_{k+1} = P_k + P_k J_{n-k} v_{n-k}^T \qquad (11.4.22)$$

with

$$P_0 = Q \qquad (11.4.23)$$

The resulting minimum cost is

$$\Phi^* = y_1{}^T P_n y_1 \tag{11.4.24}$$

After $dq = u^*$ is obtained, the actual joint variable q, to be used at this particular task point, is modified to

$$q = q^0 + dq \tag{11.4.25}$$

Positioning errors often have a scale difference which is by orders of magnitude different than orientation errors, depending on the selection of unit systems. The weight matrix Q can be chosen to relate the cost function to relative errors instead. This may be achieved by choosing a diagonal Q:

$$Q = \text{diag}(k_o, k_o, k_o, k_p / \|p\|^2, k_p / \|p\|^2, k_p / \|p\|^2) \tag{11.4.26}$$

where p is the position vector in T_n, and k_o and k_p are constants that can be selected to emphasize end-effector's positioning accuracy over orientation accuracy or vice versa.

The values of γ_k in Equation (11.4.19) may be chosen to achieve several objectives. First, the correction vector dq should also be normalized whenever the robot consists of both revolute and prismatic joints. Second, at task points that require one or more of the joint variables to be near joint travel boundaries or near robot singularities, large values of γ_k need to be selected to effectively reduce the amount of correction. A method of choosing γ_k can be:

$$\gamma_k = \delta_k \gamma / q_{k,rang} \qquad\qquad \gamma > 0 \tag{11.4.27}$$

where $q_{k,rang}$, the square of the total joint travel, is used to normalize the corrections for the case of robots with two types of joints.

Acting as a penalty coefficient, δ_k may be chosen to be inversely proportional to the minimum distance between the kth nominal joint command and its travel boundaries or singularity zone boundaries, or it could be chosen as a switching function, giving a very large value only when the above distance is smaller than a prescribed threshold value. At the limit, $\delta_k = \infty$, resulting in a zero correction of the kth joint variable. Such large values of δ_k may degrade the quality of accuracy compensation as the second terms in the cost function (11.4.19) becomes more dominant.

Assuming that $\gamma_1, \gamma_2, ..., \gamma_m, Q$ and J have been pre-computed, the DLS algorithm requires about $0.33n^3 + 0.5mn^2$ floating point operations where m is the dimension of the pose error vector (normally $m = 6$) if the Gauss elimination method is used for matrix inversion. On the other hand, the LQR algorithm needs about $7m^2n$ floating-point operations. Therefore, the DLS algorithm is slightly more efficient for cases of few degrees of freedom manipulators. In the LQR algorithm the joint corrections are determined recursively. The corrected joint commands can thus be sequentially sent to the robot actuators. If this type of compensation strategy is feasible, the LQR algorithm will be more efficient than non-recursive algorithms. The advantage of using the LQR algorithm over the DLS algorithm for highly redundant manipulators is evident.

In terms of storage requirements, the LQR algorithm requires storage of $m \times m$ matrices while the DLS algorithm needs to store both $m \times m$ and $n \times n$ matrices. As n increases, this advantage of the LQR algorithm becomes very significant.

Finally, the LQR algorithm is more convenient to program since no matrix inversion operations are required.

E.   SIMULATION RESULTS

Simulations were conducted to test the performance of the LQR technique under various conditions. The nominal PUMA geometry was chosen for simulation with Denavit-Hartenberg (D-H) link parameters listed in Table 11.4.1. The following issues were investigated:

1.   The robustness of the LQR algorithm under different link parameter perturbations.
2.   The tradeoff between position accuracy and orientation accuracy by selection of the weight matrix Q in the cost function (11.4.19).
3.   The effectiveness of the LQR algorithm in handling manipulator singularities through the choice of the control coefficients γ_k.

The remarks made in this section are also applicable to the DLS algorithm due to the mathematical equivalence of the two algorithms.

Two nominal joint configurations, termed Test points 1 and 2, are listed in Table 11.4.1.

To test the robustness of the compensation algorithm under different levels of model uncertainties, one of three types of uniformly distributed random noise processes (listed in Table 11.4.3) was added to all link parameters of the manipulator.   Model 1 simulates a relatively precise manipulator, while Model 3 represents a manipulator with large parameter deviations.

Compensation results under different levels of parameter perturbations are listed in Table 11.4.4. Obviously, the smaller the perturbations, the better the compensation results. This can be seen through the ratio of position and orientation errors before and after compensation. For instance, for orientation errors, this ratio is 483 for Model 1, 106 for Model 2, and 21 for Model 3.

The proposed compensation method can improve the accuracy of the manipulator significantly, even when parameter deviations are large.

To test the tradeoff provided by the LQR algorithm between the position and orientation errors, different values of the weight coefficients in Equation (11.4.27) were chosen.   Table 11.4.5 lists the orientation and position errors before and after compensation. Noise model 2 was chosen in this simulation.

Table 11.4.1 D-H parameters of the PUMA 560

	$\Delta\theta$(deg)	d(mm)	α(deg)	a(mm)
1	90.0	0.0	-90.0	0.0
2	0.0	149.09	0.0	431.8
3	90.0	0.0	90.0	-20.32
4	0.0	433.07	-90.0	0.0
5	0.0	0.0	90.0	0.0
6	0.0	56.25	0.0	0.0

Table 11.4.2   Nominal joint displacements of the task points

Test point	θ_1	θ_2	θ_3	θ_4	θ_5	θ_6
1	90.0	-90.0	45.0	-45.0	45.0	0.0
2	90.0	-90.0	45.0	45.0	-0.1	0.0

Table 11.4.3 Noise injected to the nominal D-H link parameter

Model Number	Pertubation in α and θ (deg)	Pertubation in a and d (mm)
1	U[-0.1, 0.1]	U[-0.4, 0.4]
2	U[-0.5, 0.5]	U[-2.0, 2.0]
3	U[-2.5, 2.5]	U[-10.0, 10.0]

Table 11.4.4   Compensation results for Test point 1
with different noise models

Model Number	Before Compensation		After compensation	
	$\|d\|$ (mm)	$\|\delta\|$ (deg)	$\|d\|$ (mm)	$\|\delta\|$ (deg)
1	1.00603	0.0042093	0.00056	0.0000087
2	5.02269	0.0210319	0.13772	0.0001994
3	24.97189	0.1044624	3.35304	0.0049828

Table 11.4.5   Compensation results for Test point 1
with different weight coefficients in the cost function

k_p	k_o	Before Compensation		After compensation	
		$\|d\|$ (mm)	$\|\delta\|$ (deg)	$\|d\|$ (mm)	$\|\delta\|$ (deg)
1.0	1.0	5.00227	0.0210319	0.13772	0.0001994
0.0	1.0	5.00227	0.0210319	9.69099	0.0001881
1.0	0.0	5.00227	0.0210319	0.07913	0.0319594

It is shown in Tables 11.4.6 and 11.4.7 that γ_k in the cost function (11.4.19) plays an important role when the robot is near a singular configuration. Noise model 2 was again chosen. The nominal joint variables were chosen according to Test point 2 in Table 11.4.2. Since $\theta_5 = -0.1$ degree, the PUMA arm is near a wrist singularity. Table 11.4.6 shows that the joint variable corrections will be exceedingly large if γ_k are all set to zero. The LQR algorithm, when γ_k is computed using Equation (11.4.19), performs slightly better than a version of the Recursive DLS algorithm which employs constant values of γ_k for all joints. For Test point 1 (results are not shown), the joint corrections, in all three cases, were of the same order of magnitude.

Table 11.4.6 Joint corrections to Test point 2
with different control coefficients as given in Table 11.4.7

	$d\epsilon_1$	$d\theta_2$	$d\theta_3$	$d\theta_4$	$d\theta_5$	$d\theta_6$
case 1	-0.07482	-0.71254	0.99695	-0.04265	-0.00313	-0.18910
case 2	-0.09596	-0.71947	1.01922	-0.28172	-0.18975	0.06491
case 3				out of range		

Note: (1) Control coefficients for different cases are listed in Table 11.4.7.
        (2) Out of range means that the joints exceed their travel limits.

Table 11.4.7 Different control coefficients
corresponding to joint corrections given in Table 11.4.6

	γ_1	γ_2	γ_3	γ_4	γ_5	γ_6
case 1	0.00026	0.00019	0.00028	0.00019	47.023	0.00002
case 2	0.0001	0.0001	0.0001	0.0001	0.0001	0.0001
case 3	0.0	0.0	0.0	0.0	0.0	0.0

Note: In case 1 the value of γ (of Equation (11.4.19)) was set to 0.01.

V.   SUMMARY AND REFERENCES

Techniques for robot accuracy compensation can be classified into two major categories: model-based and numerical fitting.   Numerical fitting techniques do not distinguish between geometric and nongeometric errors. These are particularly suitable for error compensation in 2D space.   On the other hand, model-based techniques are more effective when the dimension of the error space is large.

A key operation in a workspace mapping task is interpolation. There are numerous interpolation techniques in the literature, for instance, refer to Mortenson (1985). A good reference on the use of polynomial fitting of joint position corrections to end-effector positioning error is the paper by Shamma and Whitney (1987).

The pose-redefinition procedure was initially proposed by Veitschegger and Wu (1988). Some explicit formulas, developed by Vuskovic (1989), depend on linearization with respect to the error parameters, an unnecessary approximation, in our opinion. The algorithm may fail at task points that are near the robot workspace boundaries. Another simple strategy is based on Inverse Jacobian (i.e., Newton-Raphson) solution of the joint command updates based on estimated positioning and orientation errors of the end-effector (Veitschegger and Wu (1988) and Huang and Gautam (1989)). This algorithm breaks down if task points happen to be near nominal singular configurations of the manipulator.

The Robot Accuracy Compensation problem is closely related to the numerical solution of the robot inverse kinematics problem. As argued by Nakamura and Hanafusa (1986) and Maciejewski and Klein (1988), special accommodations need to be provided as part of a numerical inverse kinematic solution to avoid an algorithm break-down near interior manipulator singularities. The location of such a singular configuration is known for most simple manipulators (i.e. manipulators whose consecutive joint axes are either parallel or perpendicular), and can be avoided during task planning by creating suitable safety zones around singular points.

In certain critical applications, at which trajectories that pass near singularities cannot be avoided either due to a lack of precise knowledge of the singularities locations or due to special needs of the application, robust algorithms play an important role in avoiding failures. One possible solution that has been suggested by Wampler (1986) is to find a set of joint displacements that minimize position errors, while keeping joint variable corrections as small as possible. The problem can then be solved by the DLS procedure, an iterative version of which is devised for solving nonlinear least squares problems is the Levenberg-Marquardt (L-M) algorithm (Marquardt (1963)).

A recursive version of the DLS algorithm is the LQR procedure. Application of linear optimal control theory to the design of robust accuracy compensators was proposed in Zhuang, Hamano and Roth (1989) and expanded in Zhuang, Roth and Hamano (1993). In both DLS and LQR algorithms, the compensation solution exists at all poses including singular configurations, due to a particular choice of a performance index. The weight matrix in the performance index can be selected to achieve specific objectives, such as emphasizing of end-effector's positioning accuracy over orientation accuracy or vice versa, or taking into account proximity to robot joint travel limits and singularity zones.

For most industrial robots, both the LQR and the DLS algorithms have similar computational complexity, therefore either one can be used for accuracy compensation if all the corrected joint commands are computed before any of these being sent to the actuators. The efficiency of robot controllers that incorporate the LQR accuracy compensation algorithm can be significantly improved if the corrected joint commands are sent to the robot actuators sequentially. For highly redundant manipulators such as "snake robots", the LQR algorithm is much more efficient than the DLS algorithm. As to storage requirement, the LQR algorithm has an edge over the DLS algorithm as the number of degrees of freedom becomes larger. The LQR algorithm is also far more convenient to program, due to the fact that no matrix inversion operations are required.

It should be stressed that pose errors, even after compensation, cannot go all the way to zero due to: (i) the use of a linearized error model, and (ii) effects of unmodeled nongeometric errors.

Chapter 12

SELECTION OF
ROBOT MEASUREMENT CONFIGURATIONS

I. INTRODUCTION

After covering in detail the four distinct actions that constitute a robot calibration process: selection of a kinematic model, measurement of the robot end-effector pose in world coordinates, identification of the robot kinematic parameters, and accuracy compensation, it becomes clear that a measurement of robot end-effector poses is the most critical step towards successful robot calibration. As observed by many researchers, a good selection of the set of robot measurement configurations can significantly improve the quality of kinematic identification and cut down the calibration time.

Qualitatively, optimal selection of robot configurations can be stated as the problem of determining a set of robot measurement configurations within the reachable robot joint space so that the effect of measurement noise on the estimation of robot kinematic parameters is minimized.

In this chapter, the problem of optimal measurement configuration selection is first stated in Section II. Two simple configuration selection procedures are given in Section III. Selection of optimal measurement configurations using the Simulated Annealing algorithm is detailed in Section IV. Section V provides summary and references.

II. PROBLEM STATEMENT

A. PERFORMANCE MEASURES

Recall from Chapter 5 that the observability of robot error parameters is defined in terms of the rank of the Identification Jacobian matrix. The kinematic error parameters are observable if $J^T J$ is full rank. Two observability measures that have been used by robot researchers are the condition number of the Jacobian matrix and the observability index. The condition number of J is defined in terms of the maximum and minimum singular values of J:

$$\text{Cond}(J) = \sigma_{max}/\sigma_{min} \qquad (12.2.1)$$

The observability index uses all singular values of J:

$$O(J) = (\sigma_1 \sigma_2 \cdots \sigma_m)^{1/m} m_r^{-1/2} \qquad (12.2.2)$$

where m_r is the number of rows in J, and σ_i, $i = 1, 2, \cdots, m$, are the *singular values* of J. Either one of these performance measures can be used in selection of optimal or sub-optimal measurement configurations for robot calibration.

B. A GENERAL PROBLEM STATEMENT

Generally, the problem of optimal robot measurement configuration selection can be stated as follows: Determine m robot measurement configurations within the reachable robot joint space so that the effect of measurement noise on the parameter estimation errors is minimized. Note that m should be sufficiently large. For instance, whenever full poses of the robot end-effector can be measured, a necessary condition for observability is that $6m > t$, where t is the number of independent kinematic parameters of the manipulator.

C. A MORE RESTRICTED PROBLEM STATEMENT

The problem stated above is difficult to tackle since the mathematical relationship among parameter errors, measurement noise and measurement configurations is vaguely defined, as yet. However, if an error-model-based technique is adopted for kinematic parameters identification, the optimal configuration selection problem may be solved through investigation of the above performance measures since the upper bound of the parameter error norm is proportional to the condition number on the Identification Jacobian. In the following discussion, the error model based technique is used to illustrate the concept of the proposed method. The configuration selection methods are however equally applicable to other robot calibration techniques in which linear transformations relating pose measurements to the unknown kinematic parameters are available (for instance, as in the linear solution methods discussed in Chapter 6).

Let us define the problem of optimal robot measurement configuration selection as follows:

Problem A: Determine m robot measurement configurations in the reachable robot joint space such that either Cond(J) is minimized or $O(J)$ is maximized.

Problem B: Determine m robot measurement configurations from g robot measurement configurations ($m \ll g < \infty$) such that either Cond(J) is minimized or $O(J)$ is maximized.

Problem B is a special case of problem A. A solution of problem B is a suboptimal solution of problem A. If the finite set of configurations in problem B is sufficiently dense, the solutions to both problems are expected to be close. Most of the methods proposed in this chapter handle problem B only except for the Genetic Algorithm, which can also solve problem A.

III. TWO SIMPLE SEARCH ALGORITHMS

A. UNIFORM RANDOM SEARCH

Given $\mathbf{A}$, a set of robot measurement configurations, it is desired to find a subset of measurement configurations $\mathbb{B}$ such that an observability measure of J based on $\mathbb{B}$ is "reasonably good". The uniform random search procedure is simply to generate and evaluate as many random permutations as possible in the time allowed, retaining the best one found.

Example: Consider a 3 degrees-of-freedom RRP robot. A randomly generated configuration set $\mathbf{A}$ is given in Table 12.3.1. A subset $\mathbb{B}$ can be generated by randomly picking m measurement configurations from $\mathbf{A}$.

Table 12.3.1 A typical configuration set $\mathbf{A}$

Index	Joint 1	Joint 2	Joint 3
1	$\theta_{1,1}$	$\theta_{2,1}$	$d_{3,1}$
2	$\theta_{1,2}$	$\theta_{2,2}$	$d_{3,2}$
$\vdots$			
g	$\theta_{1,g}$	$\theta_{2,g}$	$d_{3,g}$

B. PAIRWISE EXCHANGE

A pairwise random search algorithm can be summarized as follows: We first generate randomly a configuration set $\mathbf{A}$. We then randomly pick m configurations from the set $\mathbf{A}$ to form an initial subset $\mathbb{B}_0$. The condition number of $\mathbb{B}_0$ is evaluated. Next we remove one arbitrary configuration from $\mathbb{B}_0$ and replace it with another configuration from $\mathbf{A}$ to form $\mathbb{B}_1$. The condition number of $\mathbb{B}_1$ is evaluated and compared with that of $\mathbb{B}_0$. If cond($\mathbb{B}_1$) is smaller than cond($\mathbb{B}_0$), $\mathbb{B}_1$ is chosen, otherwise $\mathbb{B}_0$ is retained. The process is continued until certain termination criteria are satisfied.

In another pairwise exchange procedure, rather than randomly picking up a configuration from **A** to replace an arbitrary configuration from **B**, the pairwise exchange starts from the top of the list. It generates up to $g(g-1)/2$ "adjacent" permutations by swapping the elements in each pair of positions and evaluating the resulting permutations.

IV. CONFIGURATION SELECTION USING SIMULATED ANNEALING

The robot measurement configuration selection problem may also be solved by stochastic search procedures such as a Simulated Annealing (SA) or a Genetic Algorithm (GA). In this book, only the simulated annealing algorithm is discussed. For a genetic algorithm implementation of robot configuration selection, readers are referred to (Huang, 1995; Zhuang, Wu and Huang, 1996).

A. THE SA ALGORITHM

Simulated annealing as a stochastic search approach has been introduced to allow the iterative solution to escape local minima by occasionally accepting bad points. It has been successfully applied to a variety of optimization problems.

Let F be a cost function defined over a finite state space S. Let $\{T_k\}$ be a sequence of positive numbers. $\{T_k\}$ is called the *cooling schedule*. Let $Q(S)$ be the probability density of the state, and $D(\bullet,\bullet)$ be the joint probability density of generating a new state from an old one. Both probability laws are problem-dependent. For every state $S \in S$, let $Q(S) \geq 0$ and $\sum_{S \in S} Q(S) = 1$. Similarly for every S, $S' \in S$, let $D(S, S') \geq 0$ and $\sum_{S, S' \in S} D(S, S') = 1$. The SA algorithm is summarized in the following set of generic steps for iteratively finding a "best" state X, according to a given cost function $F(X)$:

Given T_k for $k = 0, 1, 2, ...,$ and $F(\bullet)$, $Q(\bullet)$, and $D(\bullet,\bullet)$.

Step 1. Set $k = 0$;

Step 2. Generate X_0 with $Pr(X_0 = S) = Q(S)$;

Step 3. Generate Y_k with $Pr(Y_k = S' | X_k = S) = D(S, S')$;

Step 4. IF $F(Y_k) < F(X_k)$, THEN

$$X_{k+1} = Y_k, \text{ GO TO Step 6}$$

ELSE

Continue;

Step 5.   Select a uniformly distributed random number ξ_k, $\xi \in [0, 1)$;

IF

$$exp\left(\frac{-\left(F(Y_k) - F(X_k)\right)}{T_k}\right) < \xi_k$$

THEN

$$X_{k+1} = Y_k$$

ELSE

$$X_{k+1} = X_k;$$

Step 6.   Set $k = k+1$; GO TO Step 3.

In the SA algorithm, a state is a point is the solution space. For the robot configuration selection problem, a state is a set of measurement configurations that is sufficient for the purpose of robot calibration.

In the algorithm, X_0 is the initial state. In Step 5, a mechanism is provided for occasional "uphill moves". The occurrence of an uphill move depends on two factors: the size of the uphill move and the "temperature". An uphill move is less likely to occur as the size of the move becomes larger. The temperature T_k strongly influences the likelihood of an uphill move. T_k is often taken to be a monotonically decreasing function of k. Initially, at large values of T_k, greater uphill moves may be accepted. As T_k decreases, only small uphill moves may sometimes occur. Finally, as T_k approaches zero, no uphill moves are allowed. This means that the simulated annealing algorithm, in contrast to local search algorithms such as gradient-based procedures, can escape a local minimum while still exhibiting the favorable features of local search algorithms.

To fully implement the annealing algorithm, the following issues have to be settled:

1. Formulation of the cost function F;
2. Selection of initial configurations through specification of the probability distribution $Q(\bullet)$;
3. Generation of candidate states through specification of the conditional probability distribution $D(\bullet, \bullet)$;
4. Design of the cooling schedule $\{T_k\}$.

B. SELECTION OF ROBOT MEASUREMENT CONFIGURATIONS

Let q be a joint variables vector in a given robot measurement configuration. Let $S = [q_1^T, q_2^T, ..., q_m^T]^T$ be a set of distinct measurement configurations. S is then the set of all possible combinations of measurement configurations in the reachable joint space. To assure that S is finite, it is assumed that the joint variables can only take finite distinct values. This is in the domain of problem B defined earlier in Section II.C.

Example: Consider a two degrees of freedom planar revolute manipulator. Denote its joint variables by θ_1 and θ_2. Let the reachable joint space be $45^0 \leq \theta_i \leq 135^0$ for i = 1, 2. Let the candidate measurement configurations be $\theta_i = 45^0 + j\ 5^0$, for $j \in \{0, 1, ..., 18\}$. In this case, S is finite. To identify the 4 unknown kinematic parameters consisting of joint variable offsets and link lengths $\{\theta_1,\ l_1,\ \theta_2,\ l_2\}$, one needs as a minimum two distinct measurement configurations with different values of the angle θ_2 (Mooring, Roth and Driels (1991)). A typical state is therefore S = $\{\theta_{1j},\ \theta_{2j},\ \theta_{1q},\ \theta_{2q}\}$ where $j \neq q$ and $\theta_{2j} \neq \theta_{2q}$. In the SA algorithm, to simplify the program code we don't eliminate the case in which $\theta_{2j} \neq \theta_{2q}$ a priori. There are thus $19^4 = 130321$ distinct states.

a. Selection of a Cost Function

The cost function can be defined in terms of either the condition number or the observability index (in the latter case the sign of the cost function (12.2.2) needs to be reversed to form a minimization problem). The discrete probability density function $Q(S)$ of the initial state X_0 can be chosen to be uniformly distributed over S.

b. Generation of Candidate States

The generation of "good" candidate states is critical for having a good convergence rate of the SA algorithm. Fortunately, for the problem of robot configuration selection, the following method of generating candidate states is relatively simple and provides adequate results.

Randomly select a robot joint variable vector $q_i \in X_k$. It is then replaced by any other joint variable vector q_i' that is randomly selected from S - X_k. The other vectors that constitute X_k remain the same.

Since the new set of states which contain the newly selected joint variable vector is uniformly distributed over S, both the irreducibility and weak reversibility conditions of (Hajek (1988)) are satisfied. These are necessary conditions for the convergence of the algorithm.

It is clear that in each iteration, the SA algorithm performs a local search, which is similar to a gradient-type search algorithm. However, the difference is that the SA algorithm occasionally accepts uphill moves, and gradient-type algorithms do not.

The toughest part of implementing the SA algorithm is the design of a cooling schedule. Despite the large number of practical cooling schedules proposed in the literature, effective modification of these cooling schedules to fit the particular problem of robot measurement configuration selection is still an open research issue.

C. DESIGN OF A PRACTICAL COOLING SCHEDULE

A cooling schedule has to be carefully devised in order to guarantee the convergence of the algorithm. As is well known in the SA literature, a cooling schedule in the form

$$T_k = \frac{G}{ln(k+2)} \qquad (12.4.1)$$

for some constant G requires an infinite number of steps for the algorithm to converge to a global minimum (Hajek (1988)). This schedule is therefore impractical and another finite-time approach is needed. Such a finite-time realization can be achieved by letting

$$T_l = \frac{G}{ln(lk_s+2)} \qquad (12.4.2)$$

where k_s (> 0) is a constant, and

$$l = l(k) = trancate(k/L_l) \qquad (12.4.3)$$

where $trancate(k/L_l)$ takes the largest integer quotient of k/L_l, and L_l will be defined shortly. Note that temperature is changed only after every L_l transitions (or iterations).

A practical cooling schedule includes the following elements:

1. The initial temperature,
2. The equilibrium condition, and
3. The stopping criterion.

a.   The initial temperature

The initial temperature T_0 should be large enough to allow most state transitions to be accepted. Usually this can be achieved by requiring an *initial acceptance ratio* $\chi_0 = \chi(T_0)$ to be close to one, where the acceptance ratio is the ratio of accepted state transitions to all state transitions. This is the "initial heat-up". The value of G can be determined in terms of χ_0.

Let m_1 denote the number of possible transitions for which $F(Y_0) \leq F(X_0)$ and m_2 be the number of transitions for which $F(Y_0) > F(X_0)$. Evidently m_1 is the number of cost-decreasing transitions and m_2 the number of cost-increasing transitions. Furthermore, let ΔF_{ave} be the average cost difference among the m_2 cost-increasing transitions. Then the initial acceptance ratio χ_0 can be approximated by

$$\chi_0 \approx \frac{m_1 + m_2 \, exp\left(\frac{-\left(\Delta F_{ave}\right)}{T_0}\right)}{m_1 + m_2}$$

In our case, the initial state is uniformly distributed over S, therefore $m_1 = m_2$. Thus

$$\chi_0 \approx \frac{1 + exp\left(\frac{-\left(\Delta F_{ave}\right)}{T_0}\right)}{2} \qquad (12.4.3)$$

The initial temperature T_0 can then be solved from the above equation given χ_0; that is

$$T_0 = \frac{\Delta F_{ave}}{ln\left(\frac{1}{2\chi_0 - 1}\right)} \qquad (12.4.4)$$

Finally, G is determined from (12.4.2) by letting l be zero,

$$G = T_0 ln \, 2 \qquad (12.4.5)$$

b. The equilibrium condition

Let L_l denote the length of the lth state transition chain. The lth chain consists of L_l state transitions controlled by an identical temperature T_l. After L_l trials, the state vector X_l will not change much, therefore it is the time to change the temperature value. It is pointed out by Aarts and Korst (1989) that large decrements in T_l require longer state transition chain lengths in order to reach some sort of equilibrium of the state. Thus, there is a tradeoff between large decrements of the temperature parameter T_l and small state transition chain lengths L_l. One may set small increments of temperature and short state transition chain lengths or alternately use large temperature increments and long chain lengths. This is part of the SA tuning process.

c. The stopping criterion

During the entire annealing process, uphill transitions are accepted with a decreasing probability. When the temperature is decreased to a sufficiently low level, the system approaches a "freezing point" where very few uphill moves may occur. We can thus set the temperature to zero to terminate the annealing process. The SA algorithm is reduced to a standard random selection algorithm.

D. SIMULATION STUDIES

A simple two DOF all-revolute planar robot is used to illustrate the concept of simulated annealing technique for search of optimal measurement configurations. The procedure consists of (1) choice of a kinematic model, (2) derivation of the error model equation, and (3) application of the annealing algorithm to find the optimal (or suboptimal) measurement configurations.

The two DOF manipulator was modeled in the MCPC convention and its Jacobian matrix was derived using the technique given in Chapter 5. For simplicity, only the two link lengths are treated as unknown kinematic parameters. The condition number was taken as the cost function. The cooling schedule given in (12.4.2) was chosen for simulation. The constant G was empirically chosen to be 0.5 and the temperature was changed only after every m^2 steps, where m is the number of robot measurement configurations for evaluating the identification Jacobian. The configuration set S consisted of a total of 20 candidate robot configurations. To generate an initial state X_0, a set of measurements $\{q_1, q_2, ..., q_m\}$ was randomly chosen from S. After the initial state was determined, one of the elements in X_0 was randomly chosen and swapped with one of the candidates in S. The condition number of the Jacobian was evaluated, and the decision of acceptance of the new configuration was made based on the annealing algorithm. The algorithm repeated until the terminal condition given in the last section was satisfied. The optimal configurations were also selected by applying Borm and Menq's analytical solution method, which maximizes the observability measure of the identification Jacobian for an all-revolute manipulator (1991). In this case no joint limits were introduced. Note that the optimum set of configurations is in general not unique; i.e., the set of measurement configurations possesses infinite number of possibilities and each of the choices with a specified number of measurements has the same condition number for J. Table 12.4.1 illustrates the condition number versus number of measurements selected by the SA algorithm for three different cases. These were the configurations selected randomly, optimally, and with the SA algorithm, respectively.

Table 12.4.1 Condition number vs. number of measurement configurations

No. of Meas.	Randomly selected	Optimum	Selected by SA
4	4.9904	2.6180	2.9704
5	3.7135	2.8864	2.9005
6	3.4475	2.6510	2.7982

In terms of the condition number, the results obtained by the SA algorithm were consistently better than those by random selection and very close to the optimal solutions. The advantage is that joint limits can be introduced in the SA algorithm.

E. EXPERIMENTAL RESULTS

The experiment system consisted of a 6-axis Puma 560 robot, a CCD camera (Electrophysics, model number CCD1200), a 486 personal computer, an ITEX® video imaging board with its driver software, a camera calibration board, and a 3-axis Coordinate Measuring Machine (CMM) (refer to Figure 12.4.1). Note that only one camera was used for data collection, although two cameras were shown in the figure. The CMM was used to move the calibration board to different locations to enlarge the measurable robot work space.

The monocular camera was calibrated at each robot measurement configuration to provide end-effector poses of the robotic system. The SA algorithm was tested in a reduced kinematic parameter error space that consisted only of the tool and base transformations. In this case, 10 error parameters are to be identified. A total number of 40 robot configurations was recorded and 10 configurations out of 40 were arbitrarily chosen for the initial configuration set. Five other robot configurations were taken for verification. The SA technique was applied to select 10 optimal configurations. Comparisons in 3D accuracy between those randomly chosen and those obtained by simulated annealing were made. Solutions of optimal configurations were affected by the selection of initial configuration set and the cooling schedule. These two crucial factors were tested through using different initial configurations and cooling schedules in the SA algorithm. Variations in the cooling schedule included the choice of the initial temperature, the temperature decrement rule, and the maximum number of iterations. Figure 12.4.2 illustrates the condition number versus the number of iterations for a particular number of configurations. Figure 12.4.3 shows the corresponding cooling schedule. x_0 was set to 0.5 in Figures 12.4.2 and 12.4.3. Figures 12.4.4 and 12.4.5 describe the case of using a faster temperature decremental schedule. By examining Figures 12.4.3 and 12.4.5 a faster temperature schedule always forces the system into an equilibrium state with less steps.

To verify the experimental results, pose measurements corresponding to the selected set of configurations are taken. These poses are in turn used to estimate parameters of the robot. The estimated parameters are then used to compute poses at robot configurations that are different from those used for parameter estimation. The computed poses are then compared with the measured poses of the robot. Figures 12.4.6 and 12.4.7 plot the position errors versus the number of configurations for comparison between two different approaches. One is randomly chosen, whereas the other is computed by simulated annealing. By observing the verification results, one can conclude that sufficient accuracy of robot calibration can be obtained by using only a few well-selected robot postures. As the number of measurement configurations increases, the superiority of using the SA method diminishes.

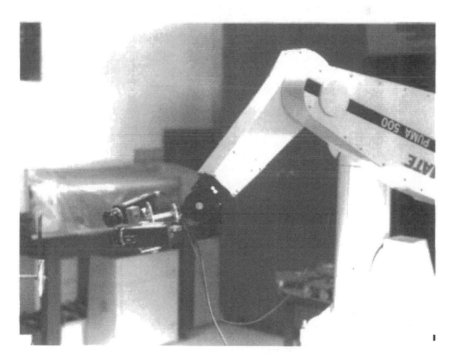

Figure 12.4.1 The experiment system setup

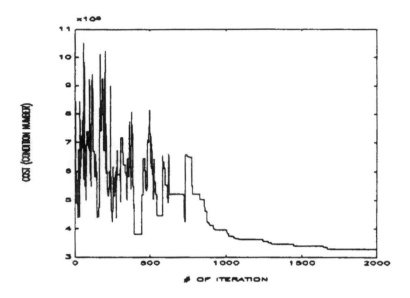

Figure 12.4.2 A cooling schedule

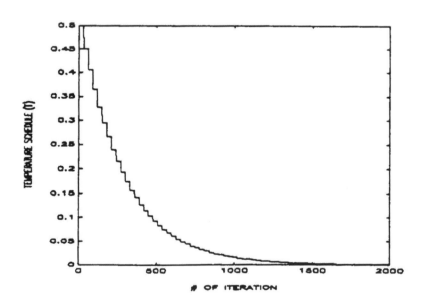

Figure 12.4.3 Condition number vs. number of iterations corresponding to the cooling schedule given in Figure 12.4.2 (Number of configurations = 10)

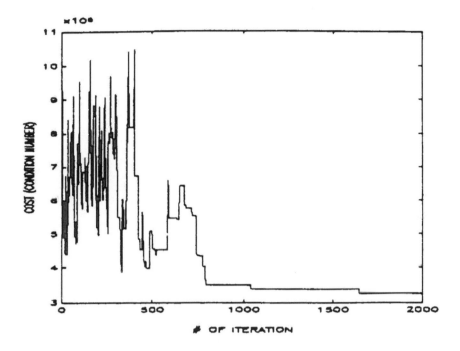

Figure 12.4.4 Another cooling schedule

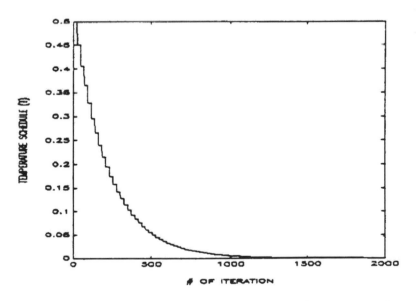

Figure 12.4.5 Condition number vs. number of iterations corresponding to
the cooling schedule given in Figure 12.4.4 (Number of configurations = 10)

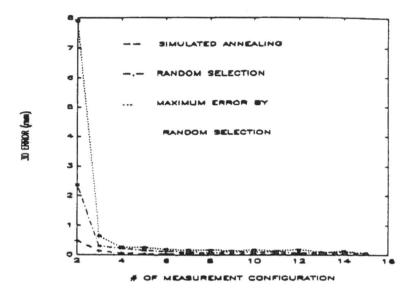

Figure 12.4.6 Comparison of 3D errors using the SA and random selection
algorithms

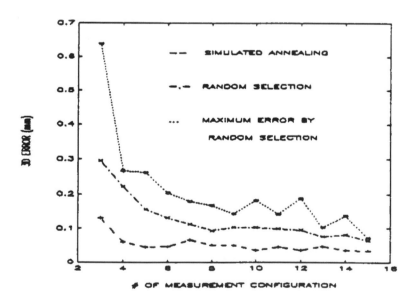

Figure 12.4.7 Subplot of Figure 12.4.6

V.   SUMMARY AND REFERENCES

Robot calibration researchers have paid much attention to optimally planning robot calibration experiments.   Driels and Pathre (1991) examined various factors affecting robot calibration performance.   Their method of selecting an optimal set of measurement configurations was based on monitoring the condition number of the Identification Jacobian.

Borm and Menq (1991) (also refer to Menq and Borm (1989)) proposed an observability index based on all singular values of the Identification Jacobian. They concluded that the number of measurements is not as important as the proper selection of measurement configurations.   Whenever the observability index increases, the contribution of geometric errors to robot positioning errors becomes dominant and the effect of measurement noise and unmodeled errors become less significant.   Although a closed-form solution was derived for an all-revolute robot, joint travel limits were not included in the analysis and hence the practical application of the method is limited.

Khalil, Gautier and Enguehard (1991) chose the condition number of the Identification Jacobian as the cost function and used a conjugate-type optimization method to search for optimal configurations.   Joint constraints were incorporated into the optimization problem.

While optimal robot configurations may be found through exhaustive off-line simulation studies using quantitative performance measures such as the observability index or the condition number of the Identification Jacobian, the computational cost can be way beyond the capacity of today's computers. This is due to the following two factors: 1) The dimension of the configuration space is generally very large. For instance, if one wants to use 10 robot measurements for the calibration of a 6 degrees of freedom manipulator, the number of variables involved in the exhaustive search is 60. 2) The cost function is normally complex.   For instance if the condition number of the Identification Jacobian is used as the cost function, each optimization step will require an estimate of the condition number. Although there exist efficient algorithms that estimate condition numbers ((Dixon (1983); Linz, Moler, Stewart and Wilkison (1979); Linz (1991); Maciejewski and Klein (1988) and Chiaverini (1993)), the accumulative computational cost is still significant.   On the other hand, conjugate-type or gradient-based algorithms may get stuck at local minima.

For most practical purposes, a simple search algorithm such as those given in Section III would be sufficient for the user to find an acceptable set of measurement configurations for robot calibration experimentation.

The Simulated Annealing (SA) approach was adopted in Zhuang, Wang and Roth (1994) to obtain optimal or near optimal measurement configurations for robot calibration.   Simulated annealing is a stochastic optimization method derived from Monte Carlo methods in Statistical

Mechanics (Kirkpatrick, Gelatt and Vecchi (1983) and Aarts and Korst (1989)). It has been successfully applied to a variety of optimization problems. Extension of the simulated annealing technique include the mean field annealing (Nobakht, Bout, Townsend and Ardalan (1990)) and the tree annealing (Han, Snyder and Bilbro (1990)). The simulated annealing algorithm is very computationally elaborate, though it often provides a better solution.

As to other type of stochastic search algorithms, Huang (1995) (also refer to Zhuang, Wu and Huang (1996) applied a genetic algorithm for the robot configuration selection problem. It was found that robot accuracy improvement by using genetic algorithms over random search is marginal, although genetic algorithms can provide a set of multiple suboptimal solutions ranked according to their performance measure value.

Chapter 13

PRACTICAL CONSIDERATIONS AND CASE STUDIES

I. INTRODUCTION

It is common to all engineering disciplines that theory must be motivated by practical engineering problems and that it is very important to verify the applicability of theoretical results through real-world applications. This is especially true for robot calibration that is basically a very practical problem. Robot calibration experiments in controlled laboratory conditions are essential for studying the feasibility of new ideas. It is a necessary step towards implementing such techniques on the manufacturing floor. This chapter examines through review of experimental results the pros and cons of some of the camera-aided calibration methods presented earlier in this book.

II. PRACTICAL CONSIDERATIONS

By trying to answer some frequently asked questions related to robot calibration experiments, we hope to steer robot users into venturing their own calibrations without taking the robots away from the manufacturing floor.

1. Which robot kinematic modeling convention is preferred?

This choice depends on the model complexity required for calibration and the amount of freedom that the user has (i.e., level of access to the robot control software). If robot accuracy is satisfactory by just calibrating the world-to-base and tool-to-camera transformations, the Denavit-Hartenberg kinematic model or any other model can be used to describe the robot internal geometry. In the case of SCARA arm or PUMA-type robots, if the user is able to choose any kinematic modeling convention, either the MCPC model or Hayati's modification to the D-H model may be used for kinematic modeling as these conventions are not singular for the standard SCARA arm or PUMA geometries. Some recent models of SCARA arms include a pitch axis as a last joint, in which case the MCPC model can still be made nonsingular. To completely calibrate a robot with an arbitrary geometry, a kinematic model that does not have model singularities such as the CPC model or Hayati's model needs to be used.

Readers are referred to Chapter 4 for a detailed account of kinematic modeling.

2. How is the maximum number of independent kinematic error parameters determined?

The upper bound on the number of independent kinematic parameters, excluding scale factors for joint variables, is $4N - 2P + 6$, where N is the number of degrees of freedom of the robot, and P is the number of prismatic joints. For instance, the maximum number of independent kinematic parameters for a PUMA arm is 30 as it is all revolute and has six degrees of freedom.

Care must be exercised, however, in determining the number of independent parameters in a particular robotic system. This is because the use of $4N - 2P + 6$ error parameters may yield a singular Identification Jacobian. For instance, as it for a typical *RRPR* SCARA arm, the upper bound on the number of kinematic error parameters is 20. However, since the end-effector has only four degrees of freedom, the maximum number of error parameters is reduced to 18. If one measures only the end-effector position, the number of identifiable parameters drops to 15. Furthermore, if the measured end-effector point lies along the roll axis, the number of independent parameters is further reduced to 13.

There is a convenient numerical way of determining the maximum number of independent parameters for a particular robot. First, a set of forward kinematic equations for the robot is developed. An Identification Jacobian can then be constructed numerically. As a rule of thumb, the number of randomly generated measurement configurations should be at least three times as much as the theoretical minimum number of required pose measurements. If the full pose of the robot can be measured, each pose measurement provides 6 scalar measurement equations. The minimum number of required pose measurements equals to $int((4N - 2P + 6)/6 + 0.5)$. Thus a recommended number of randomly generated configurations is at least $int((4N - 2P + 6)/2 + 0.5)$.

The next step is to apply a Singular Value Decomposition procedure to the Identification Jacobian. Theoretically, the number of nonzero singular values of the Jacobian provides the maximum number of independent parameters for the robot. Numerically, a suitable small threshold value can be set to separate the near-zero singular values from the other singular values.

For an analytic approach to the observability issue of robot kinematic parameters, readers are referred to Section 5.4.4.

3. How is a favorable set of kinematic error parameters selected?

It must be stressed that just by knowing the correct number of independent error parameters, it is still nontrivial to determine which specific error parameters are to be used.

Unlike the determination of the number of independent parameters of the robot, a favorable set of kinematic parameters cannot be computed by a simulation means only. In some cases, the answer is obvious. For instance, if the robot manufacturer does not provide the user any mechanisms to change internal parameters of the robot model, the sole option that the user can take is to calibrate only the world-to-base and tool-to-camera transformations. On the other extreme, if the user is allowed to use his own modeling convention within the robot controller (which is unlikely for most industrial robots), a complete set of independent parameters can then be employed to calibrate the robot. Another possibility is that the manufacturer provides the user with some flexibility in adjusting certain robot parameters such as joint offsets and link lengths. These parameters, together with the parameters in the world-to-base and tool-to-camera transformations may in this case be chosen for calibration.

After it is figured out which error parameters are available, a best set of error parameters may be determined by checking the effectiveness of accuracy compensation. Recall that the same set of data will be used repeatedly for kinematic parameter identification, with different combinations of error parameters. Moreover, the finite difference error model is handy for this purpose. The reader is referred to the last question and answer in this section for ways of conducting verification and compensation experiments.

4. How is robot pose data collected using a single hand-mounted camera?

The measurement strategy, devised by Lenz and Tsai, is reviewed again as follows. Assume that the robot is at a particular measurement pose. The world coordinates of the calibration points on the camera calibration board, together with their image coordinates are used to compute the intrinsic and extrinsic parameters of the camera model. Following this camera calibration step, the "pose" of the camera, which is just the position and orientation of the camera system in the world system, becomes known as these entities are uniquely defined from the extrinsic parameters of the camera model.

Assuming next that the robot is moved from one configuration to another while keeping the calibration board visible, the camera calibration task is to be repeated at each robot configuration, with the exception that the intrinsic parameters of the camera, once found, are treated as known quantities. This means that the poses of the camera, or equivalently the poses of the robot end-effector, are computed for each robot configuration. After the robot is moved to a sufficient number of measurement configurations, the robot kinematic error parameter vector can be estimated using the corresponding poses, which are essentially the models of the camera at different robot configurations.

Readers are referred to Chapter 4 for a detailed discussion of pose measuring with cameras.

5. Which set of measurement configurations should the user select?

It is well-known that by carefully selecting a proper set of measurement configurations, one can significantly improve on the calibration efficiency. Whenever a hand-mounted camera and a small camera calibration board are used to assist in calibration, the set of robot measurement configurations covers a relatively small portion of the entire robot joint space, leaving a very limited room for the user. One working method is to move the robot with the camera to various locations within the measurable robot workspace without conducting image acquisition and camera calibration. The main purpose of this exercise is to assure that the calibration points are fully visible from every desired robot configuration. After a sufficient number of robot joint configurations is collected, a search algorithm such as pairwise exchange or simulated annealing algorithm, can be applied off-line to select a near-optimal subset of robot measurement configurations from the collected set of joint configurations.

Chapter 12 describes several search algorithms for this purpose.

6. How can the calibrated robot workspace be enlarged?

As noted earlier, a severe problem with the camera-aided calibration approach is the limited workspace covered by calibration. Obviously, one way to remedy this problem is to use a sufficiently large calibration fixture, provided that it is not too invasive in terms of the application and costly. Another way is to calibrate the robot in several different locations. Each calibrated robot model will then be limited to the motion control of the robot for a specific region only. As a simple example let us consider a robot used for an assembly task. One can apply three different kinematic models to control the robot, one for operating near the parts feeder, another for operating within the assembly area, and yet another to be used whenever the robot operates near the conveyor that carries the outgoing assembled units. These kinematic models may be allowed to share the same internal link transformations, but will vary in terms of the world-to-base and tool-to-camera transformations.

7. How is the calibrated robot kinematic model verified?

There are two ways to check whether or not the calibrated model is valid. The first method consists of measuring the poses of the manipulator at different locations; then comparing these with the computed poses using the forward kinematics of the calibrated model. Because of applying the forward kinematics, this method is referred to as the *forward verification method*, or simply the *verification method*. The second method consists of solving the inverse kinematics of the identified model given the poses of the manipulator; then measuring, by using the output values of the inverse kinematic algorithm, the actual poses driven to. This method is defined as the *inverse verification method*, or simply the *compensation method* because it essentially solves an Accuracy Compensation problem. For techniques of robot accuracy compensation, readers are referred to Chapter 11.

Obviously whenever a forward verification method is employed, the sampled poses used for it have to be different from those used for the identification of the model. If the world coordinate system during the verification or the compensation phase is not the same as that during the identification phase, the transformation relating the new world system to the base system must be recalibrated.

8. How are pose errors compensated without modifying the kinematic model residing in the robot controller?

Robot manufacturers often provide users with the ability of redefining the world coordinate system and the end-effector coordinate system. If only the world-to-base and tool-to-camera transformations are calibrated, the user just needs to modify these transformations in the controller software before performing any motion planning. Similarly if the manufacturer allows the user only to modify certain parameters of the robot.

If the user decides to use the identified kinematic model for accuracy compensation, one can apply the so-called pose-redefinition procedure. Say that the goal pose of the robot is Location N, whose corresponding world-to-camera transformation is T_N. One can obtain a joint variable vector q_N by solving an inverse kinematic problem using the nominal kinematic parameter vector ρ_N. However, due to errors in the robot geometry, the robot will not reach Location N if the control command q_N is used. Pose T_A of the robot under this control command can be estimated by solving a forward kinematic problem using the identified kinematic parameter vector ρ_A. Let dT denote the difference between T_N and T_A ($dT = T_N - T_A$). Now, rather than using T_N as the pose of the task point N, one may redefine the pose to be $T_C = T_N + dT$. By solving another inverse kinematic problem with the *nominal* parameter vector ρ_N, one obtains q, which can in turn be used to control the robot to move to Location N. This procedure produces satisfactory results if the error parameters are small.

The compensation procedure described above can be readily implemented on any host computer. Again, readers may refer to Chapter 11 for techniques on robot error compensation.

III. CALIBRATION OF A PUMA ARM

A. THE SYSTEM SETUP

The experiment set-up consisted of a PUMA 560 robot, two CCD cameras, a PC-based image processing system and a Mitutoyo Model CX-D2 Coordinate Measuring Machine, as shown in Figure 13.3.1.

The CMM has a rated accuracy of approximately 0.1 mm and a work volume of 400x500x800 mm³. It was used to replace the robot calibration cube advocated in Chapter 4 in providing global measurements. The stereo cameras were employed to perform local measurements.

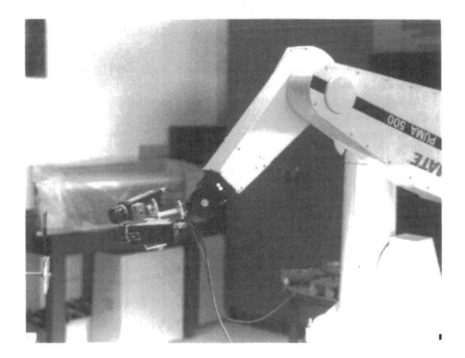

Figure 13.3.1 The measurement system set-up for PUMA calibration

The vision algorithms for camera calibration were written in Microsoft-C®. To estimate the image coordinates of a calibration point, an adaptive thresholding algorithm was devised. The algorithm first smoothed an image of the camera calibration board by applying a 3x3 low-pass mask three times. It then detected consecutively each calibration point, and computed the histogram of the image within a window surrounding each point. For example, if five calibration points were captured in the image, a typical size of each window would be about 50x50 pixels in a 512x512 image. The intersection of two Gaussian curves which fitted the histogram was chosen as the threshold value for binalization of each window area. The centroid of the image of calibration point was selected as the estimate of its image coordinates. This procedure could yield image coordinates accuracy to within 1/5 of a pixel.

B. PUMA CALIBRATION USING A HAND-MOUNTED STEREO CAMERAS

The experimental procedure was made up of the following steps:

1. Camera calibration, from which models were obtained for both cameras.
2. Pose measurement. Robot end-effector poses (or camera poses) in different robot measurement configurations were measured by utilizing the camera models.
3. Kinematic identification. By using these pose measurements, robot kinematic models of various complexities were identified.
4. Accuracy assessment. The identified kinematic models were verified.

Camera calibration was performed using the RAC-based camera calibration algorithm discussed in Chapter 2. Results for the camera calibration experiment part were already presented in Section 2.10, and are thus omitted here.

a. Robot Pose Measurement

Pose measurements were conducted using both the CMM and the stereo cameras. The CMM provided global information about robot poses, whereas the cameras measured the exact locations. The same precision board used for camera calibration was employed for robot pose measurement. This time it was mounted on the CMM. The cameras were attached to the robot end-effector.

The CMM coordinate frame was defined as the world frame. The transformation from CMM frame to the board coordinate frame (or $\{E_i\}$ as denoted earlier) were computed by using CMM readings. The transformation from the camera frame to the board coordinate frame was determined by stereo imaging. Thus the poses of the cameras in the world frame were obtained.

As the robot changed configurations, the CMM followed the robot hand so that the cameras could clearly view the precision board. This way, a sequence of robot end-effector poses was measured in terms of the world coordinate frame.

There are several implementation details that should be noted. An important issue is that of how to construct $\{E_i\}$ in the camera coordinate frame. Two lines were used to fit a cross-shaped dot pattern on the precision board. The cross-shaped pattern was obtained by masking out other dots on the board. By the principal axis method, a straight line was fitted to a set of points.   After the two line equations were obtained, their intersection point was found by a linear least squares algorithm.   This point was defined as the origin of $\{E_i\}$.   Ideally, the x and y axes of $\{E_i\}$ lie on each of the two lines are directly by the two lines, and the z axis can be obtained by a cross product operation.   However, due to measurement uncertainties, the axes of $\{E_i\}$ determined by the least squares method are rarely orthonormal.   An orthonormalization operation is therefore necessary, using a technique described in Section 3.4.D based on quaternions.

b.   Kinematic Parameter Identification
    An error model based method was applied to identify kinematic parameters of the PUMA arm.   The error model was represented in terms of the Complete and Parametrically Continuous (CPC) model.   Other error models can also be used for kinematic identification.
    The number of CPC error parameters for the PUMA robot is 30: each internal link transformation requires 4 parameters and 6 parameters account for the arbitrary assignment of the tool and world frames.   The error parameter vector dp used in the kinematic parameter identification is $[db_{0,x}, db_{0,z}, dl_{0,x}$ $dl_{0,y}, db_{1,x}, db_{1,z}, dl_{1,x}, dl_{1,y}, db_{2,x}, db_{2,y}, dl_{2,x}, dl_{2,y}, db_{,x}, db_{3,z}, dl_{3,x}, dl_{3,y}$ $db_{4,x}, db_{4,z}, dl_{4,x}, dl_{4,y}, db_{5,x}, db_{5,z}, dl_{5,x}, dl_{5,y}, db_{6,x}, db_{6,y}, dl_{6,x}, dl_{6,y}, dl_{6,z}$ $d\beta_6]^T$.   For definition of these error parameters, refer to Chapter 5.
    A nominal CPC model of the PUMA arm was determined by inspection from the robot data sheets with the exception of the *BASE* and *TOOL* transformations.   *BASE*, the transformation from the world frame to the base frame, was initially obtained using rough measurements with a simple ruler. The nominal *TOOL*, the transformation from the last link frame to the camera frame, was found from the robot nominal forward kinematic model using a single pose measurement.   This nominal CPC model of the PUMA arm obtained is listed in Table 13.3.1.

Table 13.3.1   Nominal CPC parameters of the PUMA 560

i	$b_{i,x}$	$b_{i,y}$	$b_{i,z}$	$l_{i,x}(mm)$	$l_{i,y}(mm)$	$l_{i,z}(mm)$	$\beta_i(rad)$
0	0.02786	0.99941	-0.01462	-123.973	-962.890	-320.146	3.08064
1	0	1	0	0	0	149.090	0
2	0	0	1	431.820	0	0	0
3	0	-1	0	-20.310	0	433.050	0
4	0	1	0	0	0	0	0
5	0	-1	0	0	0	0	0
6	-0.19356	0.16082	0.96782	-64.403	-93.906	323.364	3.06106

Twelve robot end-effector poses all in the PUMA right-arm configuration were collected for the kinematic identification process. Theoretically, 5 pose measurements are sufficient for identification of the 30 PUMA CPC parameters. In the experiment, a varying number (5, 6, 7, ..., 12) of pose measurements was used. The algorithm converged in 3-4 iterations in each of those cases. Table 13.3.2 lists the identified CPC parameters of the PUMA arm obtained by using 10 pose measurements. Identification results with other combinations of poses came to be very similar to those listed in Table 13.3.2.

The identified CPC models were verified through 7 additional pose measurements, also collected in the PUMA right-arm configuration. Figure 13.3.2 provides the means and standard deviations of the end-effector orientation errors (defined by $\|\delta l\|$) and position errors (defined by $\|dl\|$) against the number of poses used for kinematic identification. Because position errors from using only 5 measurements were far out of range, the result is excluded from Figure 13.3.2 (a). Evidently because the location of the nominal base frame in $\{W\}$ was measured by a ruler, pose errors were very large. These errors became smaller and smaller when more poses were used for parameter identification. As the number of poses exceeded 10, additional pose measurements did not improve the model accuracy in a noticeable way. As far as robot accuracy improvement is concerned, these results, based on data obtained with the mobile cameras, were consistent with published results obtained with other measurement techniques.

Table 13.3.2 Identified CPC parameters of the PUMA 560
with 10 pose measurements

i	$b_{i,x}$	$b_{i,y}$	$b_{,z}$	$l_{i,x}(mm)$	$l_{i,y}(mm)$	$l_{i,z}(mm)$	$\beta_i(rad)$
0	0.00606	0.99995	-0.00721	-102.155	-963.104	320.146	3.08064
1	-0.11893	0.99290	-0.00105	-7.226	61.237	149.090	0
2	-0.00100	0.00023	1.00000	430.434	37.921	0	0
3	0.09892	-0.99509	-0.00417	-21.388	-1.752	433.050	0
4	-0.12340	0.99235	0.00224	-0.337	1.280	0	0
5	0.04659	-0.99891	0.00266	0.545	-0.078	0	0
6	-0.09844	0.07576	0.99226	35.083	43.213	304.240	2.94444

With a PC-based 486 (33-MHz) vision system, it took about half a minute to grab and process images of a single robot pose. For a complete calibration of a robot manipulator, about 10 poses are sufficient. If proper communications between the robot and the vision system is established, the whole measurement process should take less than 10 minutes. The camera calibration and kinematic identification require additional time. The kinematic identification algorithm steps should take only in a few minutes on a PC-based system. The RAC-based camera calibration algorithm also takes only a couple of minutes. Therefore, in total, a complete robot calibration along the lines proposed in this book can be done in a fairly short time.

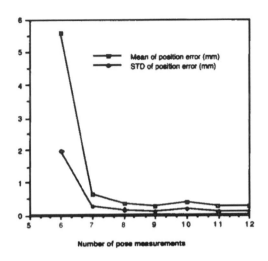

(a) Position errors

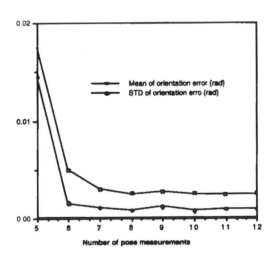

(b) Orientation errors

Figure 13.3.2   Verification results

C. PUMA CALIBRATION USING A HAND-MOUNTED CAMERA

The technique of simultaneous calibration of a robot and a hand-mounted camera presented in Chapter 7 was demonstrated on the PUMA 560 robot. Simulation as well as experimental results are both presented in this section.

a. Simulation results

In the simulation studies we chose to work with the modified complete and parametrically continuous (MCPC) model developed in Chapter 5. The main reason for selecting the MCPC model (over the CPC model) was its very compact error model. The MCPC model is singularity-free for the PUMA geometry, because for perpendicular axes of the PUMA arm, the link frames can be assigned in the way depicted in Figure 3.7.1. The nominal MCPC parameters of the PUMA 560 are given in Table 13.3.3.

Table 13.3.3 Nominal MCPC link parameters for the PUMA 560

i	x_i (mm)	α_i (deg)	y_i (mm)	β_i (deg)	z_6 (mm)	γ_6 (deg)
0	-100	90	-1000	180		
1	0	-90	-260	0		
2	431.8	0	0	0		
3	-20.3	90	149.09	0		
4	0	-90	-433.07	0		
5	0	90	0	0		
6	0	0	0	0	300	0

To simulate an actual robot, a set of randomly selected parameter deviations, uniformly distributed between ± 2 mm for translational parameters and ± 0.005 rad for orientational parameters, were added to the robot kinematic parameters. The following three issues were investigated:

1. Number of iterations required for the algorithm to converge.
2. Accuracy performance of the algorithm under different levels of noise intensities.
3. Accuracy performance comparison of the one-stage algorithm to a two-stage algorithm. The later is based on recalibration of the camera at every robot calibration measurement configuration. For the two-stage algorithm, the RAC-based camera calibration technique was employed to compute robot poses at those measurement configurations. These poses were then plugged into the robot geometry identification model.

Table 13.3.4 lists three types of noise parameters used in the simulations. U[a, b] denotes a random variable uniformly distributed in the interval [a, b]. Type I noise models a moderate noise level for both position and orientation measurements. Type II has a larger orientation noise level. Type III models a very noisy measurement system.

Table 13.3.4 Noise intensity levels used in simulation

	Position noise (mm)	Orientation noise (rad)
Type I	U[-0.1, 0.1]	U[-0.0001, 0.0001]
Type II	U[-0.1, 0.1]	U[-0.0005, 0.0005]
Type III	U[-0.5, 0.5]	U[-0.0005, 0.0005]

The Levenberg-Marquardt nonlinear least-squares algorithm was applied to identify link parameters under different simulation conditions. The initial condition for the robot link parameters was listed in Table 13.3.3. In all simulation runs, the algorithm converged within five iterations. The accuracy performance of the algorithms was evaluated using the forward verification approach described in the Chapter 11. Representative simulation results are shown in Figures 13.3.3-13.3.5. The horizontal axes in these figures represent the number of "measurements" (note that each produces two algebraic equations) for the estimation problem. The vertical axes of these figures provide the mean and standard deviation of the rotation and position estimation errors. For the one-stage algorithm, five image points were measured in each robot configuration. Similarly, five object points were measured to determine a pose in the two-stage algorithm. Determining a pose from the measured five object points can be formulated as the problem of fitting a homogeneous transformation from two sets of point measurements, and either the SVD or the quaternion based technique can be applied to find the solution.

As we study the simulation results, the following observations are due:

1. As expected, estimation error increases in an amount which is roughly proportional to the noise intensity (comparing Figures 13.3.3-13.3.5).
2. The one-stage algorithm generally outperforms the two-stage algorithm in terms of accuracy (Figure 13.3.5).
3. By adding measurements, estimation errors decrease gradually until the number of measurements becomes greater than ten, thereafter no significant improvement can be observed.

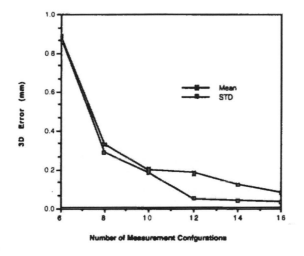

Figure 13.3.3 Performance of the one-stage approach under type-I noise

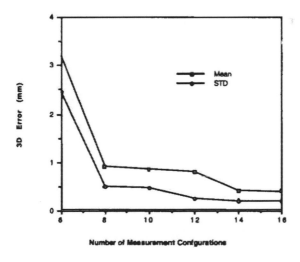

Figure 13.3.4 Performance of the one-stage approach under type-III noise

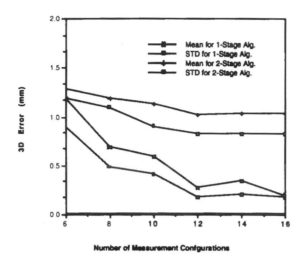

Figure 13.3.5 Comparison of the one-stage and two-stage approaches

b. Experimental results

The experiment system was the same as that described in Section 13.3.1 with the exception that only a single camera was used. Since a full-scale camera calibration was not needed, only the following steps were involved in the experimentation:

1. Measurement data collection. Images of calibration points were captured while the robot was moved to different joint configurations. Subsequently the 2D image coordinates of these points were estimated.
2. Kinematic identification. The robot and camera parameters were identified from the above measurement data.
3. Accuracy assessment. The identified kinematic model was verified. Accuracy performance was compared to that achieved by using other approaches.

The following issues were studied experimentally:

1. The performance of the one-stage algorithm with and without taking into account lens distortion.
2. Performance comparison between the one-stage algorithm and a two-stage algorithm.
3. The performance of the one-stage algorithm for various modeling complexities.

Note that in both 2 and 3, we used the one-stage algorithm with a lens distortion model. In 3, it is further assumed that in order to calibrate a particular subsystem of the robotic system, the remaining parts have been precalibrated.

The initial conditions of the robot link parameters for the Levenberg-Marquardt algorithm are listed in Table 13.3.3 with the exception that the initial condition for $BASE$, the transformation relating the robot base frame to the world frame, and $TOOL$, the transformation relating the camera frame to the robot flange frame, were found by crude measurements.

Four data sets with a total of 52 measurement configurations were recorded. Out of these four steps, one was reserved for the purpose of verification. At each measurement configuration, an average number of five calibration points of the calibration board and their corresponding image points were used for parameter identification. The transformation between the CMM and the calibration board, which was mounted on the CMM probe, can be obtained by separately moving the calibration board in the x, y and z directions of the CMM. Thus the world coordinates of the calibration dots on the calibration board can be computed following each CMM movement. The accuracy performance of the algorithms was again evaluated using the forward verification method. Figures 13.3.6-13.3.8 illustrate the experimental results.

The following observations can be made based on the experimental results:

1. The accuracy performance of the one-stage algorithm is on par with those obtained using non-vision techniques. Following the robotic system calibration process, the robot position error residuals were at the level of 0.2 mm (refer to Figure 13.3.6).
2. Inclusion of the lens radial distortion in the robot-camera model coefficient significantly improves the accuracy performance of the one stage algorithm (refer to Figure 13.3.6).
3. Experimental results confirm that the one-stage algorithm in general outperforms the two-stage algorithm (refer to Figure 13.3.7). Note that the accuracy performance of the two-stage algorithm highly depends on that of the camera calibration procedure. In our study, the ED error of the camera model came to be about 0.05 mm.
4. If the rest of the system had been calibrated in advance, partial calibration can achieve an accuracy compatible with that obtained by calibration of an entire robotic system calibration (refer to Figure 13.3.8).

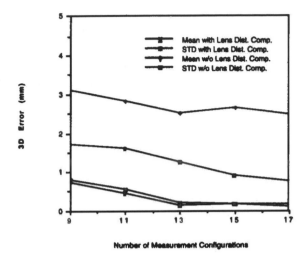

Figure 13.3.6 Accuracy performance comparison
for system calibration with or without lens distortion compensations

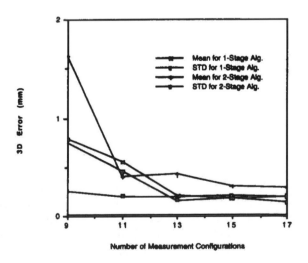

Figure 13.3.7 Comparison of accuracy achieved by the one-stage
and two-stage approaches

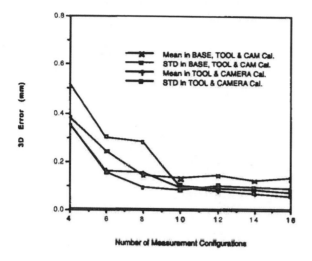

Figure 13.3.8 Compensation results of BASE, TOOL and CAMERA calibration vs. TOOL and CAMERA calibration

IV. CALIBRATION OF A SCARA ARM

A. THE SYSTEM SETUP

The experimental system consisted of a 4-DOF Intelledex robot, a hand-mounted CCD camera, a 486 (33-MHz) PC, an ITEX® video imaging board with its driver software, and a camera calibration board (refer to Figure 13.4.1).

Similar to the calibration of the PUMA arm, this experimentation also followed the following four steps:

1. Repeated Camera calibration. Because the camera optical axis was nearly perpendicular to the camera calibration board, the special calibration procedure described in Section 2.5 was adopted.
2. Pose measurement. Directly the camera models, robot end-effector poses (or camera poses) in different configurations were obtained.
3. Kinematic identification. Robot kinematic models of various complexities were identified using these pose measurements.
4. Accuracy assessment. The identified kinematic models were verified.

Readers are referred to Section 2.10.4 for details on results of the camera calibration. We now concentrate on robot calibration.

Figure 13.4.1   The measurement system set-up for SCARA arm calibration

B. CALIBRATION OF AN INTELLEDEX ROBOT

Lenz and Tsai's pose measurement strategy, outlined in Section 3.3.2, was applied to collect robot pose data using a monocular camera. Prior to performing robot calibration, we tested the repeatability of the robot/camera system. The robot was commanded to repeatedly move its hand-mounted camera to the same location, at which the camera was recalibrated. Repeatability error has been defined as the difference between the mean position and the ith measured position at each direction on the image plane. The standard deviations for position repeatability errors at the x and y directions were found to be 0.606 and 0.827 pixels, respectively.

The link coordinate systems of the Intelledex robot are shown in Figure 13.4.2. Note that because the camera was mounted on the third link of the manipulator, we were only able to calibrate the first three joints. A miniature camera is needed to be able to mount the camera on the last link of the robot. The nominal MCPC parameters used in the parameter identification are listed in Table 13.4.1. The number of identifiable parameters can be shown by the SVD method to be 14, as the motion of the last revolute joint is not measured. The calibration board was used to define the world coordinate system.

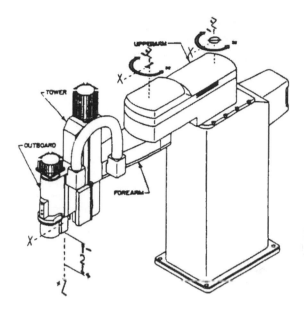

Figure 13.4.2 Link coordinates assignment for the Intelledex robot

The number of identifiable parameters reduces to 11 whenever we choose to calibrate the world-to-base and tool-to-camera transformations. Different combinations of error parameters were then tested. Results of some typical cases are presented. Table 13.4.2 lists the identification result of using 14 parameters to describe the entire robot. The identification result by using only 11 parameters is shown in Table 13.4.3. In both cases, the same four robot poses were used in the parameter identification process.

The identified kinematic models were verified at five new robot poses. The test results are shown in Tables 13.4.2 and 13.4.3. Note that the orientation of the robot was defined by the ZYX Euler angles, denoted by α, β and γ.

The following comments can be made with regard to the verification results shown in Tables 13.4.2 and 13.4.3:

1. It can be seen that the accuracy at the z direction is the lowest. This is because the accuracy of the camera measurement in this direction is inferior compared to other directions. Fortunately, for most SCARA applications, such as electronic assembly, the accuracy in the z direction is of secondary importance.

2. The accuracy performance, especially of orientation accuracy, whenever using a well-selected complete set of error parameters is better than that of using only parameters from the world-to-base and tool-to-camera transformations. This is a much anticipated result.

Table 13.4.1 Nominal MCPC Parameters of the Intelledex Robot

Trans.	α (rad)	β (rad)	γ (rad)	x (mm)	y (mm)	z (mm)
w-0	π	0	0	-29.5	58.2	17
0-1	0	0	0	30.48	0	0
1-2	π	0	0	45.72	0	0
2-3	0	0	$\pi/2$	0	0	0

Table 13.4.2 Identified MCPC Parameters (I) *

Trans.	α (rad)	β (rad)	γ (rad)	x (mm)	y (mm)	z (mm)
w-0	[3.14215]	[-0.00757]	[-0.21172]	[-32.83050]	[58.73710]	[17.48809]
0-1	0	0	0	[31.368338]	[5.25398]	0
1-2	π	0	0	[44.955517]	0	0
2-3	[-0.01457]	0.014125	[1.478590]	0	[0.39430]	[0.118382]

* The parameters whose values are shown in brackets were the ones chosen as error parameters in the parameter identification. There are 14 error parameters.

Table 13.4.3 Identified MCPC Parameters (II) *

Trans.	α (rad)	β (rad)	γ (rad)	x (mm)	y (mm)	z (mm)
w-0	[3.15034]	[-0.05602]	[-0.04648]	[-23.65510]	[53.09962]	[52.10221]
0-1	0	0	0	30.48	0	0
1-2	π	0	0	45.72	0	0
2-3	[-0.00883]	[0.13563]	[1.64791]	0	[0.20024]	[35.39887]

* The parameters whose values are shown in brackets were the ones chosen as error parameters in the parameter identification. There are 11 error parameters.

Table 13.4.4 Verification Results (I) *

	Orientation Error			Position Error		
	α (rad)	β (rad)	γ (rad)	x (mm)	y (mm)	z (mm)
Mean	0.00783	0.00534	0.00048	0.03103	0.03553	0.08284
Max.	0.00988	0.00598	0.00076	0.04664	0.05698	0.14459

* These verification results correspond to the identified model given in Table 13.4.2.

Table 13.4.5 Verification Results (II) *

	Orientation Error			Position Error		
	α (rad)	β (rad)	γ (rad)	x (mm)	y (mm)	z (mm)
Mean	0.01301	0.07910	0.00750	0.02607	0.02819	0.18146
Max.	0.01745	0.07855	0.00787	0.05376	0.05475	0.37602

* These verification results correspond to the identified model given in Table 13.4.3.

3. The performance obtained when only calibrating the world-to-base and tool-to-camera transformations is quite acceptable, especially in terms of position accuracy. This is encouraging since in this way modifications to the controller software can be kept to a minimum.

Accuracy compensation of the Intelledex robot can be carried out with the methods presented in Chapter 11. The Intelledex robot can be programmed using C code on a PC, which acts as a host computer. Whenever a robot command is executed in the PC environment, a data string is sent through a RS 232 line, which represents the command to the robot controller. The robot controller then carries out the instruction accordingly. In the case of a motion command, the corresponding joint variable vector is computed by the PC using both the nominal and calibrated kinematic models, prior to being sent to the robot controller.

IV.   SUMMARY AND REFERENCES

Experimental results on the PUMA arm using the stereo cameras were reported in Zhuang, Wang and Roth (1993). Simultaneous calibration results of the PUMA and the hand-mounted camera were cited from Zhuang, Wang and Roth (1995). Finally, the results of Intelledex arm calibration were presented in Zhuang, Wu and Roth (1995).

Experimental studies on vision-based robot calibration can also be found in (Puskorius and Feldkamp (1987); Lenz and Tsai (1989) and Preising and Hsia (1991)). A study performed by Shih, Hung and Lin (1995) deserves special attention. In this paper, a closed-form solution for identifying the kinematic parameters of an active binocular head using a single calibration point was proposed. The method was based on the CPC kinematic model. As it is more difficult to measure end-effector poses than to measure their positions only, that method is more convenient for implementation in certain applications. The same authors also conducted theoretical analysis for the estimation error, which produced a method to reduce the estimation error by controlling the factors that have great influence on the parameter estimation quality. They also presented extensive simulation and experimental results, which matched closely with the predications based on the theoretical analysis.

Returning to the experimental results presented earlier in this Chapter, the methods possess the following features:

1.   The moving camera coordinates concept allows the cameras to perform precise local measurements, as the field-of-view of the camera system can be reduced to about 50x50 mm^2, resulting in an overall accuracy of the measurement system to a level of 0.05 mm. This is accomplished using off-the-shelf cameras and lenses.   Such accuracy is at least 20 times better than the accuracy provided by conventional stationary vision-based measurement systems.

2.   The measurement system provides both orientation and position of the robot end-effector.   Using a cubic-shaped robot calibration fixture will allow the robot to be exercised in a wider range of its joint workspace.

3.   By calibrating the radial lens distortion coefficient, the accuracy performance of these procedures can be greatly improved. This is true for both camera calibration and for camera-aided robot calibration.

4.   By employing a nonlinear least-squares procedure such as the Levenberg-Marquardt algorithm, the calibration procedure converges rapidly, as a good set of initial conditions for the robot and camera parameters is usually available.

5.   These procedures can be automated if a calibration fixture is designed such that the robot can be moved to a sufficiently large portion of its feasible

work space while maintaining a view of the calibration fixture with good resolution. A development is still under way to create fully automated calibration process.

As has been mentioned in earlier chapters, if one changes the camera position with respect to the robot hand, the transformation from the camera to the robot hand has to be recalibrated whenever a hand-mounted camera system is employed to collect pose data of the robot.

Among the techniques used in the experimental studies, the first method was a two-stage approach with stereo cameras. The second method was to calibrate simultaneously the robot and the camera, and the last method, also a two-stage approach, utilizes only hand-mounted cameras. The simultaneous calibration method has the following merits:

1. The camera and robot parameters are identified simultaneously. Thus propagation errors which exist in multi-stage calibration approaches are eliminated. On the other hand, accuracy performance of the two-stage algorithm depends highly on the accuracy of the camera calibration procedure. All this is reflected in the experimental results.
2. The calibration scheme uses a monocular camera, by which the field-of-view of the camera can be larger than that of a pair of stereo cameras. Moreover, the processing speed of a single camera system is faster.
3. Only two camera parameters were added to the robot calibration model. With this addition, different levels of calibration can be accomplished under a unified framework.

The major limitations of the simultaneous calibration approach are:

1. Since only one camera is used, the system cannot recover the absolute distance information of the scene.
2. The extrinsic parameters of the camera are absorbed into the robot parameters. This is equivalent to treating the camera coordinate system as an "imaginary tool system". If an actual tool is also attached to the robot hand, the transformation from the tool to the camera has to be determined separately. This problem exists in all robotic systems equipped with a hand/eye configuration.

As to the calibration of SCARA arms, the technical difficulty that arises because of the necessity of using a camera whose optical axis is nearly perpendicular to the calibration board is overcome with a special camera calibration technique. Selection of different sets of error parameters for kinematic identification becomes a simple task when a difference error model is employed. In this chapter, we have presented some results from the calibration of three joints of a SCARA arm with the aid of a camera system.

REFERENCES

Aarts, E. and J. Korst (1989) *Simulated Annealing and Boltzmann Machines*, John Wiley & Sons.

Angeles, J. (1992) "The Design of Isotropic Manipulator Architectures in the Presence of Redundancies," *Int. J. Robot. Res.*, Vol. 11, No. 3, pp. 196-201.

Arun, K. S., T. S. Huang, and S. D. Blostein (1987) "Least-Squares Fitting of Two 3-D Point Sets," *IEEE Trans. Pattern Anal. Machine Intell.*, Vol. 9, No. 5, pp. 698-700.

Athans, M. and P. L. Falb (1966) *Optimal Control: An Introduction to the Theory and Its Application*, McGraw-Hill Book Company.

Barker, L. K. (1983) "Vector-Algebra Approach to Extract Denavit-Hartenberg Parameters of Assembled Robot Arms," *NASA Technology Paper* 2191, August.

Bennett, D. J. and J. M. Hollerbach (1989) "Identifying the kinematics of robots and their tasks," *Proc. IEEE Int. Conf. Robotics & Automation*, Scottsdale, Arizona, pp. 580-586.

Bennett, D. J. and J. M. Hollerbach (1991) "Autonomous Calibration of Single-Loop Closed Kinematic Chains Formed by Manipulators with Passive Joints," *IEEE Trans. Robotics & Automation*, Vol. 7, No. 5, pp. 597-606.

Bennett, D. J., D. Geiger, and J. M. Hollerbach (1991) "Autonomous Robot Calibration for Hand-Eye Coordination," *Int. J. Robotics Research*, Vol. 10, No. 5 pp. 550-559.

Bernhardt, R. and S. L. Albright (1993) *Robot Calibration*, Chapman and Hall: London.

Borm, J. H. and C.H. Menq (1991) "Determination of Optimal Measurement Configurations for Robot Calibration Based on Observability Measure," *Int. J. Robotics Research*, Vol. 10, No. 1, pp. 51-63.

Broderick, P. L. and R. J. Cipra (1988) "A Method for Determining and Correcting Robot Position and Orientation Errors Due to Manufacturing," *ASME J. Mechanisms, Transmissions, and Automation in Design*, Vol. 110/3, March, pp.3-10.

Brown, R. G. and P. Y. C. Young (1992) *Introduction to Random Signals and Applied Kalman Filtering*, 2nd Edition, Iohn Wiley & Sons.

Chasles, M. (1830) "Note sur les Proprietes Generales dv Systeme de Denx Corps," Bull. Sci. Math. Ferrusac, 14, pp. 321-326.

Chen, C. T. (1970) *Introduction to Linear System Theory*, Holt, Rinehart and Winston, Inc..

Chen, S. (1987) *Vision-Based Robot Calibration*, M.S. Thesis, Florida

Atlantic University, Boca Raton, Florida.

Chen, J. and L. M. Chao (1986) "Positioning Error Analysis for Robot Manipulators with All Rotary Joints," *Proc. IEEE Int. Conf. Robotics Automation*, pp. 1011-1016.

Chiaverini, S. (1993) "Estimate of the Two Smallest Singular Values of the Jacobian Matrix: Application to Damped Least-Squares Inverse Kinematics," *J. Robotics Systems*, Vol. 10, No. 8, pp. 991-1008.

Chou, J.C.K. (1992) "Quaternion Kinematic and Dynamic Differential Equations," *IEEE Trans. Robotics & Automation*, Vol. 8, No. 1, pp. 53-64.

Chou, J. K. and M. Kamel (1988) "Quaternions Approach to Solve the Kinematic Equation of Rotation, $A_a A_x = A_x A_b$, of a Sensor-Mounted Robotic Manipulator," *Proc. IEEE Int. Conf. Robotics & Automation*, Philadelphia, PA, pp. 656-662.

Chou, J. K. and M. Kamel (1991) "Finding the Position and Orientation of a Sensor on a Robot Manipulator Using Quaternions," *Int. J. Robotics Research*, Vol. 10, No. 3, pp. 240-254.

Craig, J. J. (1989) *Introduction to Robotics: Mechanics and Control*, Addison Wesley.

Dahlquist G. and A. Bjorck (1974) *Numerical Methods*, Translated by N. Anderson, Prentice-Hall.

Denavit, J. and R. S. Hartenberg (1955) "A Kinematic Notation for Lower-Pair Mechanisms Based on Matrices," *ASME J. Applied Mechanics*, pp. 215-221.

Dixon, J. (1983) "Estimating Extremal Eigenvalues and Condition Numbers of Matrices," *SIAM J. Numerical Analysis*, Vol. 20, No. 4, pp. 812-814.

Driels, M. R. and U. S. Pathre (1990) "Significance of Observation Strategy on the Design of Robot Calibration Experiments," *J. Robotics Systems*, Vol. 7, No. 2, pp. 197-223.

Driels, M. R. and U. S. Pathre (1991) "Vision-Based Automatic Theodolite for Robot Calibration," *IEEE Trans. Robotics & Automation*, Vol. 7, No. 3, pp. 351-360.

Everett, L. J., M. Driels, and B. W. Mooring (1987) " Kinematic Modelling for Robot Calibration," *Proc. IEEE Int. Conf. Robotics Automation*, pp. 183-189.

Everett, L. J. and T. W. Hsu (1988) "The Theory of Kinematic Parameter Identification for Industrial Robots," *ASME J. Dynamic Systems, Measurement, and Control*, Vol. 110, pp. 96-100.

Everett, L.J. and L.E. Ong (1991) "Solving the Generalized Sensor-Mount Registration Problem," ASME Winter Annual Meeting, DSC-Vol. 29, pp. 7-14.

Everett, L. J. and A. H. Suryohadiprojo (1988) "A Study of Kinematic

Models for Forward Calibration of Manipulators, " *Proc. IEEE Int. Conf. Robotics Automation*, Philadelphia, PA, April, 1988, pp. 798-800.

Fu, K. S., R. C. Gonzalez, and C. S. G. Lee (1987) *Robotics: Control, Sensing, Vision, and Intelligence*, McGraw-Hill.

Gill, P. E., W. Murray, and M. H. Wright (1981) *Practical Optimization*, Academic Press.

Golub, G. H., F. Charles and V. Loan (1983), *Matrix Computation*, The Johns Hopkins University Press.

Hajek, B. (1988) "Cooling Schedule for Optimal Annealing," *Math. Oper. Res.*, Vol. 13, No. 2, pp. 311-329.

Hartley, H. O. (1961) "The Modified Gauss-Newton Method for the Fitting of Nonlinear Regression Functions by Least Squares," *Technometrics*, Vol. 3, No. 2, pp. 269-280.

Hayati, S. (1983) "Robot Arm Geometric Parameter Estimation," *Proc. 22th IEEE Int. Conf. Decision Control*, pp. 1477-1483.

Hayati, S. and M. Mirmirani (1985) "Improving the Absolute Positioning Accuracy of Robot Manipulators," *J. Robotic Systems*, Vol. 2, No. 4, pp. 397-413.

Han, Y., W. Snyder, and G. Bilbro (1990) "Determination of Optimal Pose Using Tree Annealing," *Proc. IEEE Int. Conf. Robot. & Automation*, Cincinnati, OH, May 1990, pp. 427-432.

Hollerbach, J. M. (1988) "A Survey of Kinematic Calibration," *Robotics Review*, edited by O. Khatib, J. J. Craig and T. Lozano-Perez, MIT Press.

Hollerbach, J. M. and D. M. Lokhorst (1995) "Closed-Loop Kinematic Calibration of the RSI 6-DOF Hand Controller," *IEEE Trans. Robot. & Automation*, Vol. 11, No. 3, pp. 352-359.

Horn, B. K. P. (1987) "Closed-Form Solution of Absolute Orientation Using Unit Quaternions," *J. Opt. Soc. Amer. A.*, Vol. 4, No. 4, pp. 629-642.

Horn, B. K. P., H. M. Hilden, and S. Negahdaripour (1988) "Closed-Form Solution of Absolute Orientation Using Orthonormal Matrices," *J. Opt. Soc. Amer. A.*, Vol. 5, No. 7, pp. 1127-1135.

Huang, W. (1995) *Optimal Planning of Robot Calibration Experiments by Genetic Algorithm*, M.S. Thesis, Florida Atlantic University, Boca Raton, Florida.

Huang, M. Z. and A. Gautam (1989) "An Algorithm for On-line Compensation of Geometric Errors in Industrial Robots," *Proc. 2nd Conf. on Recent Advances in Robotics*, May, Florida Atlantic University, Boca Raton, FLorida, pp. 174-177.

Hunt, K.H. (1978) "Kinematic Geometry of Mechanisms," Clarendon Press, Oxford, UK.

Jiang, B. C., J. T. Black, and R. Duraisamy (1988) "A Review of Recent Development in Robot Metrology,"*J. Manufacturing Systems*, Vol. 7,

No. 4, pp. 339-357.

Kalman, R. E. and R. W. Koepcke (1958) "Optimal Synthesis of Linear Sampling Control Systems Using Generalized Performance Indices," *Trans. ASME*, Vol. 80, pp. 1820-1838.

Kanatani, K. (1993) "Unbiased Estimation and Statistical Analysis of 3-D Rigid Motion from Two Views," *IEEE Trans. Pattern Anal. Machine Intell.*, Vol. 15, No. 1, pp. 37-50.

Khalil, W., M. Gautier, and C. Enguehard (1991) "Identifiable Parameters and Optimum Configurations for Robot Calibration," *Robotica*, Vol. 9, pp. 63-70.

Kim, J.-O. and P.K. Khosla "Dexterity Measures for Design and Control of Manipulators," *Proc. IEEE/RSJ Int. Workshop on Intelligent Robot. & Syst.*, Osaka, Japan, pp. 758-763.

Kirkpatrick, S., C. D. Gelatt, and M. P. Vecchi (1983) "Optimization by Simulated Annealing," *Sci.* Vol. 220, No. 4598, pp. 671-680.

Lau, K., N. Dagalakis, and D. Myers (1988) "Testing," *International Encyclopedia of Robotics, Application and Automation*, R. C. Dorf and S. Y. Nof, John Wiley & Sons, Inc., Vol III., pp.1753-1769.

Lau, K., R. Hocken, and L. Haynes (1985) "Robot Performance Measurements Using Automatic Laser Tracking Techniques," *Robotics and Computer-Integrated Manufacturing*, Vol. 2, No. 3, pp. 227-236.

Lenz, R. K. and R. Y. Tsai (1987) "Technique for Calibration of the Scale Factor and Image Center for High Accuracy 3D machine Vision Metrology," *Proc. IEEE Int. Conf. Robotics & Automation*, pp. 68-75.

Lenz, R. K. and R. Y. Tsai (1989) "Calibrating a Cartesian Robot with Eye-on-Hand Configuration Independent of Eye-to-Hand Relationship," *IEEE Trans. PAMI*, Vol. 11, No. 9, pp. 916-928.

Levenberg, K. (1944) "A Method for the Solution of Non-linear Problems in Least Squares," Quart. Appl. Math., 2, pp. 164-168.

Linz, P. (1991) "Bounds and Estimates for Condition Numbers of Integral Equations," *SIAM J. Numerical Analysis*, Vol. 28, No. 1, pp. 227-235.

Linz, P., C. B. Moler, G. W. Stewart, and J. H. Wilkinson (1979) "An Estimate for the Condition Number of a Matrix," *SIAM J. Numerical Analysis*, Vol. 16, No. 2, pp. 368-375.

Liu, Y., T. S. Huang, and O. D. Faugeras (1990) "Determination of Camera Location from 2-D to 3-D Line and Point Correspondences", *IEEE Trans. Pattern Anal. Machine Intell.*, Vol. 12, No. 1, January, pp. 13-27.

Lou, R. C. and W. S. Yang (1990) "Motion Estimation of 3-D Object Using Multisensor Data Fusion," *J. Robotics Systems*, Vol. 7, No. 3, pp. 419-443.

Maciejewski, A. A. and J. M. Reagin (1994) "A Parallel Algorithm and Architecture for the Control of Kinematically Redundant Manipulators," *IEEE Trans. Robot. & Automat.*, Vol 10, No. 4, pp. 405-414.

Maciejewski, A. A. and C. A. Klein (1988) "Numerical Filtering for the Operation of Robotic Manipulators through Kinematically Singular Configurations," *J. Robot. Syst.*, Vol. 5, No. 6, pp. 527-552.

Marquardt, D. W. (1963) "An Algorithm for Least-Squares Estimation of Nonlinear Parameters," *J. Soc. Indust. Appl. Math.*, Vol. 11, No. 2, pp. 431-441.

Martins, H. A., J. R. Birk, and R. B. Kelley (1981) "Camera Models Based on Data from Two Calibration Planes," *Computer Graphics and Image Process*, 17, pp. 173-180.

MACSYMA® Reference Manual (1985) Symbolics, Inc.

Maybank, S. and O. D. Faugeras (1990) "A Theory of Self-Calibration of a Moving Camera," *Int. J. Computer Vision*, Vol. 8, No. 2, pp. 123-151.

Menq, C. H. and J. H. Borm (1988) " Estimation and Observability Measure of Parameter Errors in a Robot Kinematic Model," *Proc. U.S.A.-Japan Symposium on Flexible Automation*, Minneapolis, MN, July, pp. 65-70.

Mooring, B. W. (1983) "The Effect of Joint Axis Misalignment on Robot Positioning Accuracy," *Proc. ASME Int. Computers in Eng. Conf. Exhibit*, Illinois, pp. 151-156.

Mooring, B. W. and S. S. Padavala (1989) "The Effect of Kinematic Model Complexity on Manipulator Accuracy," *Proc. IEEE Int. Conf. Robot. & Automat.*, Scottsdale, Arizona, pp. 593-598.

Mooring, B. W. and G. R. Tang (1984) "An Improved Method for Identifying the Kinematic Parameters in a Six Axis Robot," *Proc. Int. Comput. in Eng. Conf. Exhibit*, August, Las Vegas, Nevada, pp 79-84.

Mooring, B. W., Z. S. Roth, and M. Driels (1991) *Fundamentals of Manipulator Calibration*, John Wiley & Sons.

Mortenson, M. E. (1985) *Geometric Modeling*, John Wiley & Sons.

Murray, R. M., Z. Li, and S. S. Sastry (1994) *A Mathematical Introduction to Robot Manipulation*, CRC Press.

Nobakht, R.A., D. E. Bout, J. K. Townsend, and S. H. Ardalan (1990) "Optimization of Transmitter and Receiver Filters for Digital Communication Systems Using Mean Field Annealing," *IEEE J. Selected Areas in Communications*, Vol. 8, No. 8, pp. 1472-1480.

Nakamura, Y. and H. Hanafusa (1986) "Inverse Kinematic Solutions with Singularity Robustness for Robot Manipulator Control," *ASME J. Dynamic Systems, Measurement, and Control*, Vol. 108, pp. 163-171.

Paul, R. P. (1981) *Robot Manipulators: Mathematics, Programming, and Control*, Cambridge, Mass., MIT Press.

Preising B. and T. C. Hsia (1991) "Robot Performance Measurement and Calibration using a 3D Computer Vision System," *Proc. IEEE Int. Conf. Robotics & Automation*, Sacramento, CA, pp. 2079-2084.

Puskorius, G. V. and L.A. Feldkamp (1987) "Global Calibration of a Robot/Vision System," *Proc. Int. Conf. on IEEE Robotics &*

Automation, Raleigh, NC, pp. 190-195.

Ravani, B. (1985) *Advanced Robots*, Lecture Notes, Florida Atlantic University.

Ravani, B. and Q. J. Ge (1991) "Kinematic Localization for World Model Calibration in Off Line Robot Programming Using Clifford Algebra," *Proc. IEEE Int. Conf. Robot. Automat.*, Sacramento, CA.

Renders, J. M., E. Rossingnol, M. Becquet, and R. Hanus (1991) "Kinematic Calibration and Geometrical Parameter Identification for Robots," *IEEE Trans. Robotics & Automation*, Vol. 7, No. 6, pp. 721-732.

Roberts, K. S. (1988) "A New Representation for a Line," *Proc. Int. Conf. Comput. Vision & Pattern Recog.*, pp. 635-640.

Roth, Z. S., B. W. Mooring, and B. Ravani (1987) "An Overview of Robot Calibration," *IEEE J. Robotics Automation*, Vol. 3, No. 5, pp. 377-385.

Shamma, J. M. and D. E. Whitney (1987) "A Method for Inverse Robot Calibration," *J. of Dynamic Systems, Measurement, and Control*, Vol. 109, No. 1, pp. 36-43.

Sheth, C. and J. J. Uicker, Jr. (1972) "IMP (Integrated Mechanism Program), a Computer-Aided Design Analysis System for Mechanisms and Linkages," *ASME J. Engineering for Industry*, Vol. 94, pp.454-464.

Shih, Hung and Lin (1995) "Comments on 'A Linear Solution to Kinematic Parameter Identification for Robot Manipulators,'" *IEEE Trans. Robotics & Automation*, Vol. 11, No. 5, pp. 777-780.

Shiu, Y. C. and S. Ahmad (1989) "Calibration of Wrist-Mounted Robotic Sensors by Solving Homogeneous Transform Equations of the Form $AX = XB$," *IEEE Trans. Robotics & Automation*, Vol. 5, No. 1, pp. 16-27.

Sklar, M. (1988) *Metrology and Calibration Techniques for the Performance Enhancement of Industrial Robots*, Ph. D. Dissertation, University of Texas, Austin, Texas.

Spangelo, J., J. R. Sagli, and O. Egeland (1993) "Bounds on the Largest Singular Values of the Manipulator Jacobian," *IEEE Trans. Robot. & Automation*, Vol. 9, No. 1, pp. 93-96.

Srinivasa Rao, K. N. (1988) " The Rotation and Lorentz Groups and Their Representations for Physicists," John Wiley & Sons, pp. 11-12, 146-152.

Stone, H. W. (1987) *Kinematic Modeling, Identification, and Control of Robotic Manipulators*, Kluwer Academic Publishers.

Suh, C. and C. W. Radcliffe (1978) *Kinematics and Mechanisms Design*, John Wiley, New York.

Sutherland, I. (1974) "Three-Dimentional Data Input by Tablet," *Proc. IEEE*, Vol. 62, No. 4, pp. 453-461.

Tan, H. L., S. B. Gelfand, and E. J. Delp (1991) "A Cost Minimization Approach to Edge Detection Using Simulated Annealing," *IEEE Trans.*

PAMI, Vol. 14, No. 1, pp. 3-18.

Tsai, R. Y. (1987) "A Versatile Camera Calibration Technique for High Accuracy 3D Machine Vision Metrology Using Off-the-Shelf TV Cameras and Lenses," *IEEE J. Robotics Automat.*, Vol. RA-3, No. 4, pp. 323-344.

Tsai, R. Y. and R. K. Lenz (1989) "A New Technique for Fully Autonomous and Efficient 3D Robotics Hand/Eye Calibration," *IEEE Trans. Robot. Automat.*, Vol. 5, No. 3, pp. 345-357.

Trivedi, H. (1991) "A Semi-Analytic Method for Estimating Stereo Camera Geometry from Matched Points," *Image and Vision Computing*, Vol. 9, pp. 227-236.

Umeyama, S. (1991) "Least-Squares Estimation of Transformation Parameters Between Two Point Patterns," *IEEE Trans. Pattern Anal. Machine Intell.*, Vol. 13, No. 4, pp. 376-380.

Vaishnav, R. N. and E. B. Magrab (1988) "A General Procedure to Evaluate Robot Positioning Errors," *Intl. J. Robotics Research*, Vol. 6, No. 1, pp. 59-78.

Veitschegger, W. K. and C. H. Wu (1988) "Robot Calibration and Compensation," *IEEE J. Robotics Automation*, Vol. 4, No. 6, pp. 643-656.

Vuskovic, M. I. (1989) "Compensation of Kinematic Errors Using Kinematic Sensitivities," *Proc. IEEE Intl. Conf. Robotics Automation*, Scottsdale, Arizona, pp. 745-750.

Wampler II, C. (1986) "Manipulator Inverse Kinematic Solution Based on Vector Formulations and Damped Least-Squares Methods," *IEEE Trans. Syst. Man Cybernetics*, Vol. 16, No. 1, pp. 93-101.

Wang, C. -C. (1992) "Extrinsic Calibration of a Vision Sensor Mounted on a Robot," *IEEE Trans. Robotics & Automation*, Vol. 8, No. 2, pp. 161-175.

Weng, J., P. Cohen, and M. Herniou (1992) "Camera Calibration with Distortion Models and Accuracy Evaluation," *IEEE Trans. Pattern Anal. Machine Intell.*, Vol. 14, No. 10, pp. 965-980.

Whitney, D. E. (1972) "The Mathematics of Coordinated Control of Prosthetic Arms and Manipulators," *ASME J. Dynamic Systems, Measurement and Control*, Vol. 122, pp. 303-309.

Whitney, D. E., C. A. Lozinski, and J. M. Rourke (1986) "Industrial Robot Forward Calibration Method and Results," *ASME J. Dynamic Systems, Measurement, and Control*, Vol. 108, Mar., pp. 1-8.

Wilkinson, J. H. and C. Reinsh, *Linear Algebra*, Springer-Verlag, 1971.

Wu, C. H. (1984) "A kinematic CAD Tool for the Design and Control of a Robot Manipulator," *Int. J. Robotics Research*, Vol. 3, No. 1, pp. 58-67.

Xu, X. (1991) *Camera Calibration Techniques*, M.S. Thesis, Florida Atlantic University.

Zhuang, H. (1989) *Kinematic Modeling, Identification and Compensation of Robot Manipulators*, Ph.D. Dissertation, Florida Atlantic University, Boca Raton, Florida.

Zhuang, H. (1995) "A Self-Calibration Approach to Extrinsic Parameter Estimation of Stereo Cameras," Robotics and Autonomous Systems, Vol 15, pp. 189-197.

Zhuang, H., F. Hamano, and Z. S. Roth (1989) "Optimal Design of Robot Accuracy Compensators," *Proc. IEEE Int. Conf. Robot. & Automat.*, Scottsdale, Arizona, pp. 751-756.

Zhuang, H., B. Li, Z. S. Roth, and X. Xie (1992) "Self-calibration and Mirror Center Offset Elimination of a Multi-beam Laser Tracking System," *Int. J. Robotics and Autonomous Systems*, Vol 9., pp. 255-269.

Zhuang, H. and Z. Qu (1994) "A new Jacobian formulation for robotic hand/eye calibration," *IEEE Trans. Systems, Man & Cybernetics*, Vol. 24, No. 8, pp. 1284-1287.

Zhuang, H. and Z. S. Roth (1991a) "A Unified Approach to Kinematic Modeling, Identification and Compensation for Robotics Calibration," book chapter in *Advances in Control and Dynamic Systems*, Vol. XXXVIII, edited by C. T. Leondes, Academic Press, pp. 71-128.

Zhuang, H. and Z. S. Roth (1991b) "Comments on 'Calibration of Wrist-Mounted Robotic Sensors by Solving Homogeneous Transformation Equations of the Form $AX = XB$,'" *IEEE Trans. Robotics & Automation*, Vol. 7, No. 6, pp. 877-878.

Zhuang, H. and Z. S. Roth (1992) "Robot calibration using the CPC error model", *Int. J. Robotics and Computer-Integrated Manufacturing*, Vol. 9, No. 3, Sept., pp. 227-237.

Zhuang, H. and Z. S. Roth (1993) "A Linear Solution to Kinematic Parameter Identification for Robot Manipulators," *IEEE Trans. Robot. & Automation*, Vol. 9, No. 2, pp. 174-185.

Zhuang, H. and Z. S. Roth (1995) "A Note on 'A Linear Solution to Kinematic Parameter Identification for Robot Manipulators,'", *IEEE Trans Robotics & Automation*, Vol. 11, No. 6, pp. 922-923.

Zhuang, H. and Z. S. Roth (1996), "A Note on Singularities of the MCPC Model," *Int. J. Robotics and Computer-Integrated Manufacturing*, in press.

Zhuang, H., Z. S. Roth, and F. Hamano (1990) "Observability issues in Kinematic Identification of Manipulators,"*Proc. ACC Conf.*, pp. 2278-2293.

Zhuang, H., Z. S. Roth, and F. Hamano (1992a) "Observability issues in Kinematic Identification of Manipulators,"*ASME J. Dynamic Systems, Measurement, and Control;* Vol. 114, June, pp. 319-322; also in *Proc. ACC Conf.*, 1990, pp. 2278-2293.

Zhuang, H., Z. S. Roth, and F. Hamano (1992b) "A Complete and

Parametrically Continuous Kinematic Model," *IEEE Trans. Robotics & Automation*, Vol. 8, No. 4, pp. 451-463.

Zhuang, H., Z. S. Roth, and F. Hamano (1993) "Optimal Design of Robot Accuracy Compensators," *IEEE Trans. Robotics & Automation*, Vol. 9, No. 6 December, pp. 854-857.

Zhuang, H., Z. S. Roth, and R. Sudhakar (1991) "Application of Quaternions to World Coordinates Identification of Robot Manipualtor," *Proc. 4th Conf. Recent Advances in Robotics*, Boca Raton, FL, pp. 145-155.

Zhuang, H., Z. S. Roth, and R. Sudhakar (1992) "Fusion Algorithms for Rotation Matrices," *J. Robotic Systems*, Vol. 9, No. 7, pp. 915-932.

Zhuang, H., Z. S. Roth, and R. Sudhakar (1994) "Simultaneous Robot/World and Tool/Flange Calibration by Solving Homogeneous Transformaiton Equations of the Form $AX = YB$," *IEEE Trans Robotics & Automation*, Vol. 10, No. 4, pp. 549-554.

Zhuang, H., Z. S. Roth, and K. Wang (1994) "Robot Calibration by Mobile Camera System," *J. Robotic Systems*, Vol. 11, No. 3, pp. 155-168; also in *Proc. ASME Winter Annual Meeting*, 1991 DSC-Vol. 30, pp. 65-72.

Zhuang, H., Z. S. Roth, X. Xu, and K. Wang (1993) "Camera Calibration Issues in Robot Calibration with Eye-On-hand Configuration," *Int. J. Robotics and Computer-Integrated Manufacturing*, Vol 10, No. 6, pp. 401-412.

Zhuang, H. and Y. Shiu (1993) "A Noise-Insensitive Algorithm for Robotic Hand/Eye Calibration with or without Sensor Orientation Measurement," *IEEE Trans. System, Man & Cybernetics, IEEE Trans. Systems, Man and Cybernetics*, Vol. 23, No. 4, pp. 1168-1174.

Zhuang, H., K. Wang, and Z. S. Roth (1993) "Error-Model-Based Robot Calibration Using a Modified CPC Model," *Int. J. Robotics and Computer-Integrated Manufacturing*, Vol. 10, No. 4, pp. 287-299.

Zhuang, H., K. Wang, and Z. S. Roth (1994) "Optimal Selection of Measurement Configurations for Robot Calibration Using Simulated Annealing," *Proc. IEEE Int. Conf. Robot. & Automation*, San Diego, pp. 393-398.

Zhuang, H., K. Wang, and Z. S. Roth (1995) "Simultaneous Calibration of a Robot and A Hand-Mounted Camera," *IEEE Trans. Robot. & Automation*, Vol. 11, No. 5, pp. 649-660.

Zhuang, W. Wu, and Z. S. Roth (1995) "Camera-Assisted SCARA Arm Calibration," *Proc. IEEE/RSJ IROS*, Vol. I, pp. 507-512.

Zhuang, H. and W. Wu (1996) "Camera Calibration with Near-parallel (Ill-Conditioned) Calibration Board Configuration," *IEEE Trans. Robotics and Automation, in press*.

Zhuang, H., J. Wu, and W. Huang (1996) "Optimal Planning of Robot Calibration Experiments by Genetic Algorithms," to appear in *Proc. of IEEE Int. Conf. Robotics & Automation*.

APPENDICES

I. SUMMARY OF BASIC CONCEPTS IN MATRIX THEORY

The summary is an adaptation of (Golub and Van Loan (1983)), restricted to real matrices.

A. EIGENVALUES AND EIGENVECTORS

The eigenvalues λ_i and eigenvectors x_i of an $n \times n$ matrix A are defined by the relationship

$$(A - \lambda_i I)x_i = 0, \qquad x_i \neq 0$$

Consequently, the n eigenvalues λ_i are the roots of the characteristic equation

$$p_A(\lambda) = \det(A - \lambda I) = 0$$

where the characteristic polynomial $p_A(\lambda)$ in λ is of degree n.

B. VECTOR AND MATRIX NORMS

For the sake of quantifying vector errors, it is convenient to associate a nonnegative scalar with any vector or matrix that in some sense provides a measure of the vector magnitude. Such measures, called *norms*, must satisfy several consistency conditions such as the triangle inequality. The most common vector norms belong to the family of L_p-norms

$$\|x\|_p = (|x_1|^p + |x_2|^p + \ldots + |x_n|^p)^{1/p}, \ 1 \leq p \leq \infty.$$

where n is the dimension of x. Two particular cases are particularly useful:
$p = 2$, the *Euclidean norm*:

$$\|x\|_2 = (|x_1|^2 + |x_2|^2 + \ldots + |x_n|^2)^{1/2},$$

$p = \infty$, the *infinity norm*:

$$\|x\|_\infty = \max \left\{ |x_i|, i = 1, 2, \ldots, n \right\}.$$

Vector norms have properties that generalize the concept of length.

Readers are referred to (Golub, Charles and Loan (1983)) for a detailed description of these properties.

If a matrix norm and a vector norm are related in such a way that

$$\|Ax\| \leq \|A\| \|x\|$$

is satisfied for any A and x, then the two norms are said to be *consistent*. For any vector norm, there exists a consistent matrix norm. In fact such an induced norm is given by

$$\|A\| = \sup\left\{\frac{\|Ax\|}{\|x\|}, x \neq 0\right\}$$

where sup($\bullet$) denotes the "lowest upper bound". If $\|x\|_\infty$ is used, then

$$\|A\|_\infty = \max\left\{\sum_{j=1}^{n} |a_{ij}|, i \in \{1, 2, \ldots, n\}\right\}$$

If $\|x\|_2$ is used, then

$$\|A\|_2 = (\lambda_{max}(A^TA))^{1/2}$$

where λ_{max} is the maximum eignenvalue of A^TA, and the superscript T denotes the transpose of a matrix.

C. SINGULAR VALUE DECOMPOSITION

If the transpose of a matrix equals its inverse, then the matrix is called a unitary matrix. Let A be an $m \times n$ matrix of rank r. There exist an $m \times m$ unitary matrix U, an $n \times n$ unitary matrix V, and an $r \times r$ diagonal matrix D, such that

$$A = U\Sigma V^T, \qquad \Sigma = \begin{bmatrix} D & 0 \\ 0 & 0 \end{bmatrix}$$

where $D = \text{diag}\{\sigma_1, \sigma_2, \ldots, \sigma_r\}$ and σ_i, which are strictly positive, are called the *singular values* of A. σ_i are the positive square roots of the eigenvalues of A^TA. Note that sometimes the zeros on the prolonged diagonal D are also called the singular values of A.

II. LEAST SQUARES TECHNIQUES

Two least squares techniques are summarized: linear least squares and nonlinear least squares.

A. LINEAR LEAST SQUARES

The objective is to fit a linear mathematical model to a given set of data. In order to reduce the influence of measurement errors, one needs to take a sufficient number of measurements to provide a number of equations greater than the number of unknowns. The resulting problem is to "solve" an over-determined system, and it can be formulated as the following *linear least squares* problem:

Given a system of linear equation

$$Ax = b \qquad\qquad (A.2.1)$$

where A and b are respectively an $m \times n$ matrix and an $m \times 1$ vector such that $m > n$, find a $n \times 1$ vector x such that $\|r\|_2$ is minimized, where

$$\|r\|_2 = (r^T r)^{1/2}, \qquad r = b - Ax.$$

The solution to the least squares problem is given by the following expression

$$x = A^+ b, \qquad A^+ = (A^T A)^{-1} A^T, \qquad\qquad (A.2.2)$$

where the $n \times m$ matrix A^+ is the *pseudo-inverse* of A.

Numerically stable algorithms for the least squares problem can be found in (Wilkinson and Reinsch (1971)).

B. NONLINEAR LEAST SQUARES

In calibration problems, it is often required to fit a vector-valued function to a set of measured data points. This type of problems can be formulated as an over-determined nonlinear system with m equations and n unknowns such that $m > n$,

$$f(x) \equiv 0$$

This is a problem of finding a local minimum of a real-valued function $\phi(x)$, where

$$\phi(x) = f(x)^T f(x)$$

By Taylor's expansion,

$$f(x_0) + f'(x_0)(x - x_0) \cong 0$$

where $f'(x_0)$ is the $m \times n$ Jacobian matrix of the vector-valued function f. Let $d = x - x_0$. d can be written as

$$d = [f'(x_0)]^+ f(x_0)$$

where $[f'(x_0)]^+$ is the pseudo-inverse of the Jacobian at $x = x_0$. The unknown vector is updated by

$$x_1 = x_0 - d$$

This iterative procedure is continued until no more improvement on the solution can be obtained. This is called the *Gauss-Newton* method.

The *Levenberg-Marquadt* (or *damped Gauss-Newton*) method may be used to improve on the robustness of the algorithm. In its simplest form, the Levenberg-Marquadt algorithm computes the adjustment d in the following way,

$$d = [f'(x_0) + \lambda]^+ f(x_0)$$

where λ, a damping factor, is normally in the range of 0.001 to 0.1. For more details of the Levenberg-Marquardt algorithm, readers are referred to (Marquardt (1963)).

III.   SENSITIVITY ANALYSIS

In (A.2.1), the vector b and matrix A correspond to measured data and uncertain system model and as such both are subject to errors. Consequently the resulting x is also subject to errors. Thus

$$x + \delta x = (A + \delta A)^{-1}(b + \delta b)$$

where δx reflects the sensitivity of the identification algorithm to errors in the model and the data characterized as δA and δb.

Let us discuss first the case that $m = n$; i.e., A is a square matrix. Assume first that $\delta A = 0$ but $\delta b \neq 0$. Then

$$\|\delta x\| \leq \|A^{-1}\| \|\delta b\|$$

Also by Schwartz inequality,

$$\|b\| \leq \|A\| \|x\|$$

Combining the above two equations and assuming that $b \neq 0$, we have

$$\frac{\|\delta x\|}{\|x\|} \leq \|A\| \|A^{-1}\| \frac{\|\delta b\|}{\|b\|}$$

$$(1.2.3)$$

where equality can occur and, therefore, the bound is tight.

Define a *condition number* $\kappa(A)$ as follows

$$\kappa(A) = \|A\| \|A^{-1}\|$$

Then (1.2.3) can be rewritten as

$$\frac{\|\delta x\|}{\|x\|} \leq \kappa(A) \frac{\|\delta b\|}{\|b\|}$$

$$(1.2.4)$$

The larger the condition number is, the large is the numerical sensitivity of the solution to data errors.

Assuming this time that $\delta A \neq 0$ but $\delta b = 0$, one can obtain the following error bound

$$\frac{\|\delta x\|}{\|x + \delta x\|} \leq \kappa(A) \frac{\|\delta A\|}{\|A\|}$$

The above results can be easily extended to the case that $m > n$. In this case, we have an over-determined system. The condition number is then defined as

$$\kappa(A) = \|A^T A\| \|(A^T A)^{-1}\|.$$

INDEX

Absolute depth measurement 131
Accuracy assessment 320, 323
Accuracy compensation 1, 6, 273
   Bilinear interpolation 276
   Damped least squares (DLS)
      algorithm 282
   Linear quadratic regulator (LQR)
      algorithm 283
   Gradient-based methods 278
   Newton-Raphson algorithm 282
   Pose redefinition 277
   Workspace mapping 273
Accuracy compensation task 274
Actual joint axis 137
Ahmad 226, 241
All-revolute 172, 174
Autonomous calibration 9, 132, 215
Axis misalignment 77

Base and tool calibration 253
   Observability 261, 262
   Position estimation 263
   Problem statement 256
   Rotation estimation 258
Base calibration 201, 245
   Experiment 252
   Position estimation 251
   Problem statement 245
   Quaternion-based 247, 254
   Rotation estimation 247
   SVD-based 251, 254
Base coordinate frame 149, 260
Base coordinate system 203
Base/tool -simultaneous calibration
   255
   Linear solution 257

   Position vector 265
   Problem formulation 256
   Rotation 260
   Simulation 266
Base transformation 77, 202, 211,
   245, 314
Basic homogeneous transformation
   67
Bennett 132
Bilinear interpolation 276
Broderick 7, 105, 199

Calibration
   Accuracy 192
   Case studies 305
   Hand/eye calibration 217
   Practical considerations 307
   Robot/camera simultaneous
      calibration 200
   World-to-base transformation 187,
      244
Calibration board 35, 54, 115, 309
Calibration error
   2D error 213
   3D error 213
Calibration fixture 108, 115
Camera aided robot calibration 10
Camera calibration 9, 11, 113, 211,
   274, 309, 312
   Basic concept 14
   Distortion 11
   Eccentricity coefficient 43, 47
   Experimentation 53, 57
   Extrinsic parameters 61
   Fast RAC-based 24
   Handling near singularity 27, 57

Image center 39, 50
Models 11
Nonlinear approach 31, 58
Perspective projection distortion 41
Radial alignment constraint (RAC) 11, 19
RAC-based 20, 53
Ratio of scale factors 35, 38, 52, 60
Self-calibration 61
Simplified RAC-based 11, 56
Simulation 50
Camera calibration fixture 126, 310, 312
Camera coordinate system 203
Camera focal length 44
Camera models
    Distortion-free model 11
    Distortion parameter 34
    Extrinsic parameters 13, 29, 34, 59, 61, 119, 203, 309
    Image center 20, 31
    Intrinsic parameters 13, 29, 34, 59, 309
    Lens distortion model 16
    Pin-hole model 12
    Radial distortion coefficient 19, 31, 203
    Ratio of scale factors 19
    Scale factors 31, 204
Camera pose 118
Camera resolution 8, 109
Case studies
    Calibration of PUMA 312
    Calibration of SCARA ARM 323
Camera-tool transformation 131
Cartesian error 156, 158
Chou 242
Cipra 7, 105, 199
Circle point analysis procedure 121
    Extension 125
Common normal 67
Compensation 311

Compensation accuracy 197, 281
Complete and parametrically continuous 6
Complete and Parametrically Continuous (CPC) model 6, 80, 82, 116, 171, 181, 314
    Convenient assignment rules 86
    Error model 157, 161
    Example 89-92
    Parameters errors 156, 158, 160
    Prismatic joint 85
    Relationship with other models 92-96
    Revolute joint 84
Completeness 75
Condition number 151, 171, 341, 291, 345
Coordinate measuring machine (CMM) 192, 245, 252, 312
Cost function 205, 231, 296
CPC error model 156
CPC model 307
CPC orientation parameter 176
CPC translation parameter 179
Craig 136

Damped least squares (DLS) algorithm 282
Degrees of freedom 71
Denavit-Hartenberg (D-H) model 68, 116, 266, 285, 307
    Common normal 77
    Distance 70
    Hayati's modification 79, 307
    Joint offset 74
    Parameters 72, 253
    Prismatic joint 74
    Revolute joint 73
    Rotation angle 70
    Twist angle 69
Depth measurement 131
D-H error model 153
Differential transformation 134
    Additive 135

Multiplicative 135
Differentiable 142, 143
Distortion parameter 34, 210
Driels 1, 104, 132

Eccentric coefficient 43, 47
Eigenvalue 177, 341
Eigenvector 177, 341
Electronic assembly 325
Encoder resolution 191
End-effector pose 61, 107, 176, 202
   Actual pose 107
   Nominal pose 107
Error model 141, 212
Errors
   Base coordinates 136
   Orientational error 147
   Rotational error 136
   Tool coordinates 136
   Translational error 136
Error modeling
   Base 163
   Hand/eye
   Robot 134, 156, 166
   Tool 165
Error parameter 277
Eular parameters 220
Everett 104, 148
Experimental study 312, 323
Extrinsic parameters 13, 29, 34, 59,
   119, 203, 309, 329

Field of view 109, 329
Finite difference approximation 140,
   169, 205
Flange 75
Florida Atlantic University Robotics
   Lab 60
Focal length 211
Forward kinematics 75, 173, 217,
   257, 308
Forward verification 311, 318, 321

Gauss-Newton algorithm 37

General rotation matrix 66
Genetic algorithm 306
Gradient-based algorithm 278

Hamano 105
Hand coordinate system 217
Hand/eye calibration 183, 201, 217,
   262
   Formulation 217
   Identification Jacobian 233
         Linear method 222
   Nonlinear method 229
   Observability 233
   Rotation 221
   Simulation 236
   Translation 227
Hand-eye coordination 7
Hand/eye transformation 211, 232
Hand-mounted camera 113, 131,
   310, 312, 324
Hand-mounted monocular camera 15
Hayati 2, 6, 79, 104
Homogeneous transformation 64,
   107, 202, 218
Hollerbach 1, 5
Hung 199

Identification Jacobian 99, 133, 141,
   291
   Finite difference method 139
   Hand/eye 232
   Robot 133
   Robot/camera 205
Image center 20, 25, 31, 60
Image coordinates 309
Identification Jacobian 147, 206,
   212, 308
   Hand/eye identification
      Jacobian 232
   Robot identification Jacobian 147
   Robot/camera identification
      Jacobian 206
Independent error parameters 309
Intrinsic parameters 13, 24, 29, 34,

59, 309
Inverse kinematics 75, 289, 311
Irreducibility 146

Jiang 5
Joint axis identification 8, 171
Joint axis misalignment 4, 77, 100,
    197, 273
Joint correction 278
Joint offset 4, 76, 77, 266, 273, 309
Joint position sensor 83
Joint space 5
Joint travel limits 290
Joint variable 71

Kamel 242
Khalil 334
Kinematic error models
    Base error model 163
    CPC error model 156, 161
    D-H error model 153, 154
    Finite difference approximation
        140, 169
    Generic error model 141
    MCPC error model 166
    Tool error model 165
Kinematic error parameters 308
Kinematic identification 1, 65, 77,
        133, 171, 202, 207, 253, 291,
        314, 320, 323
    Base calibration 253
    Direct identification 108
    Error-model-based 107, 133
    Hand/eye calibration 217
    Linear method 171
    Robot/camera simultaneous
        Calibration 201
    World-to-base transformation
        187
Kinematic modeling
    Basic concepts 165
    Completeness 63, 75
    CPC model 64, 80, 82, 136, 171,
        181

D-H model 1, , 64, 68, 136
MCPC model 64, 80
Modified D-H model 78
Parametrically continuity 63, 97,
    100
Proportionality 63, 75
S-model 105
Zero reference model 104

Laser tracking system 7, 130
Lau 5
Least squares 34, 152, 213, 226, 249
    Linear 19, 343
    Nonlinear 34, 343
Lens distortion 202, 320, 322
    Radial distortion 16-18
    Tangential distortion 17, 18
Lenz 118, 131
Levenberg-Marquardt algorithm 318,
    344
Linear least squares 19, 339, 343
    Damped least squares 282
Linear kinematic identification
    All-revolute robot 174
    All-recursive 181
    Comparison with error-model-
        based technique 197
    Hybrid linear method 188
    Numerical example 189
    Orientation parameter 176, 178,
        188
    Prismatic joint 183
    Problem statement 173
    Revolute joint 185
    Solution strategy 172
    Translation parameter 179, 189
    World-to-base 187
Linear quadratic regulator (LQR)
    algorithm 283
Linearized error model 277, 290
Link length 309
Local search algoithm 295

Manipulator Jacobian 278

Prismatic joint 281
Revolute joint 281
Manipulator singularity 285
Matrix theory 341
  Condition number 345
  Eigenvalue 341
  Eigenvector 341
  Singular value decomposition 342
MCPC error model 166
MCPC model 307
Measurement configuration selection
    291, 310
  Experimental study 300
  Genetic algorithm 306
  Pairwise exchange method 293
  Problem statement 292
  Simulated annealing algorithm
    294
  Simulation study 299
  Uniform random search method
    293
Measurement data collection 320,
    323
Measurement noise 186, 269, 292
Measurement phase 107
Measurement residual 231
Measurement techniques
  Coordinate measuring machine 5
  Laser tracking system 5
  Theodolite 5
Model singularity 97, 100
  First type 97
  Second type 98
Modified Complete and
    Parametrically Continuous
    (MCPC) model 87, 317, 324
  Assignment rules 87
  Example 90
  Singularity 99, 105
Modified D-H model 78
Monocular camera 59
Mooring 1, 6, 104
Motion and shape matrices 105, 144
Motion matrix 182

Motion planning 311
Motorized zoom camera 130
Moving camera approach 109, 114

Near parallel axes
Nominal Jacobian 278
Non-geometric errors 7
Nonlinear least squares 31, 34, 35,
    37, 38, 205, 343
  Gauss-Newton 344
  Levenberg-Marquardt 344

Norm 341
  Euclidean norm 341
  Infinite norm 341
  Matrix norm 341
  Vector norm 341

Observability 7, 147, 212, 220, 291,
    308
  Definition 147
  Hand/eye calibration 233
  Robot/camera calibration 219
Observability index 291
Observability measure 149
Offline programming 270
Orientation accuracy 197, 267, 284
Orientational error 147, 316
Orientation parameter 176, 178, 188
Orthonormalization 33, 237, 314

Parameter continuity 64, 97, 105,
    142, 143
  Definition 97, 99
Parameter estimation error 292, 318
Paul 1, 66, 142
Perspective projection distortion 41,
    47, 61
Perspective transformation matrix
    (PTM) 11, 14, 15
Pixel resolution 191
Point measurement
Polytope 39-40
Pose error 144, 233, 277

Pose measurement 1, 107, 308
  Between tool and camera 119
  Hand-mounted camera 113, 132
  Monocular camera 118, 129, 132
  Moving camera 109
  Stationary camera 110, 116, 120, 132
  Vision-based 108
Pose measurement error 192, 194
Pose redefinition 311
Position accuracy 197, 267, 284
Position equation 173
Position error 267, 274, 316
Position measurement 246
Position measurement error 111
Positional error 207, 212, 237
Practical considerations 307
Practical implementation issues 6
Prismatic joint 63, 71, 76, 308
Proportionality 75
Pseudo-inverse 343
PUMA arm calibration
  Experimental results 320
  Monocular camera 317
  Stereo cameras 313
  System setup 312

Quantization error 111
Quaternions 220, 242, 250, 258, 314, 318
Quaternion algebra 220
  Addition and subtraction 221
  Eular parameters 222
  Hamiltonian conjugate 220
  Multiplication 221
  Norm, division and inversion 221
  Pure quatenion 222
  Scalar quaternion 220
  Unit quaternion 221
  Vector quaternion 220

RAC-based camera calibration 20, 25, 53, 316

Radial alignment constraint (RAC) 11, 19
Radial distortion 16
Radial distortion coefficient 19, 31, 203
Radial lens distortion 328
Ratio of scale factors 19, 25, 60, 202
Ravani 6
Reachable robot joint space 291
Reducibility 146, 149
Redundant manipulator 285
Relative pose 217
Relative rotation angle 261
Repeatability 4, 8, 275, 324
Repeated camera calibration 59, 61
Reprogramming 1
Research issues 6
Revolute joint 63, 71, 308
Roberts 80, 105
Robot accuracy 307
Robot calibration fixture 115
Robot-camera model 202, 216
Robot/camera-simulataneous calibration 201
  Cost function 204
  Error model 208
  Experimental result 317
  Identification Jacobian 205, 210
  Implementation issue 210
  Kinematic model 202
  Stereo camera case 214
Robot control software 75
Robot localization 7, 246
Robot replacement 1
Rodrigues equation 4, 64, 103, 104
Rotation angle 66, 258
Rotation axis 66, 258, 261
Rotation axis estimation 187
Rotation equation 173
Rotation matrix 12, 28, 42, 47, 66, 81, 222, 258
Rotational error 207, 212, 237

Scale factors 13, 203, 211
SCARA arm 27, 100, 307
SCARA arm calibration 323
    Results 324
    System setup 323
Screw axis 65
Screw parameters 65
Self-calibration 192
Self-calibration of cameras 61
Sensitivity 344
Sensor coordinate frame 217
Shape matrix   105, 144
Sheth 105
Shih 199
Shiu 226, 241
Simulated annealing algorithm 294
    Candidate state 296
    Cooling schedule 297
    Cost function 296
    Equilibrium condition 298
    Initial temperature 297
    Simulation 299
Simultaneous calibration 10, 132,
    201, 255, 317
Simultaneous calibration of robot
    and camera 201
    Experimental result 320
    Monocular camera 201, 317
    Simulation result 317
    Stereo camera 202, 214
Simultaneous calibration of base
    and tool 255
Singular joint configuration 282
Singular Value Decomposition
    (SVD) 152, 188, 242, 249,
    251, 263, 308, 318, 342
Singular values 292, 342
Singularity 97, 100, 168
Singularity zone 290
Sklar 132
S-model 105
Snake robot 290
Spacial quantization 110
Statical error 181, 192

Stationary camera 110
Stereo camera setup 116, 312
Stereo vision 15, 38, 111, 113, 192
    Accuracy 112
    Quatization error 111
Stone 104, 132, 199

Tang 6
Tangential distortion 16
Theodolite 5
Tool-to-camera transformation 309,
    310, 329
Tool calibration 127, 242
Tool coordinate frame 127, 147,
    217, 260
Tool transformation 75, 76, 99, 245,
    314
Translation parameter 179, 189
Tsai 20, 60, 118, 131, 241
Twist angle 69
Two-dimensional calibration 274

Uicker 105

Verification 191, 213, 253, 327
    Forward verification 311, 318
    Inverse verification 311
Vision algorithm 312
Vision based pose measurement 108
Vision system 13, 315
Veischegger 6

Wang 59
Whitney 148
World-to-base transformation 309,
    310
Wrist singularity 288
Wu 5, 6

Xu 59

Zero position of a joint 83, 104
Zero reference model 104, 105, 116

Printed and bound by CPI Group (UK) Ltd, Croydon, CR0 4YY

23/10/2024

01778237-0012